Hydrology and Global Environmental Change

We work with leading authors to develop the strongest educational materials in geography, bringing cutting-edge thinking and best learning practice to a global market.

Under a range of well-known imprints, including Prentice Hall, we craft high quality print and electronic publications which help readers to understand and apply their content, whether studying or at work.

To find out more about the complete range of our publishing, please visit us on the World Wide Web at: www.pearsoneduc.com

Hydrology and Global Environmental Change

Nigel Arnell
Professor of Geography, University of Southampton

An imprint of Pearson Education

Harlow, England · London · New York · Reading, Massachusetts · San Francisco · Toronto · Don Mills, Ontario · Sydney
Tokyo · Singapore · Hong Kong · Seoul · Taipei · Cape Town · Madrid · Mexico City · Amsterdam · Munich · Paris · Milan

Pearson Education Limited
Edinburgh Gate
Harlow
Essex CM20 2JE

and Associated Companies throughout the world

Visit us on the World Wide Web at:
www.pearsoneduc.com

First published 2002

ISBN 0582 36984 3

British Library Cataloguing-in-Publication Data
A catalogue record for this book is available from the British Library

Library of Congress Cataloging-in-Publication Data
A catalog record for this book is available from the Library of Congress

10 9 8 7 6 5 4 3 2 1
06 05 04 03 02

Typeset in 11/12pt Garamond by 35
Printed and bound in Malaysia

To James, Fay and Rosie: thanks for letting me monopolise the computer.

Contents

Series preface

Understanding Global Environmental Change, the general title of this book series, has become a progressively high priority for all environmental scientists over the past decade. Agenda 21 and the Earth Charter that emanated from Rio de Janeiro in 1992 was accompanied by a failure of the 'Miami Group' to sign the associated Biodiversity Convention and was followed shortly thereafter by the political debacle of the Kyoto Protocol to the U.N. Framework Convention on Climate Change (1997). As concerned citizens and as physical geographers, we are impressed by the urgency of the global environmental crisis and the inadequacy of global governance mechanisms to deal with this crisis.

The distinctive role that physical geographers can play in this crisis is to point out to the community of environmental scientists that geography matters. Jeffrey Sachs and his economist colleagues at Harvard have been writing about the geography of poverty and wealth (e.g. *Scientific American*, March 2001, 70–75) and have concluded that geography matters; a realisation that has apparently escaped economists for most of the twentieth century.

We, too, believe that geography matters. Not so much in an environmental determinist global scale sense, but in the sense that global environmental change is expressed quite differently from place to place. The variable surface properties of the Earth and the socio-economic status of the inhabitants of the region make a difference to the policy implications of global environmental change and any attempts to implement those policies.

Because the original series preface was composed in 1997, the series editors proposed an updated preface to coincide with the publication of the fourth volume in this seven volume series. The series continues to be intended for use by first and second year level geography and environmental science students in universities or colleges of higher education in the UK. It is also suitable as text material for second or third year level geography and environmental science students in Canada and the USA. Our main assumption is that the reader will have completed an introductory physical geography course.

We continue to be grateful to the outstanding scientists who have committed their expertise, and acknowledge the long-suffering patience of Matthew Smith, Commissioning Editor, first with Addison Wesley Longman and latterly with Pearson Education.

We close with a quotation from Sir Crispin Tickell from an address to the Kingston University Faculty of Science, 22 June, 2001: 'Before the Rio Summit of 1992, George Bush senior tried to reassure the American people by saying that no-one was going to change the American way of life. As we approach the Johannesburg Summit, George Bush junior has maintained the same view. They are both dead wrong. North Americans must change their way of life, as we in Europe must change ours. Otherwise Nature will do what she has done to over 99% of species that have ever lived, and do the job for us.'

August 1, 2001
Olav Slaymaker, Vancouver
Tom Spencer, Cambridge.

Volume preface

At the beginning of the 21st century we are living in a rapidly changing world. Land use and land cover are changing (an extra four million houses in south east England by 2010?), climate is changing, and our demands on the water environment are increasing. All these pressures – some local, some global – are affecting hydrological systems and water resources. Meanwhile, environmental scientists are increasingly understanding the rhythms and patterns in natural climatic variability associated with, for example, El Niño and the North Atlantic Oscillation.

This book has four main objectives. The first is to provide an overview of the implications of global environmental change for rivers, lakes and groundwater, distinguishing between those changes which affect the catchment and those which affect the inputs to the catchment. The second objective is to explore how the hydrological system itself affects global environmental change through the links between catchment, atmosphere and ocean and the role of the hydrological system in biogeochemical cycling. The third is to review the effects of natural climatic variability on hydrological behaviour, and the implications for the identification of human-induced trends. The fourth is to summarise the current developments in the discipline of hydrology, looking at new measurement techniques, the interfaces between hydrology and other parts of earth system science, and the role of the increasing number of international collaborative hydrological studies. Underpinning these four objectives is a comprehensive review of the components of the water balance, hydrological processes, and the factors affecting water quality.

Hydrological science is advancing rapidly. The examples in this book are drawn from recently published or classic studies, but new examples and interpretations of patterns of hydrological behaviour are continually appearing in the literature; some of the research programmes described in the book are still ongoing. The reader is urged to use the book as a *framework* for exploring the emerging literature: Chapter 1 describes some of the key journals, and a list of web-sites is given in an appendix.

This book focuses on the hydrological effects of and contributions to global environmental change. It is not explictly concerned with the *impacts* on the economy, society or the environment of changes in hydrological systems, or about water management practices and procedures. It does not explain in detail techniques used by engineering hydrologists, such as frequency analysis and forecasting, to estimate hydrological properties as part of the water management process. However, it is hoped that this book will be of value not just to students of geography and environmental science, but also to students of engineering and water management: sustainable water management requires an understanding of hydrological processes, the consequences of environmental change, and the limits of our hydrological knowledge.

July 2001

Acknowledgements

The form of this book has evolved out of the hydrology course I now teach at the University of Southampton, but its roots lie in the decade I spent at the Institute of Hydrology in Wallingford (now the Centre for Ecology and Hydrology, Wallingford). Although my work there focused on frequency analysis, regional hydrology and modelling the impacts of global change, my hydrological education benefited enormously from working in an environment with over a hundred scientists – from a diversity of backgrounds – looking at all aspects of the hydrological cycle. My involvement in many international projects, starting with FRIEND and moving on through GCIP and the IPCC, has exposed me not only to different hydrological cultures and environments, but also to the different perspectives of all the different disciplines with an interest in the movement of water and the material it contains between air, land and sea. Bringing a diverse literature together and writing this book has been a challenge, but a highly stimulating one that has been helped significantly by the support and encouragement of my wife Hilary.

Many of the figures in the book have been prepared by Bob Smith and Linda Hall of the Cartographic Unit, University of Southampton. Terry Marsh and Felicity Sanderson of the Centre for Ecology and Hydrology, Wallingford, provided me with data from the National Water Archive, and Gwyn Rees (CEH Wallingford) provided data from the European Water Archive. Colin Neal (CEH Wallingford) sent the raw data for Figure 6.6, and Janice Lough (Australian Institute of Marine Science) provided the data for the Burdekin River used in Figure 4.14. The rest of the flow data used came from the Global Runoff Data Centre archive stored at the University of New Hampshire.

I am grateful to the following for permission to reproduce copyright material:

Plate IIb from Cambridge University Press, *Global Energy and Water Cycles* (Browing & Gurney, 1999); Plate IIc from the American Meteorological Society, *Journal of Climate*, **12** (Trenberth, 1998); Plate IV from Elsevier Science, *Global Environmental Change* **9** (Arnell, 1999); figure 1.1 from McGraw Hill, *Principles of Hydrology* (Ward & Robinson, 2000); figure 1.2 from the National Academy of Sciences, *Opportunities in the Hydrologic Sciences* (1991); figure 1.5 from Longman, *Physical Geography and Global Environmental Change*, (Slaymaker & Spencer, 1998); figure 3.2 from Routledge,

Atmosphere, Weather and Climate (Barry & Chorley, 1998); figures 3.4, 3.6 and 8.15 from the American Meteorological Society, *Journal of Climate* **6** (Brubaker *et al.*, 1993), **12** (Trenberth, 1998) and 7 (Stamm *et al.*, 1994); figure 3.8 from John Wiley & Son, *Hydrological Processes* 4 (Crockford & Richardson, 1990); figure 3.10 from IPCC, *Climate Change 1995: The Science of Climate Change* (Trenberth *et al.*, 1996); figure 3.11 from John Wiley & Sons, *International Journal of Climatology* 14 (Allen *et al.*, 1994); figures 3.13, 4.26, 6.5 and 8.4 from the American Geophysical Union, *Water Resources Research* **29** (Chanzy & Bruckler, 1993), **30** (Dracup & Kahya, 1994) **32** (Jones & Grant, 1996) and **34** (Eltahir, 1998); figures 3.14, 4.25, 5.7b and 6.11 from Elsevier Science, *Journal of Hydrology* **188–9** (Wallace & Holwill, 1997), **120** (Waylen & Caviedes, 1990), **170** (Soulsby, 1995) and **188–9** (Gash *et al.*, 1997); figure 3.18 from Elsevier Science, *Agricultural and Forest Meteorology* **98–99** (Frech & Jochum, 1999); figures 3.20, 3.26 and 3.30 from Prentice Hall, *Physical Hydrology* (Dingman, 1993); figure 3.29 from John Wiley & Sons, *Earth Surface Processes and Landforms* **10** (Gurnell *et al.*, 1985); figure 4.7 from Longman, *Global Hydrology* (Jones, 1997); figure 4.18 from the American Meteorological Society, *Bulletin of the American Meteorological Society* 74 (Guetter & Georgakakos, 1993); figure 4.19 from the International Association of Hydrological Sciences, *IAHS Publication* **221** (Arnell, 1994); figure 4.22 from John Wiley & Sons, *Global Warming, River Flows and Water Resources* (Arnell, 1996); figure 4.24 American Meteorological Society, *Monthly Weather Review* **115** (Ropelewski & Halpert, 1987); figure 4.26 from the International Association of Hydrological Sciences, *Hydrological Sciences Journal* 45 (Mosley, 2000); figure 5.6 from Allen & Unwin, *British Rivers* (Walling & Webb, 1981); figures 5.9 and 7.2 from Prentice Hall, *Our Changing Planet* (Mackenzie, 1998); figure 5.10 from Prentice Hall, *Climate and Global Environmental Change* (Harvey, 2000); figure 5.11 from Blackwell Scientific, *River Flows and Channel Forms* (Newbold, 1996); figures 6.3 and 7.8 from the Environment Agency, *State of the Freshwater Environment* (1998); figures 6.4, 6.8, 6.9 and 7.7 from the European Geophysical Society, *Hydrology and Earth System Sciences* **1** (Hudson *et al.*, 1997), **2** (Robinson, 1998), **1** (Stevens *et al.*, 1997) and **1** (Hill & Neal, 1997); figures 6.13c and d, 6.14, 7.14 and 7.19 from Routledge, *The Hydrology of the UK* (Acreman, 2000); figures 6.18c and 7.20 from Elsevier, *Global Environmental Change* **3** (Glantz *et al.*, 1993) and 9 (Arnell, 1999); figure 7.4 from Kluwer, Water, *Air and Soil Pollution* **85** (Rodhe *et al.*, 1995); figure 7.6 from Ellis Horwood, *Acid Rain and Acid Waters* (Howells, 1990); figure 7.16 from Elsevier, *Technological Forecasting and Social Change* **65** (Small *et al.*, 2000); figure 7.21 from Kluwer, *Climatic Change*, **46** (Arnell, 2000), figures 8.2, 8.5, 8.7, 8.9b and 8.16 from the American Geophysical Union, *Journal of Geophysical Research* **97** (Sellers *et al.*, 1992 and Desjardim *et al.*, 1992), **101** (Sud *et al.*, 1996, Searcy *et al.*, 1996, Shao & Henderson-Sellers, 1996); figure 8.8 from the American Geophysical Union, *Global Biogeochemical Cycles* **14** (Vörösmarty *et al.*, 2000); figure 8.9a from the American Geophysical Union, *Geophysical Research Letters* 26 (Macdonald *et al.*, 1999); figure 8.11 from University of Chicago

Press, *Journal of Geology* **91** (Milliman & Meade, 1983); and figure 8.13b from the American Meteorological Society, *Journal of Applied Meteorology* **28** (Sellers *et al.*, 1989).

Whilst every effort has been made to trace copyright holders, in a few cases this has proved impossible and the publishers would like to take this opportunity to apologise to any copyright holders whose rights we may have unwittingly infringed.

Chapter 1

Hydrology, earth system science and global environmental change

1.1 Introduction

The year 2000 saw significant flooding in many parts of the world. The largest floods for many decades struck Mozambique in February, driving 733 000 people from their homes and killing 929. Over four million people living in the Mekong Delta were affected by a series of floods between September and November – the largest for 70 years – and 865 people lost their lives. Nearly 1500 people in eastern India and Bangladesh were killed by floods in September and October, and up to 24 million people were displaced. In Europe, the largest floods for at least 50 years inundated many floodplains in northern Italy during October, and Britain experienced the most widespread flooding in October and November for more than 50 years: river levels at York rose to their highest level since 1625.

After each of these floods – and indeed after most of the largest floods during the last decade of the twentieth century – journalists, the public and politicians have speculated that various human activities may be making floods both more frequent and more severe. Was the unusually heavy rainfall that triggered many of the floods of 2000 "caused" by global warming? Were the floods exacerbated by land use change in the catchment, such as urbanisation and deforestation? Were flood heights increased because of the construction of flood banks upstream? To what extent was the increased flood damage due to increased exposure to flood loss? Was the year 2000 really any worse than previous years?

Floods are just one of the ways in which human society interacts with the hydrological system. People need water for drinking, washing and preparing food, farmers need water to irrigate crops, industry uses water both as a raw material and for cooling, and rivers are important parts of the transport network in many countries: rivers are also used to carry away waste. Pressures on the available water are increasing in many countries, as demand for water rises, and at the same time the resource is often being polluted and degraded. Polluted water can carry disease and lead to ill-health and death, and the extent and timing of many water-borne diseases – such as malaria and schistosomiasis – are very much influenced by hydrological regimes.

Activities in the catchment, such as deforestation, urbanisation, abstraction of water and the use of agricultural chemicals, are also affecting hydrological processes and regimes – and these effects have an impact on the water environment, not only in the river channel but also in wetlands, on the floodplain and along the river corridor. Human activities are also affecting the inputs to the catchment, through global warming and acid rain. The United Nations Environment Program's review of the state of the world's environment (UNEP, 1999) highlights many examples of the degradation of the water environment, as do continental-scale assessments in Europe (Stanners & Bourdeau, 1995; European Environment Agency, 1999). "Water" has been identified as the emerging critical environmental issue of the twenty-first century (Cosgrove & Rijsberman, 2000; WMO, 1997).

The aim of this book is to explore the linkages between *global environmental change* and the *hydrological system*: it examines how changes in the catchment and the inputs to the catchment affect hydrological regimes, and also assesses the role played by hydrological processes in global environmental change. It focuses on the hydrological system at the land surface, rather than over the ocean. This introductory chapter places hydrology in the context of earth system science, introduces the concept of global environmental change, and describes the changing way in which hydrology is being studied.

1.2 Hydrology and earth system science

Hydrology is broadly defined as the study of the occurrence, distribution, circulation and properties of water on the earth, and the hydrological cycle (Figure 1.1) lies at the heart of hydrological science. All the water that falls as precipitation has evaporated from the land and the oceans. Rivers transport some of the precipitated water across the land surface, to be evaporated elsewhere – from the sea or a lake, for example – and much is transported from one place to another as water vapour in the atmosphere. Water is being continuously recycled through the oceans, atmosphere, lithosphere, cryosphere (ice sheets, glaciers and permafrost) and biosphere (vegetation). Hydrology is therefore central to any understanding of the way the earth system works. Hydrological science can be seen as the component of "geoscience" that links land, vegetation, atmosphere and ocean (Figure 1.2), and which is a part of *earth system science*. In practice, hydrologists are usually concerned with freshwater (oceanographers deal with the oceans) and with water once it reaches the land surface (meteorologists and atmospheric chemists deal with water when it is in the atmosphere). They have also traditionally taken a catchment approach, treating the inputs to the catchment of water and energy as given, and focusing on the translation of these inputs to "outputs" of streamflow. Increasingly, however, hydrologists are working with meteorologists to understand the way the land surface affects the atmosphere – and hence the "inputs" to the catchment – and with oceanographers to contribute to the understanding of the way in which flows of water and material to the sea affect coastal and ocean processes. Hydrologists are also working with plant physiologists to investigate the process of transpiration, and with ecologists to understand the hydrological controls on ecosystem characteristics.

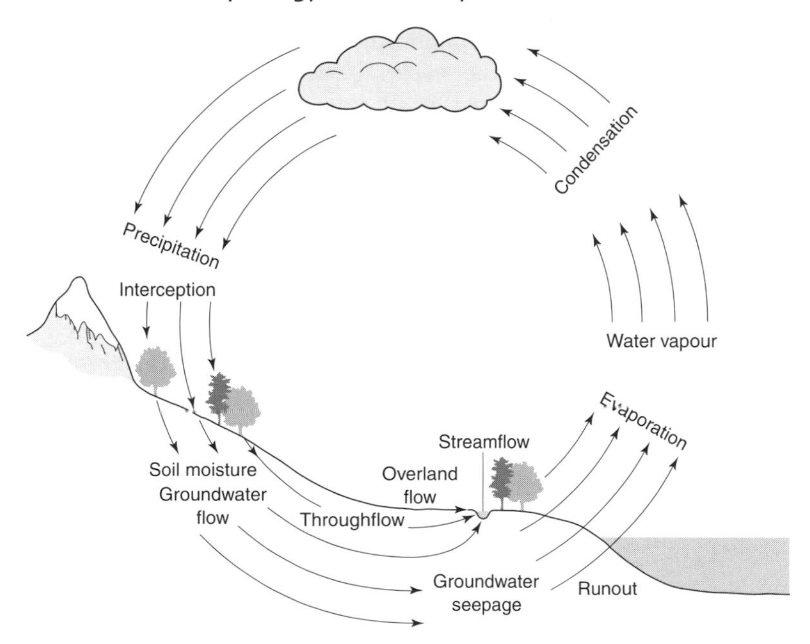

Figure 1.1 The hydrological cycle (Ward & Robinson, 2000)

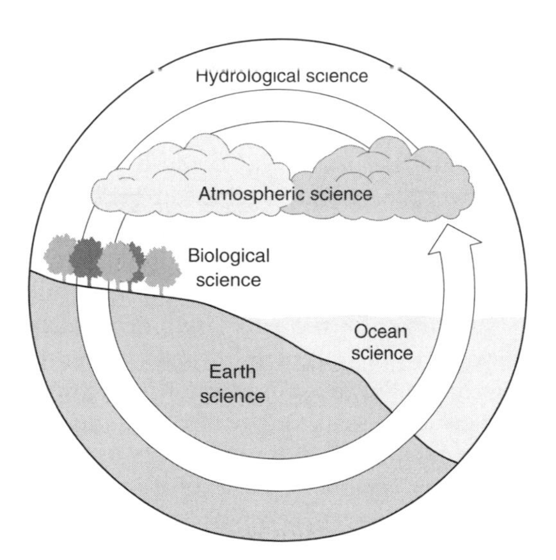

Figure 1.2 Hydrological science as a component of earth system science (modified from National Research Council, 1991)

Central to the concept of earth system science is the idea of cycles of energy and material. The hydrological cycle is an obvious example, but interest has developed over the past decade in *biogeochemical cycles* in general. A biogeochemical cycle can be defined as the cycle of any element or compound, through all its states and phase changes. Attention is often focused on the cycles of the key elements carbon, nitrogen, phosphorus, sulphur and oxygen, although it is of course possible to construct cycles for any flux of material – including sediment. Cycles can be evaluated at all scales, but only at the global scale are cycles closed with no inputs to or output from the system of interest. At smaller scales, such as the catchment or indeed continent, there will be inputs and outputs across the system boundary. Water plays a key role in all biogeochemical cycles, both as a means of transporting material from one store to another and as a medium in which transformations take place. One of water's unique features is that virtually all matter can dissolve in it, and biological and chemical transformations can also occur in water to alter the speciation of elements. Any assessment of any biogeochemical cycle must therefore consider the role of hydrological processes.

1.3 Hydrology and global environmental change

The earth is over 4.5 billion years old, and over this entire period has been subject to change. *Geological* forces include such processes as plate tectonics, metamorphism and vulcanism, which shape the broad pattern of land surface and ocean. These are internal processes, essentially driven by heat at the earth's core. Tectonic movement and metamorphism occur over time scales of millions of years, but the effects of vulcanism can be immediate and catastrophic. Three sets of *external* forces affect the earth system. The atmosphere, hydrosphere and biosphere are all driven by inputs of energy from the sun, which varies diurnally as the earth rotates, seasonally as the earth revolves around the sun, and as solar output fluctuates (solar luminosity varies by ±0.07% over the 11-year sunspot cycle (Hoffert *et al.*, 1988), and has varied more over much longer time scales). The gravitational interaction between the earth and the moon generates tides, which vary predictably from day to day, and gravitational interactions between the earth and other planets produce changes in the earth's orbit over millennial time scales which alter the amount of solar energy received. Thirdly, meteorites sometimes hit the earth, with occasionally very significant consequences. Additional to these geological and external forces are rhythms and patterns inherent in the linked atmosphere–ocean system. These rhythms, which operate on annual or decadal time scales, arise in some cases because of the different rate of response of different parts of the atmosphere–ocean system to an external forcing, and in others because the system undergoes a step change once a small progressive change pushes the system beyond a critical internal threshold. The El Niño–Southern Oscillation is an example of such a rhythm.

Superimposed onto this "natural" global environmental change, which operates at all time scales, are the effects of various human activities. The cultivation of wheat and barley began in the Middle East around 10 000 years ago, and clustered settlements began to appear at around the same time; 7000 years ago farmers in the Indus Valley and Mesopotamia were irrigating crops. Humans have therefore been impacting upon their environment for many thousands of years. However, the rate of change has been accelerating

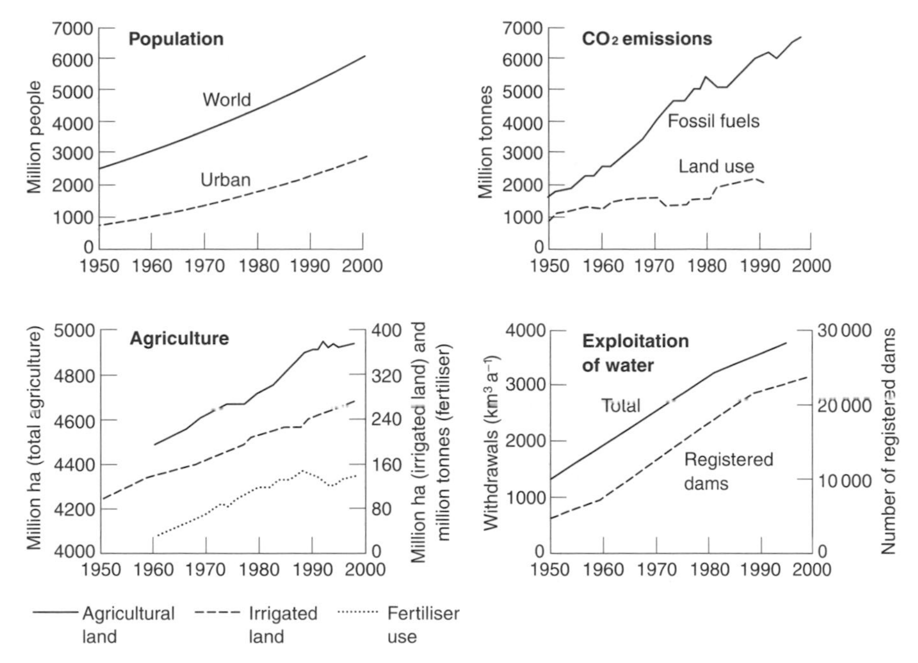

Figure 1.3 Indicators of human intervention: population, agriculture, CO_2 emissions and the exploitation of water between 1950 and 2000. Population and agriculture data from the FAO's FAOSTAT database (http://apps.fao.org), CO_2 emissions from fossil fuels from Marland *et al.* (1998), CO_2 emissions from land use change from Houghton (1999), water withdrawals from Shiklomanov (1998), and numbers of registered dams from the ICOLD database (www.icold.org). The CO_2 emissions data are available from http://cdiac.esd.ornl.gov

dramatically, and Figure 1.3 shows changes in a number of indicators of human activity since the middle of the twentieth century. These interventions are affecting hydrological systems in many catchments.

As Turner *et al.* (1990) explain, there are two components of human-induced global environmental change.[1] *Systemic* changes are operating at global scales. They include climate change due to the increasing concentration of greenhouse gases in the atmosphere, acid rain and depletion of stratospheric ozone. *Cumulative* change is the net effect of many different changes occurring at different spatial scales. Land cover change, for example, may be occurring in different places for many different reasons, although some of the economic drivers – such as globalisation – may be common. Figure 1.4 conceptualises the implications of global environmental change for the hydrological system. Changes in catchment land cover, the use of water in the catchment, and the

[1] It is important to note here that although the term "global environmental change" is rarely explicitly defined, in practice it is used in two different ways. One use of the term embraces *all* environmental change, for whatever reason, whilst the other restricts the term to cover just change caused by human activities. This book adopts the broader definition, because human-induced global environmental change must be seen in the context of "natural" environmental change.

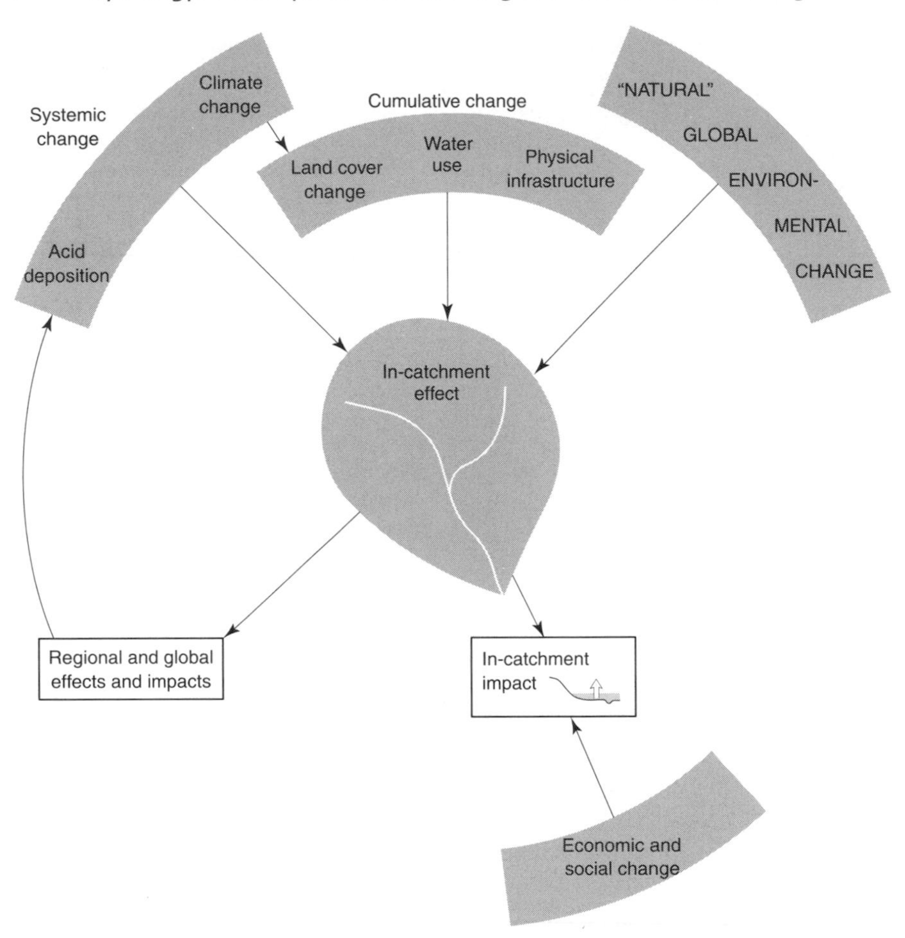

Figure 1.4 Global environmental change and the hydrological system

development of physical infrastructure in the catchment (such as regulating reservoirs and flood embankments) can be seen as cumulative global environmental change. Changes to the inputs to the catchment – in terms of the amount of water, energy and dissolved material – are due to systemic global environmental change. These two types of change affect hydrological processes within the catchment, but also feed through the linked earth system to affect other catchments. Changes in soil moisture, for example, due to changes in land cover, will affect the amount of water available to be evaporated back up into the atmosphere, and hence rainfall downwind. Changes in the output of water and material from a catchment to the sea affect coastal and possibly deep ocean processes, potentially affecting regional and global climate. *Global environmental change should not therefore be seen as just affecting the catchment: what happens in the catchment affects global environmental change.*

Both natural and human-induced global environmental change affect the hydrological processes and regimes within a catchment. The *impacts* of these changes, however, depend on the characteristics of the human use of the

Table 1.1 Example impacts of adverse hydrological change in the catchment

"Hydrological" change	Potential impact
Increase in flood magnitudes or frequency	• Flood damage and disruption • Increased channel erosion and downstream deposition • Increased risk of flushes of pollutant • Increased ill-health and disease
Reduction in low flows	• Increased risk of water shortage: drought • Reduced dilution of effluent • Smaller habitat area • Reduction in wetland extent • Siltation of river bed, with implications for ecosystems
Degradation of water quality	• Impacts on freshwater ecosystems • Contamination of water source/increased cost of water treatment • Ill-health and disease
Increased concentration of nutrients	• Eutrophication
Increased sediment concentration	• Deposition: siltation of river bed and blocking of navigation channels
Lowered groundwater levels	• Desiccation of wetlands • Reduction in flows (see above) • Saline intrusion into coastal aquifers

catchment and the ecological characteristics of the water environment. Table 1.1 summarises some of the potential human and environmental impacts of hydrological change: the list is not exhaustive. Many of these impacts are affected not only by hydrological change, but also by economic and social changes altering the exposure to environmental hazard. In the broadest terms, "hazard" is a function of both the physical environment and the human environment: *an increase in the impact of water-related hazards, such as flood, is therefore not necessarily due to global environmental change altering the physical environment.* It may be entirely due to increasing exposure to the hazard.

1.4 Hydrology in practice

Traditionally, hydrologists have taken a catchment perspective, seeing precipitation and energy as inputs and focusing on the characteristics of the output of streamflow. This largely reflects hydrology's disciplinary tradition as an applied science, aimed at developing techniques to estimate hydrological characteristics at a site as part of the process of water management. Much hydrology has therefore concerned itself with issues such as flood frequency estimation and the stochastic generation of data, central to the sound design of engineering works, and until at least the late 1970s most was done by hydrologists who trained as engineers and who were working to solve engineering problems.

Since the early 1980s the centre of gravity of hydrology has shifted towards the geosciences, and it has been forcefully argued that the role of hydrology is to solve the global water balance and understand how the hydrological system works (Dooge, 1986; Klemes, 1988; Chahine, 1992). This shift has been driven partly by a need to understand hydrological processes in order to manage some aspects of water more effectively (particularly water pollution, which requires an understanding of flow pathways), and partly by a need to understand how changes in the catchment – and global environmental change – affect hydrological systems. Hydrologists have therefore worked with meteorologists, atmospheric chemists, physicists, plant physiologists, ecologists, soil scientists, environmental chemists, geomorphologists, geologists and oceanographers – indeed the whole range of environmental scientists (including physical geographers).

Hydrological knowledge has advanced through a combination of field observation and experimentation, theoretical analysis and more recently computer modelling. The distinction between field observations and field experimentation is that in the latter the investigator intervenes to make something happen, in a controlled way, whilst in the former the investigator observes and records without intervention. Field experiments therefore tend to be very small-scale – perhaps just a few square metres – although there are some notable examples of catchment-scale experiments looking at the effects of different types of land cover change. Field observations are undertaken at scales ranging from a few square metres to several thousand square kilometres, although at this larger scale the measurements are taken by remote sensing and the definition of "field" becomes rather stretched. For approximately 100 years hydrologists have used research catchments to observe and study hydrological processes and regimes, and during the 1980s a new type of field study – known as meso-scale field experiments[2] – began to be undertaken. These meso-scale field experiments focus on the links between land surface and atmosphere, and the first few were not designed to measure streamflows at the surface: they were designed by atmospheric scientists, who did not think in catchment terms. Later experiments have involved close collaboration between atmospheric and "surface" scientists, enabling all aspects of the water and energy balance to be measured.

Theoretical analysis has concentrated on the development of equations describing the flow of water through saturated and unsaturated media, and the process of evaporation. These equations are based on the physics of the movement of mass and energy, and were generally based not only on theoretical reasoning but also in practice on detailed field observations and experiments. Prominent examples include Darcy's Law, which describes the rate of flow through a porous medium, and the Penman equation, which calculates the rate of evaporation from meteorological data.

Since the 1980s hydrologists have increasingly used computer simulation models both to understand how the hydrological system works and to predict

[2] These studies are not actually controlled experiments in the classical sense: they undertake observations in a coordinated and consistent manner.

what would happen if either the inputs to the catchment or the characteristics of the catchment itself were to change. A model can be defined as a simplified representation of reality, usually in numerical form, and in the broadest sense engineering hydrologists have been using "models" to estimate hydrological characteristics at sites with no data for decades. However, these engineering models have generally been empirically-based, constructed from relationships developed from observed data. Whilst these models may produce good estimates within the range of their calibration data, they do not necessarily give reliable predictions outside this range or if conditions in the catchment change. Neither do they provide insights into the way that hydrological system actually functions: empirical relationships do not imply causation. Hydrologists have therefore developed two different types of hydrological model (Beven, 2001). *Conceptual models* are based on a conceptual representation of a catchment, usually as a series of stores. The magnitudes of these stores and rate of outflow are determined by model parameters (Box 1.1), which generally must be calibrated using observed data. A "lumped" conceptual model treats the catchment as a single unit, with the same inputs and catchment characteristics across the

Box 1.1 Some basic modelling concepts

All models have parameters, which control model behaviour and which vary from catchment to catchment. Model parameters are estimated by calibrating the model with some observed data. This involves adjusting the parameters until the simulated model output approximates the observed data, either manually or by minimising an objective function which quantifies the difference between simulated and observed. In principle, a calibrated model should then be checked – or validated – against an independent set of data from the same catchment to assess the robustness of the estimated model parameters. In practice, several different parameter combinations can give equally good fits to a given set of observed data, and so techniques have been developed to estimate confidence intervals around model predictions based on the number of different feasible parameter sets (e.g. Freer *et al.*, 1996).

In broad terms, there are three main sources of error in model simulations. First, the model may be conceptually unsound. Second, the model parameters may be inaccurately estimated, perhaps because many different parameter sets produce similar simulations. Third, the input data may be inaccurate, or available at the wrong spatial or temporal scale. Attempts to simulate flood peaks on small catchments from monthly total rainfall over a large region would, for example, be doomed to failure. This last issue is very important for hydrological modelling: the spatial and temporal scale of the model must be consistent with the available input data, the amount of spatial variability in process, and the intended use of the simulations.

whole catchment, whilst a "distributed" conceptual model breaks the catchment into discrete units. *Physics-based models* are based on the theoretical equations of the movement of water through the catchment, and in principle have parameters which can be measured in the field and do not need to be calibrated (e.g. SHE: Abbott *et al.*, 1986 and TOPMODEL: Beven, 1997). Physics-based models are distributed, because the parameters of the underpinning theoretical equations vary rapidly over space, but in practice this variability is often greater than the feasible resolution of field data, so parameters are often calibrated. During the late 1990s many models were designed to be applied over a large geographic domain. These so-called "macro-models" were intended partly to simulate hydrological behaviour in many catchments at once, and partly as a contribution to global climate models, but all have the common characteristic that their parameters must be based on readily-available spatial databases (e.g. Abdulla *et al.*, 1996; Arnell, 1999a; Vorosmarty *et al.*, 1996).

The 1980s and particularly 1990s also saw a large increase in the use of remote sensing in hydrological science (Engman & Gurney, 1991). Remote sensing can be defined as the measurement of the attributes of a surface by a device separate from the surface. Air photography is an early example of remote sensing, but most hydrological applications are now based on the digital radiation and electromagnetic data collected by airborne and space-borne platforms. Remote sensing has three uses in hydrology: estimating the spatial extent of surface properties, estimating the magnitudes of hydrological stores, and estimating the fluxes of water between stores (Box 1.2). Remote sensing techniques

Box 1.2 Remote sensing and hydrological science

Some of the main contributions of remote sensing to hydrology are summarised below: more details can be found in later chapters.

- *Surface properties*: vegetation type and physiological properties can be inferred through remote sensing, and where vegetation is sparse it is also possible to infer soil and geological characteristics.
- *Store contents*: remote sensing has been used successfully to measure snow cover extent and the amount of water held as snow and the area of standing water. It is also possible to monitor levels of large lakes and rivers via satellite altimetry. Soil moisture contents can be inferred in some circumstances where vegetation is sparse, but it is relatively straightforward to estimate in qualitative terms the variation in soil moisture contents across space.
- *Hydrological fluxes*: weather radar is routinely used to estimate rainfall, and it is possible under some conditions to infer rainfall rates from satellite observations of cloud brightness or temperature. Evaporation can be inferred indirectly from measurements of surface temperature, again under some circumstances.

are still in the early stages of development, but offer the potential for measuring the components of the water balance, over space and over large geographic domains.

Yet another development during the last two decades of the twentieth century was a change in the way much hydrological research was conducted. Although the International Hydrological Decade between 1965 and 1975 stimulated many catchment studies in many countries, most of these studies were conducted in isolation, and usually exclusively by hydrologists. During the 1980s and 1990s a number of major international collaborative initiatives were developed, which were broad in concept, global in scale and multidisciplinary in nature. In hydrological terms, the two most important initiatives are the GEWEX project and the IGBP: these are summarised in Box 1.3, and Appendix 1 gives a list of acronyms. Unesco has coordinated hydrological research since the International Hydrological Decade through its rolling International Hydrological Programme (IHP). This mostly takes the form of review groups brought together to review progress in particular areas (such as the hydrology of the humid tropics), and Unesco has a strong capacity-building programme aimed at increasing hydrological expertise in the developing world. One very significant Unesco project, which unlike most of the others involves

Box 1.3 International hydrological research: GEWEX and the IGBP

GEWEX is the Global Energy and Water Cycle Experiment, set up under the World Meteorological Organization's World Climate Research Programme (WCRP). It has components dealing with clouds, atmospheric simulation models and oceans, together with a number of initiatives with a strong hydrological component. GEWEX Continental-Scale Experiments (including GCIP, MAGS, GAME and others) are studies looking at the water balance over a very large spatial scale. GEWEX studies are funded by individual countries, but are coordinated by international scientific steering committees to an agreed science plan.

The International Geosphere-Biosphere Programme (IGBP) is an initiative of the International Council of Scientific Unions, aimed at studying in a coordinated way the relationship between the geosphere and the biosphere. The part with the greatest hydrological component is BAHC (Biospheric Aspects of the Hydrological Cycle), which is concerned with the role of vegetation, particularly in influencing evaporation, but the IGBP also has programmes on land cover change (LUCC), past environmental changes (PAGES) and interactions between the land and sea (LOICZ) which have strong hydrological components. Individual research groups are funded by their own funding agencies, but as is the case with GEWEX, international teams work to an agreed science plan and are overseen by scientific steering committees.

international coordinated research, is FRIEND (Flow Regimes from International Experimental and Network Data), which is concerned with understanding regional hydrological behaviour and variability. It began in north west Europe, and there are now FRIEND projects in parts of every continent. The World Meteorological Organization (WMO) also has a hydrological research programme, but unlike Unesco's this is geared towards *operational* hydrology, as used by water managers. The WMO's hydrological programme therefore undertakes such activities as reviews of hydrological models or forecasting techniques. At the national level, NASA's Earth System Science research programme has stimulated multi-disciplinary research into the water cycle in the USA, as have the TIGER and LOIS programmes in the UK.

At the more individual and catchment level, the last decade of the twentieth century saw much closer collaboration in many countries between hydrologists and freshwater ecologists (e.g. Baird & Wilby, 1999), reflecting increasing concerns over the link between hydrological change and environmental degradation.

1.5 The structure of the book

The core of this book is divided into seven chapters, the first four of which lead in to the remaining three. Chapter 2 introduces some basic hydrological concepts and explores the global water balance. Chapter 3 looks at the components of the water balance, starting with precipitation, and following the water through interception, evaporation, infiltration into the soil, movement through the soil, recharge to groundwater, and the generation of streamflow. This chapter discusses in some detail the measurement techniques available to hydrologists, as these underpin enhanced understanding of the hydrological system. Chapter 4 reviews the variability of hydrological behaviour – primarily streamflow – over both space and time. Understanding this variability is essential before attempting to assess the effects of human changes to the hydrological system. Chapter 5 summarises the processes which determine the physical and chemical characteristics of rivers and groundwater, again emphasising the considerable variability in these characteristics over both time and space.

Chapter 6 explores the implications of changes in the catchment – cumulative global environmental change – for hydrological processes and regimes in the catchment. These changes are divided into changes in land use and land cover, changes in the use of water within the catchment, and changes in the physical characteristics of the river network and water storage. Chapter 7 examines the implications of changes in the inputs to the catchment – systemic global environmental change – focusing on acid deposition and climate change due to increasing concentrations of greenhouse gases. Both of these chapters concentrate on changes to hydrological regimes within the catchment. Chapter 8 looks at the way hydrological processes affect the atmosphere above the catchment and the sea downstream of the catchment, together with the implications for the atmosphere and coastal zone of hydrological changes within the catchment.

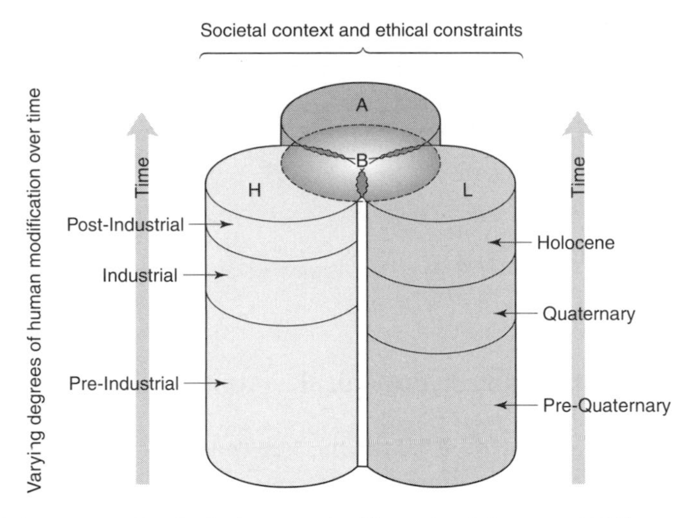

Figure 1.5 A structure for physical geography (Slaymaker & Spencer, 1998)

Figure 1.5 shows Slaymaker & Spencer's (1998) conceptualisation of a physical geography oriented around global environmental change and the concept of biogeochemical cycling. This book focuses on the hydrological column (H in Figure 1.5), but touches on the atmosphere (A: meteorology and climatology), the lithosphere (L: geomorphology) and the biosphere (B: ecology, plant physiology). It also encroaches into oceanography (subsumed within H in Figure 1.5), and draws a great deal from environmental chemistry (which cuts across H, A, B and L). The book contains many references to original research, most of which come from the key international hydrological journals (*Journal of Hydrology*, *Water Resources Research*, *Hydrological Sciences Journal*, *Hydrological Processes*, *Hydrology and Earth System Sciences*, and *Nordic Hydrology*): many are taken, however, from journals centred in other disciplines, ranging from *Marine Chemistry* to *Plant, Cell and Environment*. Many of the international programmes mentioned in the text have web-sites, and Appendix 2 gives a list of current (2001) web addresses.

Finally, it is important to define here what the book does *not* cover. The emphasis is very much on the hydrological aspects of global environmental change. These hydrological effects have consequences or impacts, for users of water and the water environment, as summarised in Table 1.1: they will not be discussed further in this book, and the reader is referred to texts such as Newson (1997). Whilst the effects of some management interventions to counter the impacts of human activities on the water environment are summarised briefly in Chapters 6 and 7, the bulk of the book does not discuss ways of managing water resources and ecosystems. The final chapter, however, concludes by discussing the broad-scale implications of global environmental change for the management of the water environment.

Chapter 2

The global water balance

2.1 Introduction to the hydrological system

The hydrological cycle (Figure 1.1) is the key concept at the heart of understanding the water balance. This chapter considers the global water balance, looking at global water fluxes and stores, and outlines the geographical variability in water fluxes. First, though, two introductory concepts need to be introduced.

The first is that of the *catchment* (also known as basin or watershed – although in British English watershed actually means the divide between catchments). A catchment is defined as the area from which rainfall flows into a river. Although there are no hard and fast rules, the term basin is usually applied to a "large" catchment. Catchments are conventionally defined in the field or on a map on the basis of topography, following the dictum that water flows downhill (Figure 2.1). Digital elevation models are increasingly used to define automatically catchment boundaries. However, this simple rule may not always be appropriate. *Groundwater* catchments are defined on the basis of the height of the water table – and the highest part of the water table may not coincide with the highest part of the surface topography. The true catchment area of many catchments in the chalk of southern England or northern France, for example, may be very different from the topographic catchment area. Patterns of regional groundwater flow may mean that water that infiltrates to an aquifer is transported for many kilometres underground before emerging in a river well away from the catchment that the rain originally fell into. The second exception to the general rule occurs where there may be isolated pockets of internal drainage within a catchment: areas into which water drains, but does not drain out. These tend to occur in dry regions, and the water flowing into the internal drainage basin is evaporated: the internal drainage basins may have been lakes in earlier, wetter, periods. The implication of these exceptions to the general rule is that it may be difficult to define in practice the area over which to calculate a water balance.

The second introductory concept is that of the *conservation equation*, which essentially defines the water balance in formal terms. In the most general terms, this states that the time rate of change of the mass of a conservative quantity stored is equal to the difference between the inflow rate and the outflow rate. Water is a conservative quantity, and given that the density of water can be taken as approximately constant, the conservation equation can be expressed

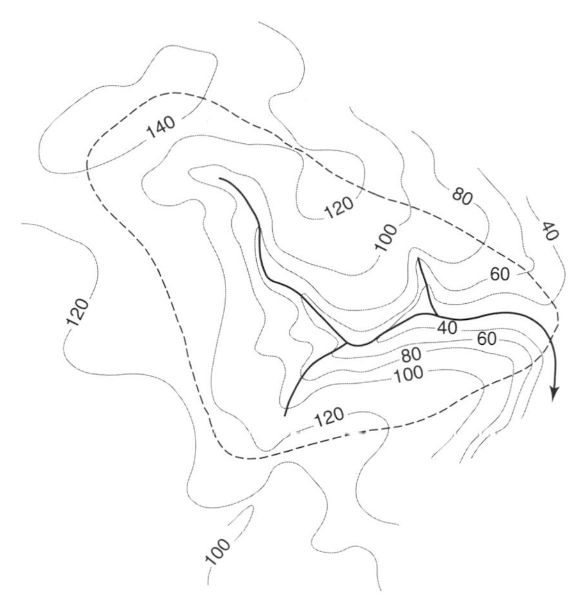

Figure 2.1 The catchment boundary

in terms of volumes of water. In mathematical terms, the general conservation equation can be stated as:

$$\frac{\mathrm{d}V}{\mathrm{d}t} = I - O \qquad \textbf{2.1}$$

where V is volume, I is inflow, O is outflow and t is time, and in hydrological terms this can be written as:

$$\frac{\mathrm{d}S}{\mathrm{d}t} = P - E - Q \qquad \textbf{2.2}$$

where S is the volume of water in a store, P is precipitation falling onto that store, E is evaporation from the store and Q is discharge from the store. In practice, a catchment has a large number of stores – introduced in Chapter 3 – and the water balance equation gets considerably more complicated.

2.2 The global hydrological cycle: budgets and fluxes

Table 2.1 gives estimates of the volumes of water held in the various stores, at the global scale. By far the largest volume of water is held in the oceans, and three-quarters of freshwater is held in ice (largely in Greenland and the Antarctic). Most of the rest is held in groundwater, and only a small portion of this is actually close to the surface and therefore accessible by plants and people. Virtually all groundwater, at whatever depth, is "meteoric", or in other words composed of rainfall that has percolated downwards, often in periods in

Table 2.1 Estimated storages of water

Store	Volume (thousand km^3)
Oceans	1 320 000–1 370 000
Atmosphere	13–14
Land	
Ice caps and glaciers	24 000–29 000
Groundwater	8000–60 000
Freshwater lakes	91–125
Saline inland seas	85–104
Soil water	16.5–85
Water stored in permafrost	300
Rivers	1.2–2.1
Swamps and marshes	11.5
Biological water	0.6–1.1
Total land	*36 000–84 400*

Source: Data tabulated by Gleick (1993)

the past wetter than the present. Only a very small proportion of groundwater is derived from "connate" water, or sea water trapped in sediments at the time of their deposition (Ward & Robinson, 2000). Far less water is stored in freshwater lakes – including such large water bodies as the Great Lakes of North America, Lake Baikal and the lakes of east Africa – than is held in groundwater. "Biological water" is that stored by vegetation.

Despite the apparent precision of some of the numbers in Table 2.1, these estimates are very uncertain because of a lack of widespread measurements. The volume of water held in ice caps, for example, is calculated from the area and thickness of ice, but the thickness of ice caps is not well measured. More significantly, the estimated volume of groundwater storage is very uncertain, because it is difficult both to measure water storage at depth at any point and to extrapolate from observation boreholes: estimates range from between 8 and 60 million km^3, although most are at the lower end of this range.

Figure 2.2a shows the average annual fluxes of water from one store to another, *at the global scale*. Like the storage estimates, these values are subject to considerable measurement error, and in fact some of the terms have been inferred from the others or adjusted to complete the water balance. Precipitation over land has been estimated from networks of raingauges (but see Section 3.3), and estimates of precipitation over the sea have been derived from very sparse monitoring networks. Runoff from the land to the sea has been determined from observed streamflow data from major river basins. This runoff includes all water that reaches the river channel, either rapidly or slowly after filtering through aquifers, and also groundwater that discharges directly to the sea. The other fluxes are inferred: total evaporation from the land is the difference between precipitation and runoff, and the evaporation from the ocean must be equal to ocean precipitation plus runoff from the land. The net flux

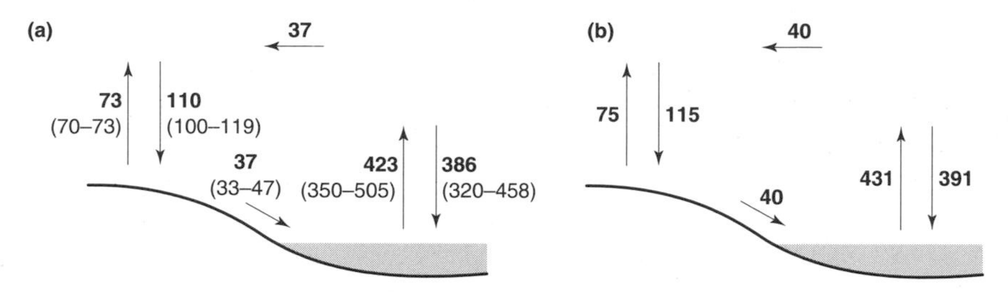

Figure 2.2 Average annual water fluxes (a) adapted from Berner & Berner (1987), with ranges in parentheses from data compiled by Gleick (1993); (b) adapted from Oki (1999)

in the atmosphere from ocean to land must be equal to total runoff from the land to the ocean, but can also be inferred from estimates of water vapour fluxes in the atmosphere made by aerological measurements taken from rawinsondes (unmanned weather balloons), as suggested by Rasmussen (1977) and done by Bryan & Oort (1984). Note that the estimates for the fluxes over the ocean have the greatest uncertainty.

Since the late 1980s, however, it has been possible to refine these global water balance elements in two main ways. First, satellites have been used to estimate precipitation over both land and sea (Arkin & Xie, 1994; Section 3.3). Second, numerical weather prediction (NWP) models have been used to simulate the components of the energy and water balance, including precipitation, evaporation and atmospheric moisture fluxes. These simulations are, of course, only as good as the numerical weather prediction model, but some of the model limitations can be tackled by combining simulated and observed data. This process (known as 4-dimensional data assimilation, or 4DDA) involves feeding observations made at fixed points in time and space into the model as it simulates weather, effectively in order to re-initialise it in real-time: the observed data constrain the model from straying too far from reality, although the results are still sensitive to the characteristics of the NWP and how the data are averaged (Trenberth & Guillemot, 1995: Chapter 8 looks at how hydrological processes are incorporated into weather prediction models). Oki (1999) used a combination of observed precipitation data (from ground-based and satellite measurements) and model output to estimate the global-scale water fluxes over the period 1989 to 1992. His estimates are shown in Figure 2.2b, and are close to the values estimated by Berner & Berner (1987).

There are, of course, very considerable variations in these fluxes across the world. Table 2.2 gives one estimate of average annual continental precipitation, evaporation and runoff (and, by inference, the net continental flux in the atmosphere is equal to runoff). The precipitation is estimated from networks of precipitation gauges, runoff is estimated from data observed at the downstream ends of major river basins, and evaporation is calculated as the difference between them.

Colour plate Ia shows average annual precipitation over land, averaged over the period 1961 to 1990. There are very considerable variations over the continents, with particularly high values in South America, equatorial Africa

Table 2.2 Estimated continental water balances

	Precipitation ($mm\ a^{-1}$)	Runoff ($mm\ a^{-1}$)	Evaporation ($mm\ a^{-1}$)	Runoff as a percentage of precipitation
Africa	740	153	587	21
Asia	740	324	416	44
Australia	791	280	511	35
Europe	790	283	507	36
North America	756	339	418	45
South America	1595	685	910	43

Source: Adapted from Korzun (1978)

and south east Asia, as well as in areas with high altitude. Plate Ib shows estimated average annual runoff across the land surface of the world. Average runoff varies from less than 25 mm a^{-1} in parts of North America, southern South America, Africa, Australia and central Asia, to over 2000 mm a^{-1} in South America and south east Asia. It is very difficult to measure evaporation routinely (Section 3.4), and Plate Ic plots evaporation over land as the residual between annual precipitation and annual runoff. Except in circumstances where water is imported to a wetland or lake from upstream (Chapter 8), or where deep-rooted vegetation is planted which taps stored water (Chapter 6), the annual evaporation from land is always less than the annual precipitation. Colour plate Id shows the percentage of annual precipitation that becomes runoff: the ratio varies from less than 10% to over 90%.

The spatial distribution of precipitation, evaporation and runoff can also be estimated by analysis of *atmospheric* moisture data. For a vertical column of atmosphere above a defined portion of the earth's surface, the change over time in the amount of water vapour stored in the atmosphere (dW/dt) is equal to the amount of water evaporated from the surface (E) minus the precipitation (P) out of the atmosphere, minus the net horizontal flux of water out of the column in the atmosphere (denoted by $\nabla\vec{Q}$ – pronounced "div Q" – where $\vec{Q}$ is the vertically integrated two-dimensional water vapour flux, and ∇ is a mathematical operator used in vector geometry: $\nabla\vec{Q}$ essentially measures the difference between the amount of water entering the column from upwind and the amount leaving it downwind):

$$\frac{dW}{dt} = E - P - \nabla\vec{Q} \qquad \textbf{2.3}$$

$\nabla\vec{Q}$ is termed the vapour flux divergence. A positive $\nabla\vec{Q}$ denotes that the region is a source of moisture, whilst a negative $\nabla\vec{Q}$ shows that the region is a sink of moisture: it is a region of *convergence*, where more moisture enters via the atmosphere than leaves. Of course, this surplus water must go somewhere, and it must either be stored in snow, ice or groundwater, or leave the region as river runoff. If there is no net change in these stores, $-\nabla\vec{Q}$ is therefore numerically equal to runoff, as can be seen by comparing equations 2.2 and 2.3. A map of $-\nabla \vec{Q}$ (or vapour flux *convergence*) would be equivalent to a map of runoff. Vapour flux divergence/convergence can be measured from rawinsondes (Starr & Piexoto, 1958) or simulated by numerical weather prediction models (e.g. Trenberth & Guillemot, 1995 and Oki *et al.*, 1995). Maps of vapour flux convergence have been produced at global (Oki, 1999) and smaller scales (Oki *et al.*, 1995), although validation exercises where runoff data are available have found very large differences in some areas, reflecting both errors in the numerical weather prediction model used with the observed data and problems associated with averaging over the time and space scales of the model.

Colour plate II shows the estimated components of the water balance over the entire globe. Precipitation (Plate IIa) is estimated from a combination of ground observations and satellites (as compiled by the Global Precipitation Climatology Project), and can be very much higher over oceans than land. Precipitation is in excess of 3000 mm a^{-1} across the equatorial Pacific and Indian oceans, and is also high in the western Pacific (off Japan) and the western Atlantic (off eastern North America). Vapour flux convergence (Plate IIb) is calculated from NWP estimates of vapour fluxes (Oki, 1999): positive values in Plate IIb represent an excess of precipitation over evaporation, and negative values an excess of evaporation over precipitation. The vapour flux convergence should be positive everywhere over the land surface (except in small areas as noted above). The maps of evaporation (Plate IIc) are calculated from NWP estimates (Trenberth, 1998), and show very high values in the tropical Atlantic and Indian Oceans in particular.

Figure 2.3 shows the meridional distribution (i.e. averaged over latitude bands) of annual average precipitation, annual average vapour flux convergence and annual average evaporation (calculated by subtracting precipitation from vapour flux convergence). Over land, the vapour flux convergence is equal to runoff. There are a few zones where "runoff" or evaporation is negative, and

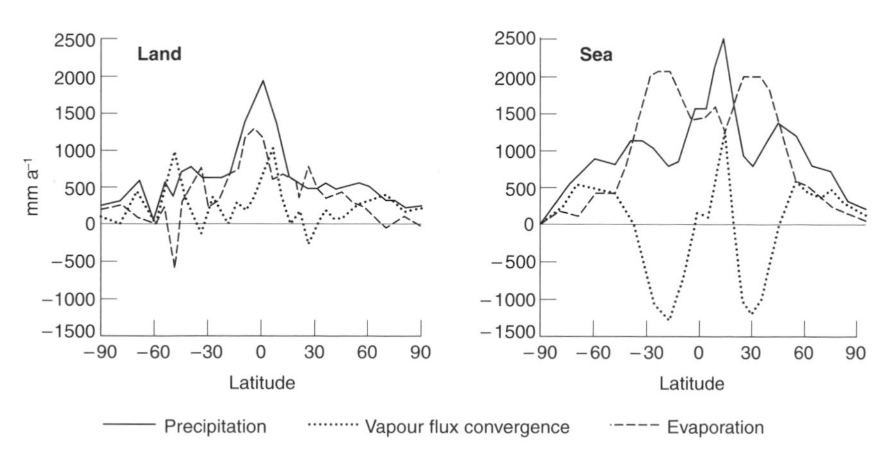

Figure 2.3 Meridional distribution of precipitation, vapour flux convergence (runoff) and evaporation: 1989–92 (Xie & Arkin, 1996; Oki, 1999)

this reflects model errors. The highest runoff is at around 5°N, corresponding to the Amazon basin and south east Asia. The fluxes over the oceans show a smoother progression from south to north, and vary much more than those over land. The oceans are a net source of moisture to the atmosphere in the subtropics (between 40 and 10°S and 10 and 40°N), but receive more in precipitation than they provide in evaporation in mid latitudes and around the equator.

2.3 An overview

This chapter has introduced the broad components of the water balance. The oceans cover 79% of the world's surface, but generate around 85% of the world's precipitation: there is therefore a net movement of water in the atmosphere from sea to land, which is balanced by river, groundwater and glacier discharge to the sea.

However, the magnitudes of both the fluxes and the stores of water are not well quantified, largely due to major gaps in data-observing networks, particularly over the oceans: satellite remote sensing is plugging some of these gaps. A clear challenge to earth system science is therefore to improve estimates of the components of the global energy and water cycle, at global and continental scales, and indeed this is the key aim of the World Meteorological Organization's GEWEX (Global Energy and Water Cycle Experiment) programme.

Chapter 3

Components of the water balance

3.1 Introduction

The previous chapter has introduced some of the patterns and dynamics of the global water cycle. It has touched upon some measurement issues. This chapter explores the components of the water balance, but can only provide a generalised introduction to the key concepts: more details and examples can be found in the cited papers. Understanding of the catchment water balance is largely based on field observations, and it is therefore important to appreciate how these observations are made and where the major measurement errors and problems lie. Measurement techniques are therefore described in boxes.

3.2 Precipitation

Types of precipitation

Precipitation is defined as the condensation of water vapour into liquid water droplets and ice particles that fall to the earth's surface. *Rain* is precipitation that is formed by water vapour condensing at temperatures above freezing or by ice crystals thawing before they reach the ground: *snow* is formed when ice crystals do not thaw. Rain and snow are the most important types of precipitation in hydrological terms. *Hail* is created when raindrops are carried further aloft by updrafts within a cloud into air below 0 °C and freeze: although hailstorms are damaging, the hydrological effects are very similar to those of heavy rainfall. Other types of precipitation include *dew*, where water vapour condenses directly onto a cold surface, and *fog-drip* (or occult precipitation), where water is "harvested" directly from fog or low-lying clouds by vegetation. Dew is important for some plants but not significant hydrologically, but fog-drip can introduce water into the soil and may, under some circumstances, generate streamflow (as shown by Gurnell (1976) in the New Forest, England). This section focuses on rain and snow. Box 3.1 summarises methods used to measure and estimate precipitation.

The formation of precipitation

The amount of water in the air is measured by the *vapour pressure* (technically, the partial pressure exerted by the molecular motion of water vapour), and the

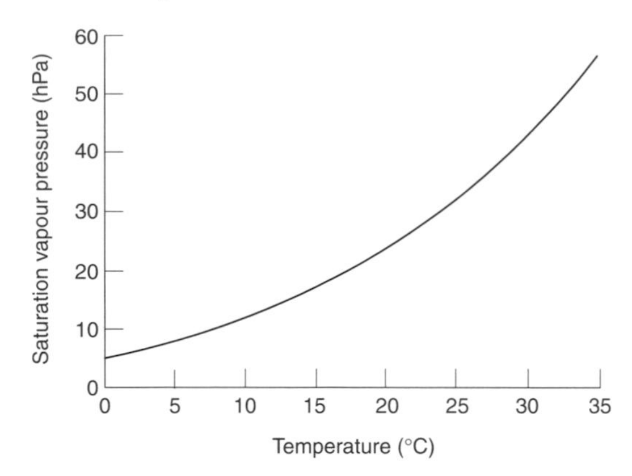

Figure 3.1 Relationship between temperature and saturation vapour pressure

maximum amount of water vapour – the *saturation vapour pressure* – that can be held in a parcel of air is a function only of temperature. Warm air can hold more water than cold air. The relationship between temperature and saturation vapour pressure is shown in Figure 3.1, and can be approximated by the following equation:

$$e_s = 6.11 \exp\left(\frac{17.3T}{T + 237.3}\right) \qquad \mathbf{3.1}$$

where T is temperature in degrees Celsius and e_s is saturation vapour pressure in hectopascals (hPa). The temperature to which a parcel of air with a given vapour pressure must cool in order for it to become saturated is termed the *dew point*. If the air cools below dew point, condensation occurs around particles in the air known as *cloud condensation nuclei*. These particles can be dust or ions, and major sources are wind-blown dust, volcanic material, smoke particles and sea salt: sulphur and nitrogen ions can also produce compounds that act as cloud condensation nuclei (and trigger acid precipitation: Chapter 7). In principle, if there are no cloud condensation nuclei, there will be no condensation and air can become "supersaturated". In practice, however, this is rare. At the other extreme, water molecules are especially attracted to some nuclei (such as sea salt), termed hygroscopic nuclei, and condensation can occur when the air is not saturated.

Condensation alone, however, does not necessarily generate precipitation because the droplets are too small (Barry & Chorley, 1998). Some mechanism must therefore generate drops large enough to overcome updrafts in the cloud and fall to earth. There are two main possible mechanisms, one operating in "cold" clouds and the other in warmer clouds. The Bergeron–Findeisen theory (Barry & Chorley, 1998) is based on the observation that the saturation vapour pressure over a water surface is higher than over an ice surface (and in fact Figure 3.1 shows the temperature/saturation vapour pressure relationship over water). If a cloud is cold enough to contain ice crystals, then water will condense onto these crystals at temperatures above dew point and these crystals will continue to grow. As they collide, ice crystals aggregate to produce still larger crystals. Central to the formation of such ice crystals is the presence of *freezing nuclei.* Although there is a good deal of uncertainty, very fine soil particles are probably the major source, but biogenic aerosols emitted by decaying plant litter may also contribute. Ice crystals fall to the ground once they reach a certain size, and if they melt before reaching the ground become rain. The second means by which raindrops can form is through coalescence. This occurs as large raindrops fall towards the ground, sweeping smaller drops into their wake. The process requires large condensation nuclei (such as salt particles) and can occur when cloud temperatures are above freezing.

Precipitation, therefore, is largely triggered by cooling of an air mass containing water vapour. There are three main ways in which air can rise and therefore cool: convergence, convection and orographic uplift. When different air masses converge from several directions, air is forced to rise. *Frontal* (or *cyclonic*) *convergence* occurs at the boundaries of air masses with different temperature (and hence density) and humidity, and is characteristic of mid latitudes. Figure 3.2a (Barry & Chorley, 1998) summarises the typical development of a depression (also termed an extratropical cyclone), forming at the boundary between a warm moist air mass and a colder, drier, air mass. As the wave develops, the warmer air rises and pressure at the centre of the depression falls. Figure 3.2b shows the pattern of precipitation associated with the two fronts in a depression (but note that under some circumstances the warm air may actually sink relative to either frontal surface, producing much less precipitation). Mid-latitude depressions are typically 1500–3000 km in diameter, and have a life span of 4 to 7 days (Barry & Chorley, 1998). The main regions of frontal development are the polar fronts (between 35 and 50°N and 30 to 45°S) – the Atlantic Polar Front produces the depressions bringing precipitation to western Europe and the Pacific Polar Front creates depressions bringing precipitation to the north west of North America – and the Arctic fronts, associated with snow and ice margins at high latitudes. The Mediterranean Front is a section of the Atlantic Polar Front which develops only during winter bringing frontal precipitation to the Mediterranean.

Non-frontal convergence occurs in the tropics within a mass of warm, moist air. The zone where the northern and southern hemisphere wind systems converge, and air rises, is termed the Inter-Tropical Convergence Zone (ITCZ). The ITCZ moves through the year, following the thermal equator, so is at its northern-most extent in July and at its southern limit in January. It produces the equatorial band of heavy precipitation clear in Plate Ia, and its movement defines the duration of wet and dry seasons. Non-frontal convergence also occurs in tropical cyclones and hurricanes. These intense cyclones are not associated with weather fronts, and are initiated by a small low-pressure disturbance in a maritime tropical air mass: they are fed by evaporation, and the condensation of the evaporated water gives them energy. Like the frontal cyclones, tropical cyclones produce precipitation over a very large area.

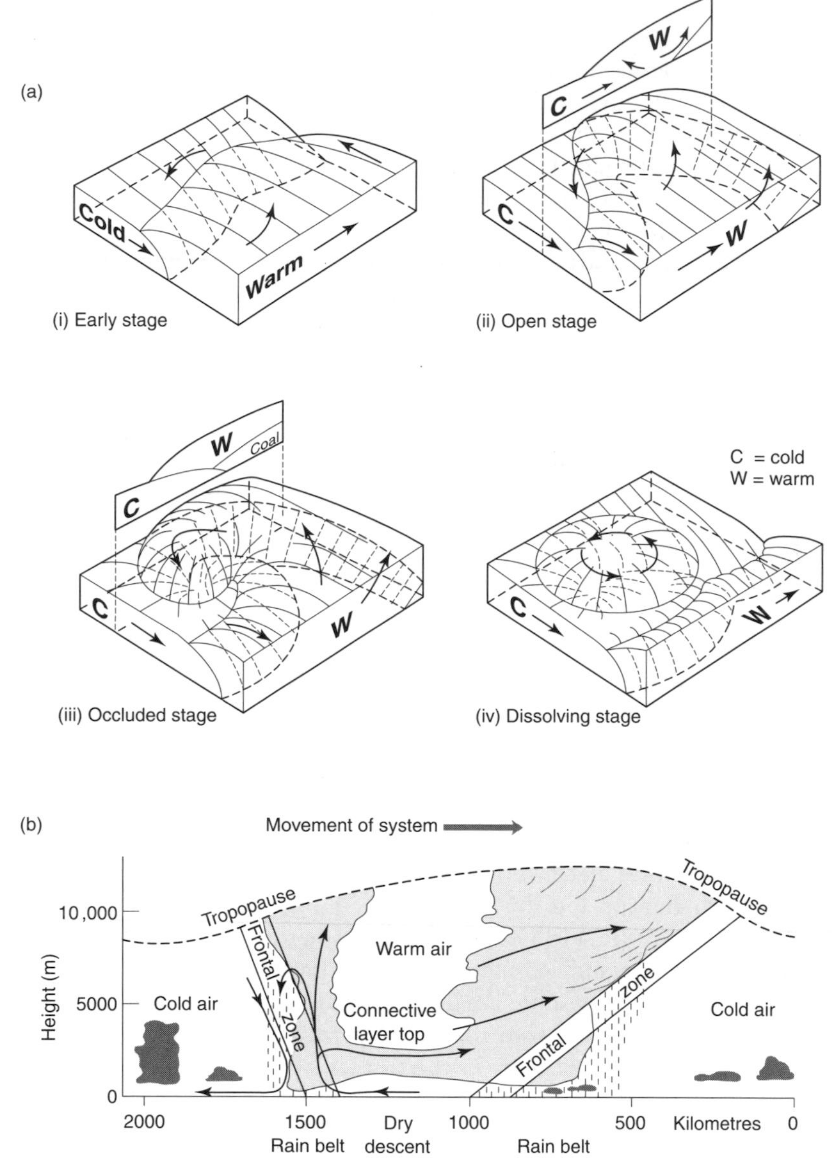

Figure 3.2 (a) Stages in the development of a mid-latitude depression (Barry & Chorley, 1998). (b) Cross-section of a mid-latitude depression, where the warm air is rising relative to each front (after Barry & Chorley, 1998)

Convective cooling occurs when a column of air is heated from below and rises. As air rises, it expands, and heat energy per unit volume – and hence temperature – falls. This is termed adiabatic cooling: heat has not been taken away from the air, but simply distributed amongst a greater volume of air. Dry air cools at a rate of 1 °C per 100 m as it rises (the dry adiabatic lapse

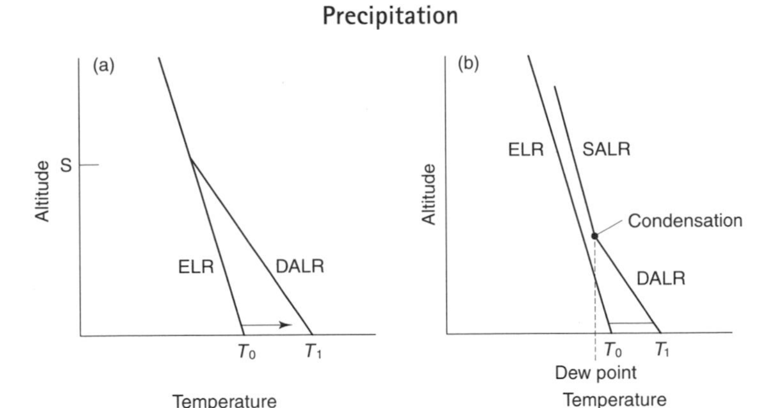

Figure 3.3 The generation of convective precipitation

rate, DALR), but moist air cools at a lower rate of around 0.5 °C per 100 m (the saturated adiabatic lapse rate, SALR), as latent heat is released by condensation and slows the rate of cooling. The environmental lapse rate describes the actual reduction in temperature *of still air* with altitude, and this varies depending on local air mass conditions. Figure 3.3 summarises what happens when the air at the surface is heated by conduction from temperature T_0 to T_1, and rises: the equilibrium lapse rate is shown by the line ELR. The air first cools, at the DALR. If the air contains little moisture, it may continue to cool at the DALR until it reaches the temperature of the surrounding air at S, and ceases to rise and cool further: the atmosphere stabilises (Figure 3.3a). There is no precipitation under these circumstances. If, however, the air reaches dew point before it reaches the temperature of the surrounding air, the rate of cooling will reduce to the SALR, and the air will continue to rise (Figure 3.3b). Uplift and condensation will continue, often resulting in significant precipitation. The amount of convective precipitation that falls is therefore dependent on the amount of warming at the surface, the amount of water in the atmosphere, and the local environmental lapse rate. Convective precipitation is localised (typically only covering a few square kilometres) and often intense. In some parts of the world – such as the tropics and also the Great Plains of the USA – clusters of convective storms may develop.

Orographic cooling occurs when an air mass is forced to rise by a topographic obstacle, such as a mountain range. In practice, of course, orographic effects are a result of convective and convergence processes interacting with topography. The "pure" orographic effect results in precipitation falling at higher elevations when it is not lower down. If precipitation frequency is the same at higher elevation, but intensities are higher, then the orography is enhancing precipitation triggered by other mechanisms. In western Britain, the number of days with rainfall varies little between the coast and high altitude, implying that orography enhances, and does not initiate, rainfall. Browning & Hill (1981) calculated the mean orographic enhancement of precipitation over England and Wales, over several days with different directions of weather systems. They showed that the effects varied with the direction of the air mass, with the greatest effects caused by upland areas perpendicular to the direction of the air mass. Orographic effects enhanced precipitation in an event by between 2 and 4 mm h^{-1} in upland Wales, and even by up to 2 mm h^{-1} over the South Downs in southern England (only around 150 m above sea level). Similarly in Scotland, Weston & Roy (1994) showed how orographic enhancement of rainfall differed between different wind directions. Both the "pure" and enhanced orographic effects account for much of the spatial variability in event and longer-term average rainfall. Dingman *et al.* (1988), for example, found that elevation accounted for 78% of the variation in average annual precipitation across New Hampshire and Vermont. Singh & Kumar (1997) showed a very strong association between elevation and rainfall in the western Himalaya. Rainfall gradients on the windward side were around 106 mm per 100 m, but only 13 mm per 100 m on the more sheltered leeward side. In south western British Columbia, however, precipitation increases with altitude up to around 400 to 800 m, and then remains reasonably constant (Loukas & Quick, 1996), implying that the increase with altitude is not consistent.

Spatial structure in precipitation patterns

From the above, it is clear that different mechanisms of precipitation generation produce different spatial patterns of precipitation. Frontal rainfall, for example, tends to occur over a large geographical area, following the path of a depression. Within the broad area of rainfall, however, there are smaller-scale patterns. Some of these may be orographically-induced (as shown in south east England by Browning & Hill (1981) and Kitchen & Blackall (1992) for example), but others reflect the characteristics of the rainfall process. Although frontal rainfall, for example, falls over several tens of square kilometres over a few hours or days, in detail it tends to occur in short-lived, localised higher-intensity cells (Austin & Houze, 1972). These cells may cover one or two square kilometres, and last for as little as 10 to 20 minutes (Shaw, 1983). A rainfall event at a particular location, therefore, may consist of several "pulses" of rainfall. Models that simulate rainfall patterns over space and time do so by simulating the locations, spatial extent, duration and total amount of rainfall of such cells (e.g. Amorocho & Wu, 1977; Cowpertwait, 1991; Khaliq & Cunnane, 1996; Onof & Wheater, 1996). More localised convective rainfall events also show small-scale variability, unrelated to topography and determined by processes within the cloud generating the precipitation. Goodrich *et al.* (1995), for example, studied variation in rainfall over a 4.4 ha area in south east Arizona, showing variations in total event rainfall of between 4 and 14% over distances as small as 100 m. Individual events covered areas as small as 300 × 300 m.

Where does the moisture come from?

For precipitation to occur, the atmosphere obviously needs to contain moisture. Until the late 1930s, it was widely believed (according to Benton *et al.*,

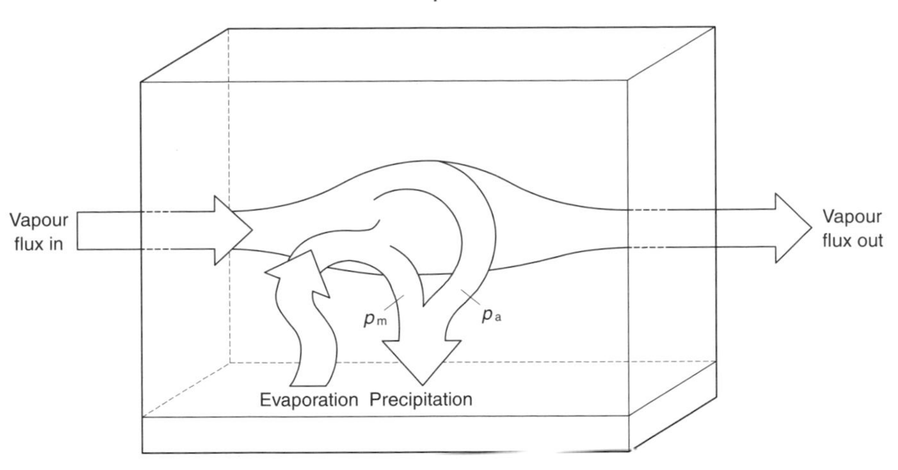

Figure 3.4 Conceptual model of atmospheric moisture recycling (Brubaker *et al.*, 1993). P_a is the "imported" precipitation, and P_m is the locally-generated precipitation

1950) that water precipitating over a continent was directly derived from evaporation from the continent (although it has long been realised that moisture over the maritime fringes of a continent – such as the British Isles – comes from upwind oceans). Benton *et al.* challenged this perception, noting that moisture added to the atmosphere by evaporation may travel large distances before being precipitated. They estimated that in fact only around 10% of the precipitation falling in the Mississippi basin came from within the catchment, and that the remaining 90% came from elsewhere. Budyko (1974) calculated that only 10% of the precipitation in the European part of the former Soviet Union was of local origin, ranging from 4% in October to 18% in May. On the other hand, some land regions can be significant sources of moisture in some seasons (Starr & Piexoto, 1958; Chapter 8).

Figure 3.4 (Brubaker *et al.*, 1993) summarises in a diagrammatic form atmospheric moisture fluxes over a land region. Moisture can be advected in from upwind (and from more than one direction), and can be generated by local evaporation. This moisture falls as precipitation over the region, or is advected out of the region. Figure 3.4 is in a sense a graphical representation of equation 2.3, with $-\nabla\vec{Q}$ equal to advection in minus advection out.

The proportion of precipitation that is generated within a region can be estimated using real or modelled atmospheric moisture data. Figure 3.5 shows the estimated average proportion of precipitation generated in each month from local evaporation in four regions (Brubaker *et al.*, 1993). On average, 11% of the precipitation of European Russia is generated from local evaporation, compared to 23% in the Mississippi basin (broadly defined) and 24% over the Amazon: 31% of the precipitation in the west African region is locally generated. There are clear variations through the year in each region, largely reflecting the amount of water available for evaporation at the land surface: again, this will be returned to in Chapter 8.

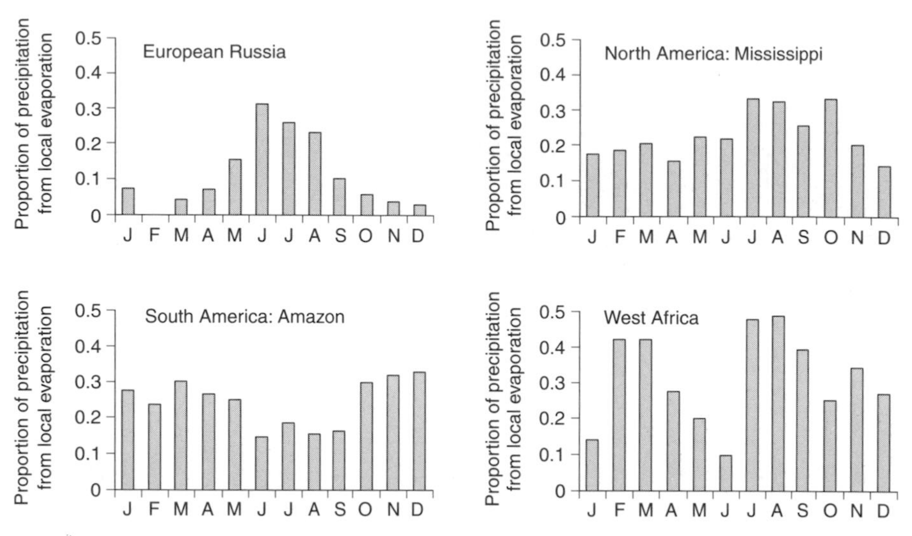

Figure 3.5 Proportion of monthly precipitation generated from within-region evaporation, for four large regions (Brubaker *et al.*, 1993)

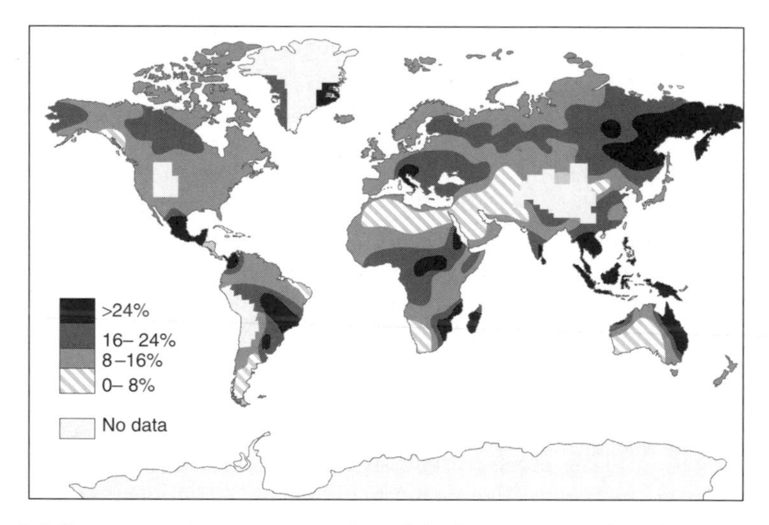

Figure 3.6 Percentage of average annual precipitation generated from local evaporation, 1979–95, over regions of 1000 × 1000 km (Trenberth, 1998)

Figure 3.6 (Trenberth, 1998) shows the global pattern of the percentage of precipitation generated by local evaporation, for regions 1000 × 1000 km (1 million km^2, or approximately 10° longitude × 10° latitude). On average, approximately 15% of land precipitation at this spatial scale was generated from local sources, with higher percentages in South America south of the Amazon basin, eastern Siberia and south east Asia (although large parts of the 1000 × 1000 km regions here are sea). In western Europe, it appears that less than 10%

of precipitation is locally generated. Brubaker *et al.*'s (1993) estimates for the four sample regions are slightly higher than those implied in Figure 3.6, and this is because their regions are two to three times the size of the 1000 × 1000 km regions used to produce the global map. As a general rule, the larger the region, the greater the proportion of precipitation that is locally generated. Trenberth (1998) showed, for example, that at a spatial scale of 500 × 500 km (or 0.25 million km^2), approximately 9% of precipitation over land regions was generated locally, and only 5% of Amazon precipitation and 7% of Mississippi precipitation was of local origin.

The advected moisture which generates most precipitation in most parts of the world largely comes from the oceans. European precipitation, for example, largely derives from the Atlantic Ocean, whilst moisture in North America mostly comes from the Gulf of Mexico (Brubaker *et al.*, 1994), with smaller influxes from the Pacific North West.

Box 3.1 Measurement of precipitation

Rainfall is, in principle, perhaps the easiest component of the water balance to measure, and raingauges have been used to record rainfall totals for close to 2000 years. Designs have obviously evolved over this time, and different national meteorological organisations use raingauges of different sizes and designs: unfortunately, this can make comparisons between data from different countries rather difficult. The simplest *non-recording* raingauge is a cylindrical container of defined height with an orifice of a defined size, emptied and measured at fixed intervals (usually daily, and generally at 9am in the morning local time). Such gauges therefore measure accumulated rainfall over the collecting period. *Recording* raingauges automatically measure rainfall at temporal resolutions of 1 minute or even shorter. There are three main types of recording gauge: weighing gauges (which continuously weigh the water held in the gauge), float and siphon gauges (where the depth of water held in the gauge is continuously measured using a float resting on the water, and the gauge is emptied automatically once a certain amount of water is stored), and tipping-bucket gauges (in which rain water fills and tips a small bucket of known capacity – often around 0.2 mm – and the times of the tips are recorded). Virtually all raingauges, however, are subject to measurement error, the two main sources of which are evaporation from the gauge and, more importantly, turbulence around the gauge. The higher the raingauge sticks up off the ground, the greater the turbulence induced and the likelihood of the gauge under-catching rainfall, particularly at high windspeeds. One response to this is to shield the gauge, either by some form of fence or turf wall, or to bury the gauge at ground level (ensuring that water does not run across the land surface into the gauge by placing it at the centre

(continued)

(continued)

of a grid). Another response is to use an aerodynamically-shaped raingauge, designed to minimise turbulence. Comparisons between different gauge types (e.g. by Rodda (1967), Essery & Wilcock (1991), Sevruk (1987), Rodda & Smith (1986)) suggest that standard raingauges can underestimate rainfall by up to 20%.

Snowfall is measured on the ground in two main ways. One measures the snow caught and melted in a gauge, but snow is even more prone to underestimation than rain as it is more easily blown around a gauge. Standard raingauges do not estimate snow well, so specialised gauges, usually with a larger orifice and shielding, are used in snow-affected regions. The other way of estimating snowfall is to measure the amount of snow lying and convert it into volume of water. This can be done by snow pillows, which continuously weigh the snow lying on top of them, or by manually measuring snow depth at defined locations. Snow density, however, varies over time, so the conversion of snow depth to water requires frequent calibration.

In practice, hydrologists need information on the rain and snow falling over a catchment area, not just at a point. It is therefore necessary to interpolate between ground-based point measurements, and there are two key issues. The first relates to the location and density of the sample sites (see Bras & Rodriguez-Iturbe (1976) and Pardo-Iguzquiza (1998) for overviews of statistical techniques which can be used to tackle the "network design" problem). As a general rule, the greater the accuracy of estimate required and the greater the spatial variability in rainfall, the denser the network needs to be. In practice, however, network density is usually constrained by practical considerations. More important, therefore, is the siting of the individual gauges, which should ensure that the full range of spatial variability is covered and that an accurate estimate of total catchment precipitation can be made. One approach often used in designing networks for experimental catchments (see Section 6.2) is to divide the study area into domains which are expected to show similar behaviour and adopt a stratified network design sampling each domain (e.g. Clarke *et al.*, 1973). The second key issue relates to the techniques used to interpolate between the observations. Techniques range from manually drawn isohyets plotting contours of equal rainfall, through Thiessen polygons (e.g. Damant *et al.*, 1983) and objective statistical interpolation methods (such as trend surface analysis and inverse distance interpolation) to kriging (e.g. Carroll & Cressie, 1996; Phillips *et al.*, 1992). Dirks *et al.* (1998) compared different interpolation techniques on Norfolk Island (35 km^2) in the South Pacific, showing that inverse distance interpolation was marginally most efficient over this small area. The strong relationship between precipitation and elevation suggests that elevation needs to be considered explicitly in interpolation schemes,

(continued)

(continued)

particularly where the area is large or variation in elevation is great (Chua & Bras, 1982; Goovaerts, 2000).

Two newer precipitation measurement techniques – radar and satellite estimation – however, make spatially distributed precipitation estimates. Radar can, in principle, provide estimates of rainfall at time resolutions as low as 5 minutes and spatial resolutions as fine as 1 km^2, up to 10 000 km^2. Radar estimation works by developing an empirical relationship between radar reflectivity and rain rate:

$$R = aZ^b$$

where R is rainfall rate, Z is reflectivity and a and b are parameters. There are several technical problems with radar, relating to beam attenuation, but most important is the fact that the relationship between reflectivity and rainfall depends on factors such as drop size and whether the precipitation is falling as rain or snow. Consequently, operational weather radar systems tend to involve real-time calibration using data from a network of recording raingauges (see Andrieu *et al.* (1997) and Seo *et al.* (1999) for some recent examples).

Like radar, satellites do not directly measure precipitation. However, precipitation rates can be inferred from two indicators, namely cloud top brightness (measured by visible light sensors) and cloud top temperature (measured by thermal infrared (TIR) sensors). The brighter the top of the cloud, or the lower the temperature, the greater the likelihood that precipitation is falling. Geostationary satellites such as Meteosat are most useful as they provide imagery at high temporal resolution and spatial resolution of a few kilometres. As with radar, empirical relationships can be constructed between one or both of these cloud indicators and precipitation at the ground. Dugdale *et al.* (1991) and Grimes *et al.* (1999) give examples using the "TAMSAT" method in west Africa. Rain over a 10-day period is estimated from the number of hours that a given pixel has a cloud top temperature below some defined threshold. The threshold and the parameters of the relationship between rainfall and "cold cloud duration" are estimated by calibration. Tsintikis *et al.* (1999) describe a different method using both cloud top temperatures and cloud top brightness, applied in the Nile basin.

Radar and satellite estimation offer many advantages over ground observations, but it is important that procedures are locally calibrated. Radar is better for high-frequency, fine spatial resolution estimates (covering a research catchment, for example), and satellite estimation is more appropriate over a larger spatial domain where very fine spatial resolution is not necessary. Also, radar estimation obviously requires an expensive ground installation.

(continued)

Table 3.1 Throughfall, stemflow and evaporation as a percentage of gross rainfall

Vegetation type	Location	Throughfall	Stemflow	Evaporation	Reference
Boreal and temperate evergreen					
Scots pine	Norfolk, England	67	2	31	Gash & Stewart (1977)
Scots pine	Highland Scotland	58		42	Gash *et al.* (1980)
Sitka spruce	Upland Wales	73		27	Gash *et al.* (1980)
Sitka spruce	Northern England	68		32	Gash *et al.* (1980)
Sitka spruce	Highland Scotland	69	3	28	Johnson (1990)
Black spruce	Manitoba, Canada	76	1	23	Price *et al.* (1997)
Temperate broadleaf					
Beech	Hampshire, England	82	4	14	Neal *et al.* (1993)
Oak	Netherlands	74		26	Dolman (1987)
Mixed hardwood	Ontario, Canada	77	4	19	Carlyle-Moses & Price (1999)
Mediterranean and sub-tropical					
Maritime pine	SE Australia	73	9	18	Crockford & Richardson (1990d)
Maritime pine	Central Portugal	83	0	17	Valente *et al.* (1997)
Maritime pine	SW France	87	1	12	Gash *et al.* (1995)
Scots pine	Pyrenees, Spain	76		24	Llorens *et al.* (1997)
Eucalyptus	SE Australia	85	4	11	Crockford & Richardson (1990d)
Eucalyptus	Central Portugal	87	2	11	Valente *et al.* (1997)
Laurel forest	Canary Islands	57		43	Aboal *et al.* (1999a)
Gorse	NW Spain	58		42	Soto & Diaz-Fierros (1997)
Sub-tropical and tropical broadleaf					
Montane forest	Puerto Rico	59	2	39	Scatena (1990)
Montane forest	Panama	62	1	37	Caudier *et al.* (1997)
Tropical evergreen	Ivory Coast	91	0	9	Hutjes *et al.* (1990)
Tropical evergreen	Brazilian Amazon	89	2	9	Lloyd *et al.* (1988)
Tropical evergreen	Colombian Amazon	82–87	1	12–17	Tobon Marin *et al.* (2000)
Tropical evergreen	Indonesia	87	1	11	Asdak *et al.* (1998a)
Tropical evergreen	Brunei	82		18	Dykes (1997)
Savanna forest	Venezuela	76	1	23	San Jose & Montes (1992)
Mixed subtropical forest	Brazil	85		15	Fujieda *et al.* (1997)

balance: how different is the evaporation of the intercepted water from the evaporation that would have taken place if the water had reached the ground surface? The evaporation of the intercepted water is explored in detail in the next section: this section concentrates on the ways in which precipitation is intercepted and reaches the ground.

Table 3.1 summarises estimates of the proportions of precipitation at the canopy that go to evaporation, stemflow and throughfall: measurement techniques are reviewed in Box 3.2. There is clearly considerable variability, reflecting different vegetation and climatic conditions.

Box 3.2 Measurement of the components of interception

To complete the "interception balance" it is necessary to measure the precipitation above the vegetation (gross precipitation), throughfall, stemflow, evaporation of intercepted water, and the amount of water in store. The measurement of evaporation is discussed in Box 3.4: this box focuses on the other components, looking particularly at interception by a forest canopy. In practice, the evaporation of intercepted water is usually inferred from measurements of gross precipitation, throughfall and stemflow.

There are two ways of measuring gross "top of canopy" precipitation (de Laine, 1969). One is to site raingauges in an area with no trees – perhaps in a clearing – but this may not of course be feasible if the forest is extensive. The other approach is to raise raingauges to the canopy top on towers. This is obviously inconvenient and expensive. As with ground raingauges (Box 3.1) there are problems of undercatch due to turbulence.

The major problem with measuring throughfall is that it is spatially very variable, and raingauges just a few metres apart may record very different totals (Crockford & Richardson, 1990a; Durocher, 1990). One solution to this problem is therefore to install a dense network of gauges, ideally located along transects (Lloyd & Marques, 1988). Moving the raingauges around from time to time may give a more representative estimate of total throughfall: Scatena (1990) found, however, that fixed and "roving" gauges gave very similar results in a lowland tropical rainforest. Raingauges under vegetation are, fortunately, less prone to undercatch due to turbulence. A second approach is to dig trough raingauges (Lloyd & Marques, 1988) which collect water and channel it to a recording device. Trough gauges sample a larger area than point gauges, but can be expensive and impractical to install. A third approach is to cover part of the forest floor with plastic sheets (Calder & Rosier, 1976), again draining to a recording device. Such gauges, however, may be impractical – as they smother surface vegetation – and will undercatch if holed.

(continued)

(continued)

Stemflow is rather easier to measure, using a collar to collect water running down the trunk. The seal must be tight, but a bigger concern is over the extrapolation of results from the sample trees to an area of forest. Aboal *et al.* (1999b) compared three methods – scaling up based on the projected crown area, scaling up based on the area of the trunk, and the use of empirical relationships between gross precipitation falling on the tree and stemflow – finding in their study laurel forest in the Canary Islands that extrapolation on the basis of trunk area gave the best estimates.

The amount of water stored on a vegetation canopy at a point in time is very difficult to measure directly, but there are two groups of approaches. The first is based on the idea that radioactive gamma rays (Calder & Wright, 1986) or microwaves (Bouten *et al.*, 1991) passing through vegetation are attenuated by stored water: the greater the volume of water stored, the greater the attenuation. This is of course expensive, and may not be practical to operate over a long period. The other approach essentially involves continuously weighing sample trees (Roberts, 1977; Lundberg, 1993), which is done by cutting a tree and placing it on a weighing device. This gives continuous measurements, but only of course for the sample trees. Studies concerned with the interception of snow tend to measure snow storage rather than compare gross snowfall with throughfall, because measurements of snow are particularly prone to error (Box 3.1; Lundberg *et al.*, 1998).

Table 3.1 shows the long-term average breakdown between the proportion of incoming precipitation that is intercepted and subsequently evaporated, and the proportion that reaches the surface by throughfall and stemflow. There are two main sets of factors determining this variation: vegetation characteristics (particularly interception storage capacity) and precipitation characteristics.

Influence of vegetation properties on interception

The storage capacity of the vegetation is clearly a major influence on the proportion of the incoming precipitation that actually reaches the ground surface by throughfall or stemflow. Storage capacities can be estimated in a variety of ways, including by scaling up from observations at leaf and branch level and by determining the amount of precipitation which occurs before throughfall and stemflow begin. Table 3.2 gives some estimates for different vegetation types: note that the interception capacity of trees is not necessarily larger than that of lower vegetation such as heather.

Somewhat surprisingly, leaves represent only a relatively small proportion of the storage capacity. Liu (1998), for example, found that only around 15%

Table 3.2 Some estimated canopy storage capacities, expressed as equivalent depth of water across the projected canopy area of the vegetation

Vegetation	Storage capacity (mm)	References
Oak	0.8–1	Dolman (1987), Carlyle-Moses & Price (1999), Rutter *et al.* (1975)
Sitka spruce	2–3	Calder & Wright (1986)
Eucalyptus	0.2–0.6	Aston (1979)
Douglas fir	2.4	Klaassen *et al.* (1998)
Cypress	0.5–0.8	Liu (1998)
Slash pine	0.38–0.5	Liu (1998)
Heather	1.1	Hall (1985)
Lowland tropical	0.74–1.15	Schellekens *et al.* (1999), Lloyd & Marques (1988), Jackson (1975), Dykes (1997)

of total storage capacity in cypress was in the foliage, with the remaining 85% evenly divided between the stems and branches. Cypress has a flaky, soft bark with high storage potential, and in slash pine the three components of storage were much more closely matched. Crockford & Richardson (1990c) compared the different components of storage in three different types of eucalyptus tree: the proportion of storage in the foliage ranged from just 3.5% to 24%. Storage capacity is therefore a function not only of the size of the vegetation, but also the shape and position of the leaves and the characteristics of the bark. There is a large contrast in the proportion of precipitation intercepted between coniferous and deciduous trees (Table 3.1). Coniferous trees tend to intercept a larger proportion of precipitation, partly because the storage capacity is greater but also partly because the smaller leaves allow greater circulation of air and hence greater evaporation.

Storage capacity varies over time. Firstly, it varies through the year as the vegetation changes, with the greatest variation of course in deciduous species. Liu (1998), for example, showed that the canopy storage capacity of (evergreen) slash pine varied over a year between just 0.46 and 0.5 mm, whilst the storage capacity of (deciduous) cypress ranged from 0.75 mm in winter to 0.89 mm in summer. Secondly, the amount of water that can be stored on the canopy varies with windspeed. The higher the windspeed during a precipitation event, the lower the effective canopy capacity (Klaassen *et al.*, 1996).

Precipitation that falls through or drips off one level of the canopy may be intercepted at lower levels. Gash & Stewart (1977), for example, estimated that around 7% of the precipitation falling from a Scots pine canopy in Thetford, Norfolk, was intercepted by a layer of bracken. Price *et al.* (1997) found that 23% of the precipitation that fell from a black spruce canopy in Manitoba was intercepted by, and subsequently evaporated from, the thick layer of feather moss covering the forest floor. The litter layer at the ground surface can also intercept water, with the storage capacity proportional to the depth of the litter layer – which varies depending on the type of tree. Putuhena

& Cordery (1996), for example, found that the litter layer at the bottom of a pine forest had an interception capacity of around 2.8 mm, compared with a capacity of around 1.7 mm under eucalyptus.

The characteristics of the precipitation event

For a given type of vegetation, both the long- and short-term proportion of precipitation that is intercepted and evaporated depend on precipitation characteristics. As a general rule, the smaller the amount of precipitation, the greater the proportion of that precipitation that is intercepted and the smaller the proportion that reaches the ground surface. This occurs because in low precipitation events, stores may not be filled: when the canopy is saturated, most precipitation will drain to the ground. Figure 3.8 shows the proportion of incoming precipitation that is intercepted by two forest types in south east Australia (Crockford & Richardson, 1990d). There is a difference between the two different vegetation types (reflecting their storage capacities) but in both cases the proportion of precipitation that is intercepted declines rapidly. The pine forest, for example, intercepts and subsequently loses by evaporation around 30% of incoming precipitation in events smaller than 10 mm, but intercepts only around 10% in events with more than 30 mm of rainfall. Eucalyptus forest intercepts only around 12% of 10 mm events, and around 5% of events of 30 mm or more. Figure 3.8 shows that, for a given vegetation type and event precipitation total, a greater proportion of precipitation is intercepted if the event is discontinuous. This happens because the intercepted water can evaporate during the gaps in the event, with subsequent precipitation replenishing

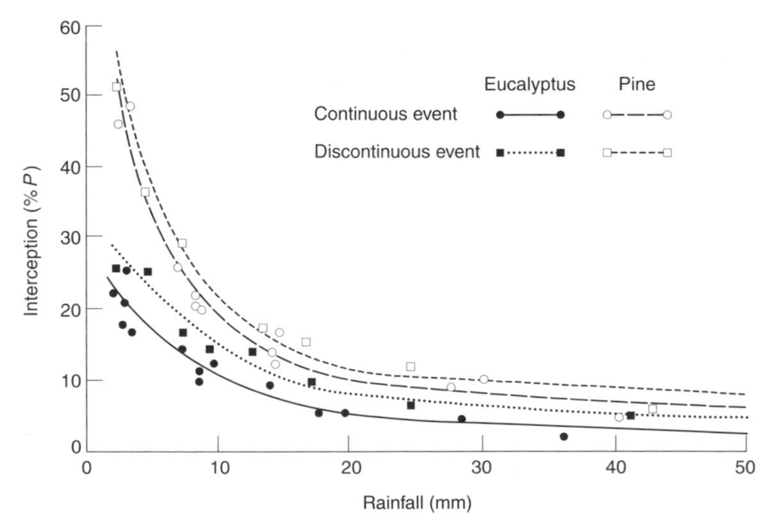

Figure 3.8 The proportion of incoming precipitation that is intercepted under two different forest covers (Crockford & Richardson, 1990d)

the storages. Similar graphs to Figure 3.8 have been produced for many different vegetation types and climatic conditions. Calder (1990) developed a simple daily-scale model to estimate interception from precipitation:

$$I = \gamma[1 - \exp(-\delta P)] \qquad \mathbf{3.2}$$

where I is the amount of precipitation intercepted, and γ and δ are parameters. Parameter values of $\gamma = 6.91$ and $\delta = 0.099$ provide a good fit to data from coniferous forests in Britain, implying that 43% of a daily precipitation total of 10 mm would be intercepted, 22% of 30 mm, and only 14% of a daily fall of 50 mm.

Many of the differences shown in Table 3.1 for a given vegetation type can be partially explained in terms of differences in precipitation climate. In the most general terms, interception will be a smaller proportion of long-term precipitation, the greater that precipitation is concentrated into short, intense events. Under these circumstances, the canopy quickly becomes saturated and only a relatively small proportion of precipitation is intercepted: much drips off the saturated canopy. If the precipitation is less intense, the canopy tends to be partially wetted frequently and stays wetter for longer, and therefore a greater proportion can be evaporated. Small raindrops also tend to lead to greater interception than large raindrops (Calder *et al.*, 1986a). The low proportion of precipitation that reaches the ground surface in the subtropical mountain forests of Puerto Rico and Panama, for example (Scatena, 1990; Caudier *et al.*, 1997) is attributed to the fact that much of this precipitation occurs as low-intensity drizzle. The low proportion in most tropical forests, however, is attributed partly to the fact that precipitation tends to occur in short-duration, high-intensity events, partly due to the large raindrops (which are not very efficient at wetting small leaves), and partly because many tropical trees have leaves which allow water rapidly to drain off.

As a general rule, a greater proportion of precipitation is intercepted when it falls as snow rather than rain. This is partly because snow falls more slowly, but largely because vegetation can store more snow than rain. Lundberg *et al.* (1998), for example, measured around 20 mm of water stored as snow on Sitka spruce near Aviemore, Scotland. Some of this intercepted snow will subsequently fall to the ground, but a proportion will be evaporated, depending on climatic conditions. As an example, 75% of a total snowfall of 19.3 mm (water equivalent) in March 1985 near Aviemore was intercepted by Sitka spruce (Lundberg *et al.*, 1998): most of this snow was subsequently evaporated. In colder conditions, however, a greater proportion of this snow would remain on the ground to melt during spring. Hedstrom & Pomeroy (1998) showed how "unloading" of snow from the canopy was important in environments where snow cover persisted for much of the winter season.

Stemflow

Table 3.1 indicates that stemflow tends to represent a small proportion of gross precipitation – and often it is not measured at all. The largest proportion in the table is close to 9%, from a pine forest in south east Australia. However, although the proportions are small, stemflow concentrates rainfall into a small area. This concentration of water may encourage recharge to groundwater (Taniguchi *et al.*, 1996), generate infiltration-excess overland flow (Section 3.7; Herwitz, 1986), provide soil moisture for plants, and also account for a large proportion of the chemical input into the soil (Crabtree & Trudgill, 1985). Durocher (1990) and Chang & Matzner (2000) showed how soil moisture was substantially increased around trees due to stemflow.

For a given amount of rainfall, the main controls on the amount of stemflow generated on a tree are the area of the crown, the smoothness of the bark, the position of the tree relative to its neighbours and the configuration of the branches, with crown area generally being most important (Crockford & Richardson, 1990b; Durocher, 1990; Navar, 1993; Aboal *et al.*, 1999b). Crown area is in a sense a surrogate for the storage capacity of the tree and the number of branches down which stemflow occurs. Figure 3.9 summarises observations of stemflow from six different species of tall Mediterranean shrub in the Canary Islands (Aboal *et al.*, 1999b). The trees are all over 10 m tall, but have quite different canopy structures. Stemflow per unit crown area varies from around 17 l m^{-2} in the myrtle to around 40 l m^{-2} in tree heath. The vertical lines in the graph, however, indicate the range between different individuals in each species, and the range for some species is clearly very large. The funnelling ratio measures the extent to which the tree focuses water collected over the canopy area to a point. It is the ratio of the stemflow running down the trunk to the rainfall which would have fallen over an area equal to the space occupied by the trunk: the water reaching the base of a Canary holly is around 40 times more than would have reached that part of the land surface if the tree was not there. Figure 3.9 shows variation between species and also, importantly, within each of the six species.

For a given tree, the amount of stemflow is a function of the amount of precipitation. Figure 3.9 shows for the six species the amount of rainfall necessary before stemflow occurs, and the range between 1 and 5 mm spans estimates from many other species. The last panel in Figure 3.9 shows the percentage of the weekly gross precipitation that becomes stemflow. There is variation between the species (very much in line with that for annual streamflow), but the greater the amount of precipitation, the higher the proportion of that precipitation that becomes stemflow. The relationship between precipitation and the proportion that goes to stemflow is non-linear, and the stemflow proportion tends to reach a maximum value.

Cloud water interception

Cloud water interception occurs when wind blows fog or low cloud into vegetation: fine airborne droplets of water are blown onto the vegetation, and stick there. It was initially believed that this was primarily a horizontal process (hence the term horizontal interception), and therefore most important at the edge of forests or patches of vegetation, but Shuttleworth (1977) showed that vertical air movement within a forest also blew droplets onto vegetation, and

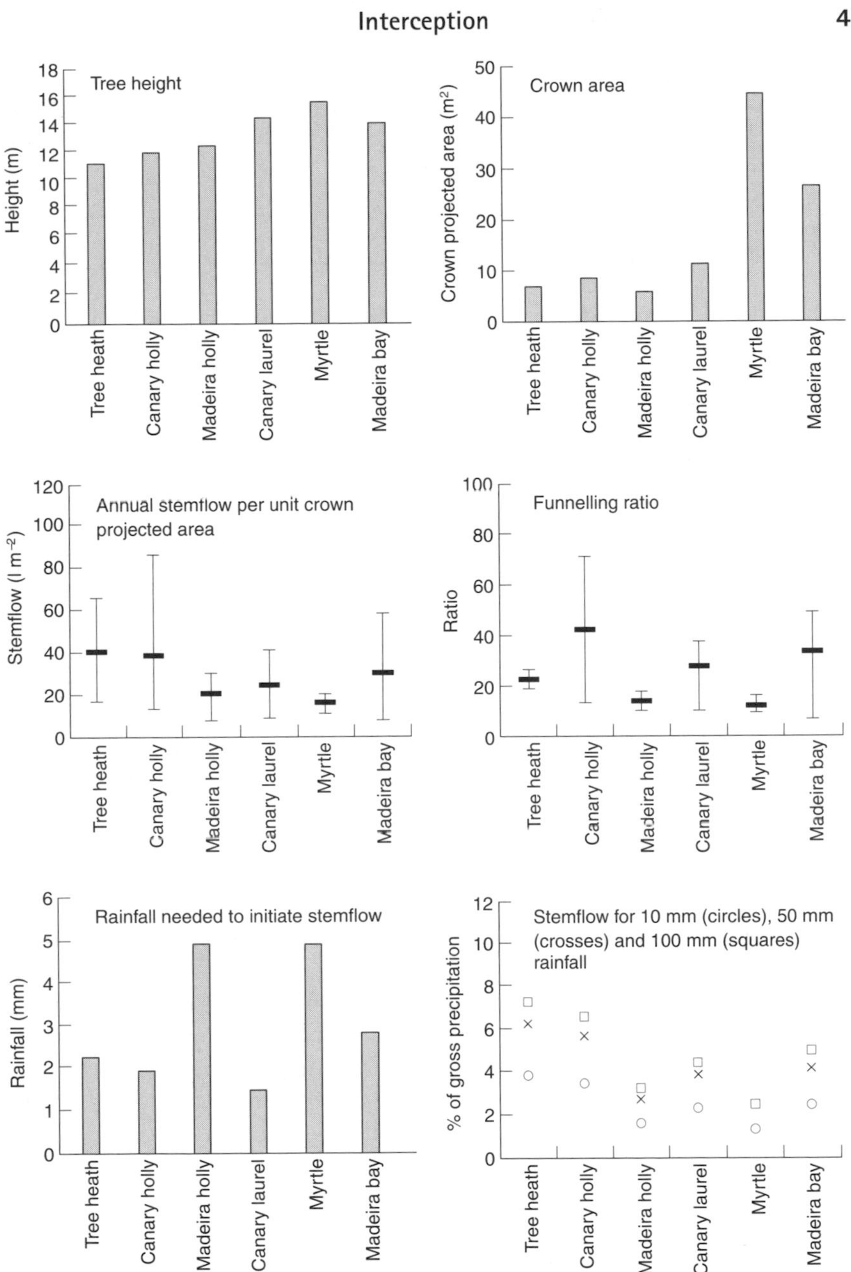

Figure 3.9 Stemflow characteristics for six Mediterranean tall shrubs: Canary Islands (adapted from Aboal *et al.*, 1999b)

that cloud water interception was not limited to edges. In some environments, such as along the Chilean coast or in tropical cloud forests, cloud water interception is an important source of moisture to vegetation – although how much of it reaches the ground as throughfall is unclear.

Cameron *et al.* (1997) measured cloud water deposition rates in tall tussocky grassland at high altitude in New Zealand. Values ranged between 0.02 to 0.26 mm h^{-1}, with a median of 0.05 mm h^{-1}, and cloud water deposition provided around 4% of annual precipitation.

3.4 Evaporation and transpiration

Evaporation is the conversion of a liquid to a vapour, and is arguably the most important "hydrological" process in the global climate system: it provides moisture to the atmosphere and uses energy. Evaporation requires energy to excite water molecules so that they leave the water surface. The energy required is termed the latent heat of vaporisation, and is slightly dependent on temperature:

$$\lambda = 2.501 - 0.0002361T \text{ MJ kg}^{-1} \qquad \textbf{3.3}$$

where T is temperature in degrees Celsius and λ, the latent heat of vaporisation, is in MJ kg^{-1}. In other words, at 10 °C it takes around 2.5 million joules to evaporate a kilogram of water. Evaporation, however, can only occur if the air above the surface is not saturated, and the size of the *vapour pressure deficit* (the difference between saturation and actual vapour pressure) can constrain the amount of evaporation. The rate at which air is moved away from the evaporating surface is also an influence. Still air soon becomes saturated. The humidity and turbidity of the air therefore together control the rate at which water vapour can diffuse into the atmosphere. Transpiration can be defined as that part of the total evaporation which enters the atmosphere from the soil through plants (Shuttleworth, 1993). The term "evapotranspiration" is sometimes used to denote both evaporation from soil and water and transpiration from plants – but is falling out of favour as it is too clumsy.

The rate of evaporation from a surface is a balance between the amount demanded by the atmosphere and the amount available for evaporation. The demand from the atmosphere can be loosely defined as *potential evaporation* (but see Box 3.3 for a discussion on the various definitions of potential evaporation), and that which actually occurs, given the amount of water that is available to be evaporated, can be, rather tautologically, termed *actual evaporation.*

The actual physics of evaporation (use of energy and diffusion of vapour) are the same regardless of the evaporating surface, whether it be water, soil, a plant or the water intercepted by and stored on a leaf. However, the type of surface does affect the *balance* of different influences. This section considers first the atmospheric controls on the rate of evaporation. It then examines evaporation from different land covers, and attempts to make some quantitative comparisons. Box 3.4 summarises the methods which can be used to measure the rate of evaporation, and Box 3.5 reviews techniques to measure transpiration. In practice, evaporation is difficult to measure, so towards the end of this section, procedures developed to *estimate* evaporation from meteorological data are described.

Box 3.3 Potential evaporation

Although potential evaporation is loosely conceptualised as the amount of evaporation demanded by the atmosphere, there are in practice several operational definitions. Thornthwaite (1948), for example, defined potential evaporation as "the water loss which will occur if at no time there is a deficiency of water in the soil for use of vegetation". Penman (1956) further restricted the term to mean "evaporation from an extended surface of short green crop, actively growing, completely shading the ground, of uniform height and not short of water", effectively defining potential evaporation to be the evaporation from a well-watered grass surface. Shuttleworth (1993), however, termed this "reference crop evaporation" (a term generally favoured by agriculturalists), and restricted potential evaporation to mean the evaporation from an extensive free water surface. Lhomme (1997) proposed a "rational" definition of potential evaporation, suggesting that it be calculated using a particular estimation procedure with defined parameters. Cain *et al.* (1998) argued, however, that since the concept of potential evaporation is imprecise – though conceptually simple and helpful – it is misleading to attempt to define it precisely.

The atmospheric demand for evaporation is driven by energy (the "thermodynamic" factors) but influenced by humidity and turbulence ("aerodynamic" factors). Actual evaporation is constrained by the amount of water available.

Energy for evaporation

The earth's climate system is driven by energy received from the sun. Figure 3.10 summarises the global mean energy fluxes. The global average incoming solar radiation is 342 W m^{-2}, which virtually all occurs in short wavelengths from ultraviolet (0.1–0.4 μm) through visible (0.4–0.7 μm) to near infrared (0.7–4.0 μm). Approximately 31% of this incoming radiation is reflected back to space, from clouds (23%) and the earth surface (8%). The remaining 69% is absorbed, 49% by the surface and 20% in the atmosphere (much of the ultraviolet radiation is absorbed by ozone in the stratosphere, and some longer wavelength radiation is absorbed by clouds). In order to maintain thermal equilibrium, energy equivalent to this 69% of the incoming radiation must be re-radiated back to space. However, the energy goes through a complicated recycling process. The earth itself emits radiation, but at a longer wavelength than the sun: the wavelengths of emitted radiation are inversely proportional to the temperature of the emitting body. On average, the earth emits 390 W m^{-2} of long-wave radiation, at wavelengths between 4 and 50 μm. A small proportion of this long-wave radiation escapes directly to space, but the largest

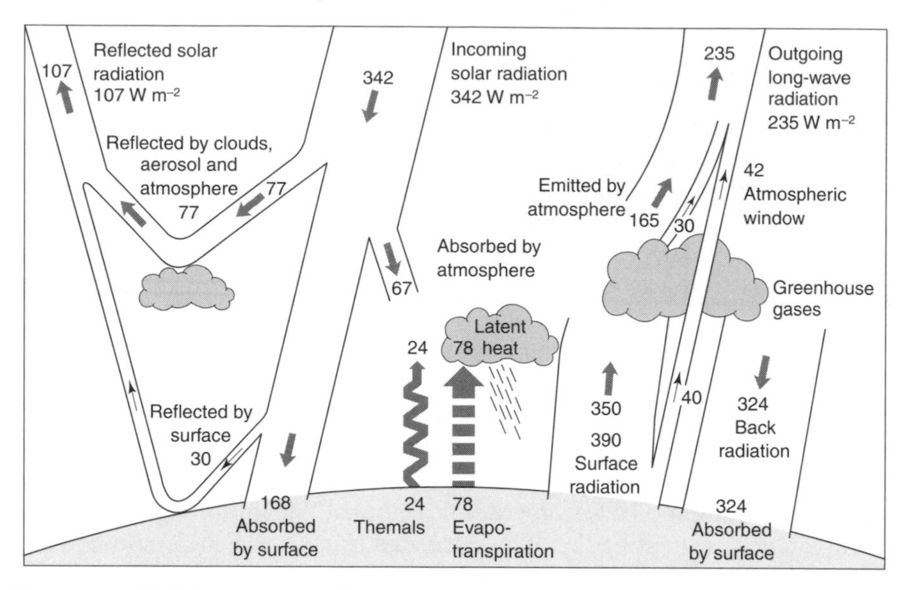

Figure 3.10 Global mean energy flows between the surface and the atmosphere (Trenberth *et al.*, 1996)

proportion is absorbed by trace gases in the atmosphere. This is combined with the incoming short-wave energy absorbed by the atmosphere, and is emitted by the atmosphere. Some goes to space, but the largest proportion (324 W m^{-2}) goes back towards the earth surface as downward long-wave radiation. The amount of energy available at the earth surface is therefore equal to:

$$R_n = S_t(1 - \alpha) + L_d - L_u \qquad \textbf{3.4}$$

where R_n is net radiation, S_t is the incoming short-wave radiation at the surface (i.e. after reflection by and absorption in the atmosphere), α is albedo, L_d is the downward long-wave radiation and L_u is upward long-wave radiation. On a global average, the net radiation at the surface is therefore 163 + 324 – 390 = 102 W m^{-2}. There is, of course, very considerable variation in net radiation over both space and time (between –50 and +170 W m^{-2} over land), largely due to variations in the net short-wave radiation. Net long-wave radiation is usually negative; in other words there is a net flux from earth to atmosphere. It varies with temperature and cloud cover. Net short-wave radiation is a function of the incoming short-wave radiation at the top of the atmosphere (which varies with time and latitude), the reflectance and absorption of radiation in the atmosphere (largely related to cloudiness), and the albedo at the earth's surface. Albedo varies with surface characteristics and, for some surface types, the angle of the sun. Table 3.3 gives some typical albedo values for different surfaces. Open water has a low albedo, so a high proportion of the incoming short-wave radiation is absorbed. New snow, on the other hand, has a very high albedo so up to 80% of the short-wave radiation is reflected back into space. Net radiation is generally higher over the oceans

Table 3.3 Typical values of albedo for different surface types (adapted from Shuttleworth, 1993)

Surface type	**Albedo**
Snow and ice	0.20 (old) to 0.80 (new)
Open water	0.08
Bare soil	0.10–0.35
Tall forest	0.11–0.16
Grass and pasture	0.20–0.26

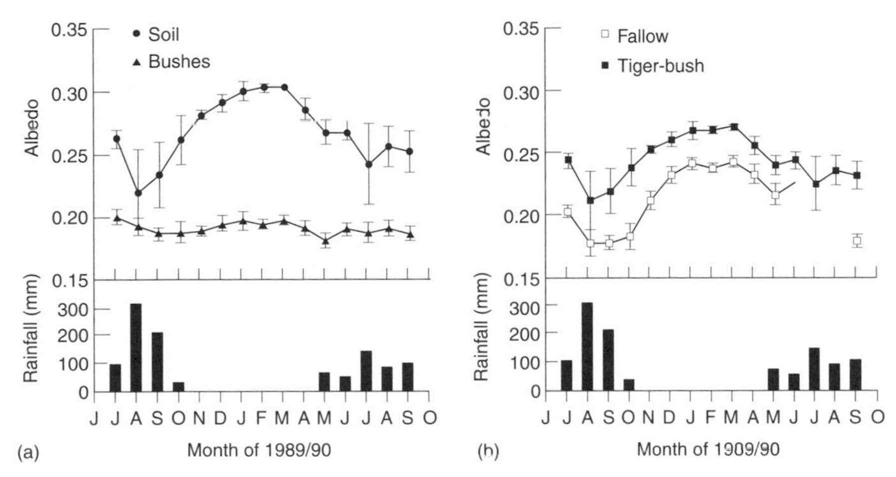

Figure 3.11 Monthly mean albedo (a) under soil and vegetation, and (b) under two different vegetation types, Sahel (Allen *et al.*, 1994)

than the land, and this is because the albedo of the sea is lower than that of the land: less of the incoming short-wave radiation is therefore reflected back into the atmosphere. Figure 3.11a plots the seasonal variation in albedo in the semi-arid Sahel region of west Africa, for two land cover types (Allen *et al.*, 1994). The albedo of the bare soil is considerably larger than that of the bush and, significantly, varies more through the year. During the wet season (May to September), the soil is wetter and so darker than during the dry season: its albedo is lower, and a greater proportion of the incoming short-wave radiation is absorbed by the surface.

The annual cycle of net short-wave radiation, net long-wave radiation and net radiation for four sites is shown in Figure 3.12. Net long-wave radiation is consistently negative, and shows little seasonal variability at any of the sites. Net short-wave radiation, however, shows a very strong seasonal cycle at the English, Canadian and Australian sites: net short-wave radiation is always positive at each of these sites, and the negative net radiation in winter at the first two is due to the negative net long-wave radiation offsetting the incoming short-wave radiation. Net short-wave radiation, and hence net radiation, shows less seasonal variability at the Tanzanian site, close to the equator.

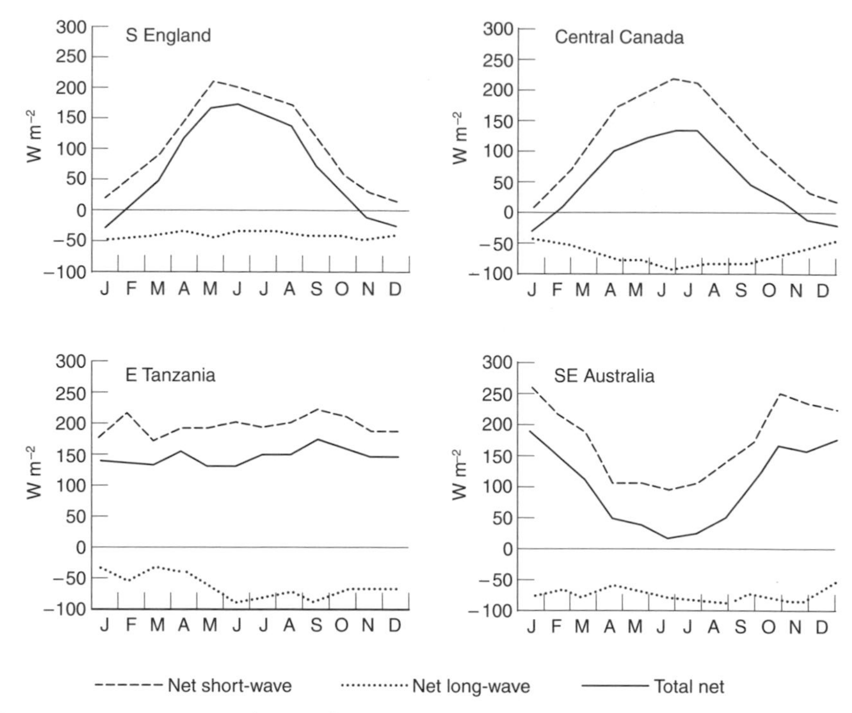

Figure 3.12 Annual cycle (in 1988) of short-wave radiation, long-wave radiation and net radiation at four locations (data from Darnell *et al.*, 1992)

Net radiation is the energy at the earth's surface, potentially available for evaporation. In practice, of course, not all the available energy is used to evaporate water, and the full energy budget for an area is given as (Shuttleworth, 1993):

$$R_n + A_d = \lambda E + H + G + P + S \qquad \textbf{3.5}$$

A_d is the net energy advected in by horizontal air movement. λE is the energy used in evaporation (it is written in this way because E usually denotes evaporation in millimetres of water: multiplying by the latent heat of evaporation λ converts evaporation into "units of energy used". It is sometimes written as LE). H is the outgoing sensible heat flux, where the energy at the ground warms the air by conduction. G is the outgoing heat flux into the soil (by conduction) or water body (by both conduction and convection). This can be ignored over the long term, but is an important component of the energy budget at daily, weekly and seasonal scales, particularly for water bodies. The last two terms in equation 3.5 are relevant if the surface is vegetated: P is the energy absorbed by biochemical processes in plants, typically around 2% of the net radiation (Stewart & Thom, 1973), and S is the energy temporarily stored in the air in the vegetation immediately above the surface. Both these terms are small, and are ignored in hydrological assessments of the energy balance. The ratio of sensible heat H to latent heat used in evaporation λE is known as the Bowen ratio.

Aerodynamic factors: humidity and turbulence

Net radiation provides the energy for evaporation, but evaporation will only take place if the air above the evaporating surface can accept moisture: if the air is saturated, there will be no evaporation. The *vapour pressure deficit* is the difference between the saturation vapour pressure e_s (equation 3.1) and the actual vapour pressure, and for a given amount of energy, evaporation will be proportional to the vapour pressure deficit.

If the air above the evaporating surface is still, then it will soon become saturated and evaporation will cease. In practice, of course, the air is moving, and the greater the velocity of the air, the greater the rate at which evaporated water is moved away from the evaporating surface. Dalton (1802) was the first to propose that, other things being equal, the rate of evaporation was a function of windspeed and vapour pressure deficit. In mathematical terms, this can be summarised as:

$$E = f(u)(e_s - e) \qquad \textbf{3.6}$$

where $f(u)$ is a function of windspeed. Water vapour is actually transported from the surface through turbulent diffusion, and the greater the degree of turbulence the greater the rate of diffusion. This turbulence is dependent not only on windspeed, but also on the roughness of the surface (which is very dependent on vegetation type – as shown below).

Box 3.4 Measuring evaporation

Evaporation is the most difficult component of the water balance to measure. The simplest approach is to use an *evaporation pan*, filled with water, in which the depth of water is regularly measured. The amount of evaporation during the recording period is the difference in the volume of water stored, minus the amounts added by precipitation or the observer. Evaporation pans are widely used by meteorological and hydrological monitoring agencies, but there are some obvious problems. Water may splash into the pan, or be blown out. The difference in depth from one period to another may be very small and therefore difficult to assess accurately. Most importantly, however, the characteristics of the pan itself may lead to it giving misleading estimates. The colour and materials of the pan may alter the local energy balance, and energy may be stored in the water and used later for evaporation. A pan rising above the ground may trigger turbulence, which encourages evaporation. *Pan coefficients* have been developed to correct pan evaporation to "real" evaporation, although even then evaporation pans essentially only give an index of evaporation from small ponds.

Lysimeters work on a similar accounting principle. They are isolated blocks of soil. The inputs – precipitation – are measured, as are outputs – drainage out of the soil – and the volume of water stored in the

(continued)

(continued)

soil block can be determined in a number of ways (see Box 3.11): some lysimeters are continually weighed to monitor changes in storage. By growing different types of vegetation on the soil block, it is possible to determine actual evaporation under different conditions, and by ensuring the soil block is well supplied with water it is possible to measure potential evaporation. Lysimeters, however, can be very expensive to install and are generally only used as research instruments.

Since the late 1980s, however, two micrometeorological approaches have become technically feasible: both rely on the ability to take precise and accurate measurements (Shuttleworth, 1993). The first – *energy balance method* – uses the Bowen ratio, or the ratio of sensible heat flux to latent heat flux. It is assumed that the ratio between temperature and humidity differences at two heights above the surface is proportional to the ratio between sensible and latent heat fluxes:

$$\beta = \frac{\gamma \Delta T}{\Delta e}$$

where β is the Bowen ratio, ΔT is the difference in temperature (degrees Celsius), Δe is the difference in vapour pressure (in kPa), and γ is a constant necessary to account for units. The energy used in evaporation can therefore be determined using β together with an estimate of net radiation R_n and ground heat flux G:

$$LE = \frac{R_n - G}{1 + \beta}$$

where evaporation LE is in the units of the energy terms. The second micrometeorological approach is regarded as more accurate, and involves measuring vertical windspeeds and specific humidity: it is known as the *eddy correlation* method (Shuttleworth *et al.*, 1988). Near the surface the mean vertical wind is zero. Turbulent eddies in the moving air, however, mean that occasionally the air moves vertically, with velocity w' (positive upwards, negative downwards). There are similar turbulent fluctuations in the specific humidity of the air, fluctuating around the average specific humidity. If both the vertical wind velocity at a small distance above the ground is positive and the specific humidity at that same point is above average, then moisture is being moved away from the ground by evaporation. Given measurements of w' (in m s^{-1})and q' (the difference between current and average specific humidity, in kg of water/kg of air), then the evaporation rate E (in mm d^{-1}) can be calculated from:

$$E = 86\,400\,\rho_a \overline{w'q'}$$

where ρ_a is the density of moist air, in kg m^{-3}. The method requires very precise, co-located, measurements, made by recording devices which do not themselves create turbulence. An example instrument is the Hydra

(continued)

(continued)

(Shuttleworth *et al.*, 1988), which is increasingly used in research studies into evaporation. Accuracy is generally of the order of 5 to 10% (Shuttleworth, 1993).

All the methods summarised so far give point estimates of evaporation: as with precipitation, hydrologists generally need measurements of evaporation over an area. One new ground-based technique uses a large aperture scintillometer to measure sensible heat fluxes across a transect (Meijninger & de Bruin, 2000). The instrument works by measuring the turbulent intensity of the refraction index of air between a transmitter and receiver of an electromagnetic pulse. The refraction index is a function of, amongst other measurable factors, sensible heat flux. The energy used in evaporation can then be determined using measurements of net radiation and soil heat flux.

Remote sensing, however, offers another way of estimating areal evaporation (Kustas & Norman, 1996). In general terms, there are two broad approaches. One calculates net radiation and vapour pressure at the surface, and then uses a conventional evaporation equation to estimate evaporation: the other measures sensible heat flux and net radiation, and determines evaporation from the energy balance. Granger (2000) gives an example of the first approach. Net radiation is determined by first measuring albedo, applying this to incoming short-wave radiation, and using an empirical relationship to estimate net long-wave radiation from net short-wave. The vapour pressure deficit at the surface is determined from satellite estimates of surface temperature using an empirical relationship. Bastiaanssen (2000) gives an example of the second approach. This uses estimates of surface temperature and roughness to determine sensible heat flux.

Evaporation from open water

Evaporation from a water surface is, almost by definition, unconstrained by the amount of water available. Energy availability is the most important control, which is largely a function not only of net radiation but also of heat stored within the water body, but windspeed may also be important. Heat storage is related to the size and depth of the water body, with the greatest potential for storage in large, deep water bodies. In Lake Superior, for example, the time with the highest evaporation rates is not summer, when net radiation is highest, but autumn when the lake is at its warmest. Sene *et al.* (1991) found that evaporation from Lake Toba, Indonesia, was largely determined by the lake heat storage and windspeed (affecting the rate of removal of saturated air from the lake surface). Heikenheimo *et al.* (1999) found similar results from two lakes in eastern Sweden: in the larger, but shallower, of the two, high evaporation resulted from high windspeeds, but in the smaller, more sheltered, deeper lake similar rates of evaporation occurred due to the larger amounts of heat stored in the lake. Heikenheimo *et al.* noted that evaporation from the lake had a much smaller diurnal cycle than that from surrounding land – and could occur at night.

Evaporation rates are also affected by salinity, with the rate declining as salinity increases (because if a substance is dissolved in a liquid the motion of the liquid molecules is restricted: the saturation vapour pressure over saline water is lower than over open water). The effect is, however, small, and the rate of evaporation from sea water is typically only 2 to 3% less than that from freshwater (Ward & Robinson, 2000).

Evaporation from snow on the ground

Evaporation from snow is generally very low, largely because snow tends to be lying when radiation inputs are low and air is cold (and therefore the vapour pressure deficit is low). Very little of a snow pack is therefore lost through evaporation, and melt is far more important. The evaporation of intercepted snow is discussed below.

Evaporation from bare soil

In principle, evaporation from bare soil can operate at the same rate as evaporation from open water, as the water held between the soil particles is evaporated. In practice, however, there are two differences. Firstly, the characteristics of the soil can affect the albedo (Figure 3.11), and secondly the amount of water available, particularly at the soil surface, may constrain the rate of evaporation. Albedo varies over time for a given soil, primarily with moisture content (Idso *et al.*, 1975), but also varies between different soil types. Darker soils obviously have a lower albedo than lighter soils, and thus there is more energy available for evaporation.

It is the moisture content at the surface of the soil which largely controls the rate of evaporation from soil, because the rate of upward movement of water by capillary action (see Section 3.5) tends to be slow. Different types of soil can hold different amounts of water, varying primarily with soil texture, and some soils drain more quickly than others. In general, coarse soils can hold the smallest volume of water and drain most quickly, so over time the total evaporation from a coarse soil can be expected to be lower than from a finer soil. Figure 3.13 shows the ratio of actual to potential evaporation plotted against soil water content in the top centimetre of soil for three different soil types (Chanzy & Bruckler, 1993). The rate of decline is exponential. The effect of soil water content on the partitioning of energy into latent heat and sensible heat fluxes is shown in Figure 3.14, which plots energy fluxes over a bare soil in the Sahel on a wet and a dry day (Wallace & Holwill, 1997): a considerably greater proportion of the available net radiation is used for evaporation on the wet day.

In very hot dry soils, evaporation may occur *beneath* the soil surface, at the base of the dry surface layer (Yamanaka *et al.*, 1998). The thickness of the evaporation zone depends on soil texture, and is greatest in fine-textured soils.

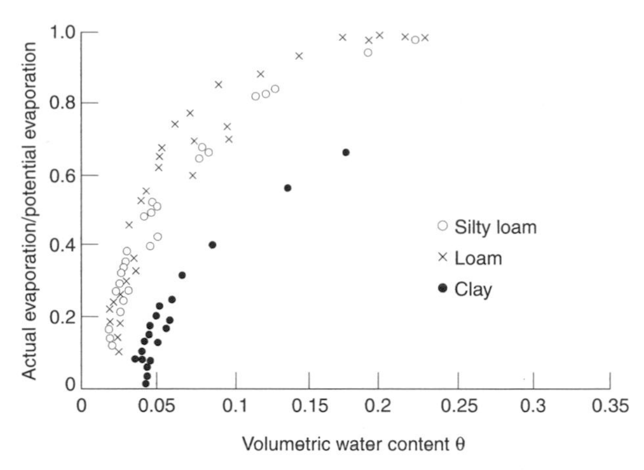

Figure 3.13 Effect of surface soil moisture content on the ratio of actual to potential evaporation (Chanzy & Bruckler, 1993)

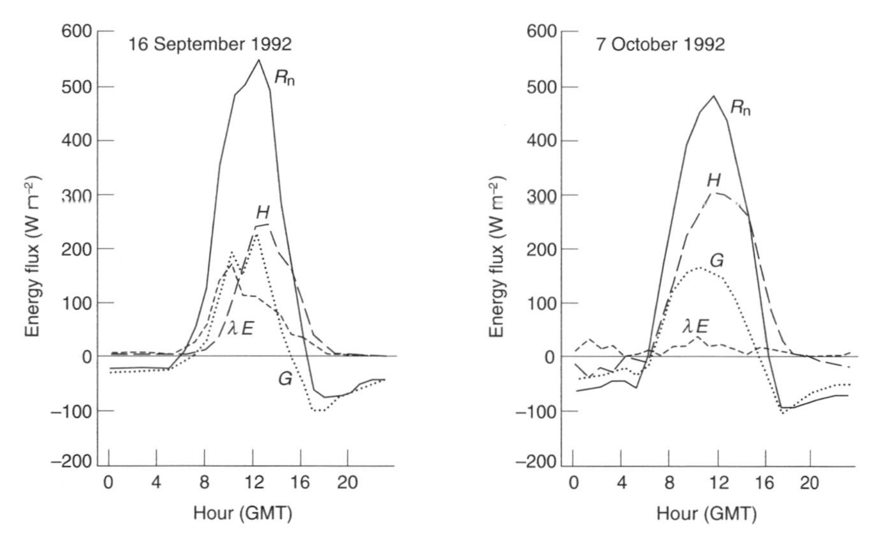

Figure 3.14 Energy fluxes over bare soil in the Sahel during a wet day (16 September) and a dry day (7 October). The graphs show net radiation (R_n), evaporation (λE), sensible heat flux (H) and the soil heat flux (G) (Wallace & Holwill, 1997)

Vegetation and evaporation

There are two types of evaporation from vegetation: transpiration and the evaporation of intercepted water. Transpiration is the process by which water evaporates from a plant. Essentially, water is evaporated through small holes (known as *stomata*) in the leaves, and this draws water up through the plant (in microscopic tubes termed *xylem*) from the soil. This "transpiration stream" brings water to the plant, to be used in photosynthesis, to produce carbohydrates, to maintain turgidity (rigidness) in the cells and tissues, and to bring dissolved minerals from the soil. Very little of the water is actually used in photosynthesis, and most of the water sucked up from the soil is evaporated through the stomata, whose primary purpose is actually to exchange carbon dioxide and oxygen with the atmosphere.

Both transpiration and the evaporation of intercepted water are driven by the energy available, humidity and turbulence, but vegetation has two main influences. Firstly, it alters the "driving forces" of evaporation, and secondly plants exercise some control over their rate of transpiration: plants are not simply wicks through which water is evaporated. It is important to remember, however, that transpiration is a physical, not a metabolic, process.

Vegetation effects on thermodynamic and aerodynamic driving forces

Vegetation alters the thermodynamic and aerodynamic influences on the rate of evaporation in two main ways. First, albedo, and hence the amount of energy available for evaporation, varies with vegetation cover. Table 3.3 shows that tall forests (which tend to be dark green) have a lower albedo than short vegetation, and therefore more energy is available for evaporation. Figure 3.11b shows the variation in albedo through the year for two vegetation types in the Sahel (Allen *et al.*, 1994).

The second influence is through turbulence. Vegetation induces turbulence in an air stream, with the stiffest, tallest vegetation having the greatest effect. The most widely used model for the transport of water from surface to atmosphere uses the analogy of an electrical current: the flux of water is proportional to the difference in concentration divided by the resistance. The *aerodynamic resistance* r_a characterises the resistance to turbulent diffusion. It is inversely proportional to windspeed and varies with the height of the vegetation (Shuttleworth, 1993):

$$r_a = \frac{\ln[(z-d)/z_{om}]\ln[(z-d)/z_{ov}]}{(0.41)^2 U_z}\ \mathrm{s\,m^{-1}} \qquad \textbf{3.7}$$

where U_z is windspeed (m s^{-1}) at height z (m) above the vegetation, d is the zero plane displacement and z_{om} and z_{ov} are the roughness heights for momentum and vapour fluxes respectively. The zero plane displacement is the height at which there is no air movement. It can be estimated from $d = 0.67\ h$, where h is the vegetation height in metres, z_{om} can be estimated from $z_{om} = 0.123\ h$ and z_{ov} can be estimated from $z_{ov} = 0.0123\ h$. Grass has a typical height of 0.1 m, agricultural crops 1.0 m and forest 10 m. Figure 3.15 shows the variation in aerodynamic resistance with windspeed and vegetation type (the curve for water is calculated using a slightly different equation). Note the logarithmic scale: aerodynamic resistance varies considerably with windspeed and the aerodynamic resistance of forests is approximately a seventh of that of grass,

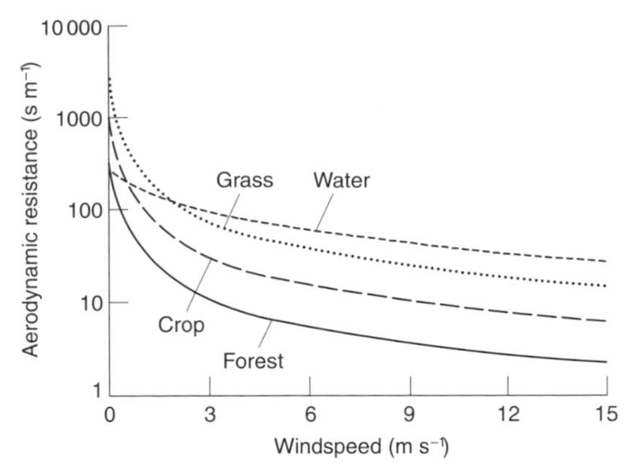

Figure 3.15 Variation in aerodynamic resistance with vegetation type and windspeed

meaning that turbulent diffusion is much easier from trees than short grass – and hence rates of evaporation from trees are likely to be higher. The reciprocal of aerodynamic resistance is known as atmospheric conductance.

Plant controls on transpiration

Transpiration is the evaporation of water through stomata on the plant leaves, and is driven by the available energy, turbulence and the vapour pressure deficit. As with aerodynamic resistance, it is possible to conceive of a *stomatal resistance* to the movement of water away from the leaf. Often, plant physiologists use instead the concept of *stomatal conductance*, the reciprocal of stomatal resistance. Stomatal conductance varies between species and also, importantly, varies over time. Table 3.4 shows typical maximum values of stomatal conductance and resistance for a number of vegetation types. The total transpiration from a plant (or indeed a canopy) is equal to the sum of the losses from all the

Table 3.4 Typical values of maximum stomatal and canopy resistance and conductance (Kelliher *et al.*, 1995; Kabat *et al.*, 1997)

Vegetation type	Stomatal		Leaf area index	Canopy	
	Conductance ($mm\ s^{-1}$)	Resistance ($s\ m^{-1}$)		Conductance ($mm\ s^{-1}$)	Resistance ($s\ m^{-1}$)
Temperate grass	8	125	2.1	17	59
Cereals	11	90	2.9	32.5	31
Conifers	5.7	175	3.7	21.2	47
Deciduous	4.6	217	4.5	20.7	48
Tropical rainforest	6.1	164	2.1	13	77
Patterned savanna woodland				40	25
Fallow savanna				10	100

individual leaf surfaces. If the leaf area index (*LAI*) is the surface area of all the leaves in a unit area of land, then the *canopy* (or bulk) *resistance* r_c can be calculated from:

$$r_c = \frac{\overline{r_l}}{LAI} \qquad \textbf{3.8}$$

where r_l is the average leaf stomatal resistance. The leaf area index varies between vegetation types, and also, for deciduous plants at least, varies through the year.

The values shown in Table 3.4 are *maximum* values of stomatal and canopy conductance. Stomatal conductance varies with five factors (Stewart, 1988): light intensity, CO_2 concentration, the leaf–air vapour pressure difference, leaf temperature and leaf water content. CO_2 concentration does not vary over short time scales, so can be ignored here (but it is potentially very important when assessing the effects of climate change on evaporation: Chapter 7). Stomata are closed when it is dark – so transpiration does not take place at night – but are fully open with only small amounts of light. The other three controls are much more important over the short term, and several studies (e.g. Stewart, 1988; Hanan & Prince, 1997; Granier *et al.*, 2000) have shown how stomatal conductance for different vegetation types responds differently to vapour pressure deficits, temperature and soil moisture deficit.

Figure 3.16 shows such relationships for a pine forest in England (Stewart, 1988). The graphs show stomatal conductance relative to the maximum stomatal conductance. Temperature has a clear effect, with maximum stomatal conductance occurring at around 18 °C. Stomatal conductance is unaffected by soil water content (a surrogate for leaf water content) until a deficit of around 60 mm is built up, and then stomatal conductance falls dramatically: the trees are lowering stomatal conductance in order to conserve water. Kelliher *et al.* (1998) found a factor of five change in daily *canopy* conductance over

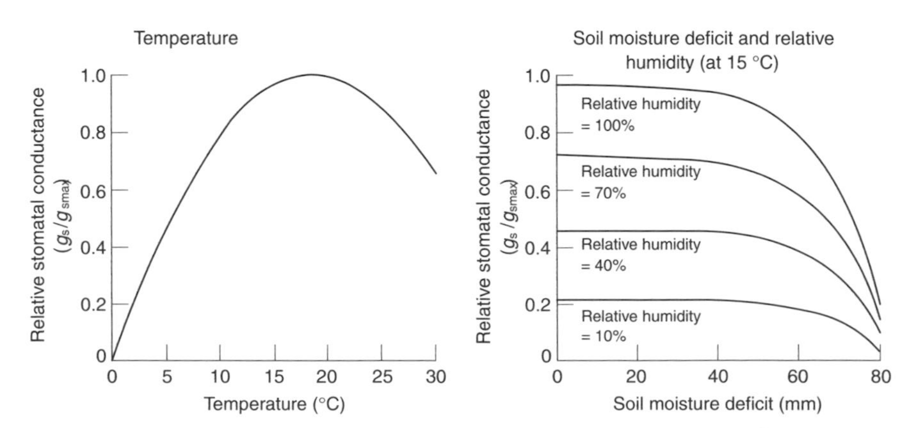

Figure 3.16 Effect of meteorological controls on plant stomatal conductance (from Stewart's (1988) model with parameters appropriate for a pine forest in south east England, and a temperature of 15 °C)

18 days in a Siberian pine forest, with 79% of this variation explained by variations in soil moisture content. Stomatal conductance in different plants, however, responds differently to given soil moisture states. Plants also lower their stomatal conductance when the air is dry. In the example in Figure 3.16, the stomatal conductance in humid conditions is up to five times as high as when the air is dry (with a relative humidity of 10%). Kabat *et al.* (1997) showed that the response to vapour pressure deficit of the stomatal conductance of two different savanna vegetation types differed, due to different ways in which the plants used CO_2 in photosynthesis. Variation through the day in vapour pressure deficit (due to variation through the day in temperature) means that stomatal conductance often has a diurnal cycle.

Box 3.5 Measuring transpiration

Transpiration of individual plants can be estimated by measuring "sap flow", the rate of movement of water up through the xylem. This can be done through injecting tracers into the roots of the plant – such as deuterium (Calder *et al.*, 1986b) – or by measuring the conductance of heat through the plant stem: this is proportional to the rate of sap flow (Smith & Allen, 1996). The major problem, as with measuring stemflow (Box 3.2) lies in the extrapolation from the measured plants to the larger scale.

A plant acquires water through its roots, and the density of the root network is therefore an influence on the total amount of transpiration and the rate at which stomatal conductance declines in response to water shortage: other things being equal, trees will have access to more water than shorter plants. Indeed, in some environments eucalyptus trees "mine" deep groundwater (Le Maitre & Versfeld, 1997). However, a denser root network alone does not guarantee more access to water, and the ability of the plant roots to "suck" water from the soil may be as important.

Given the diversity in plant controls on the rate of transpiration, it is therefore somewhat surprising that seasonal and daily transpiration rates from different forest species are very similar (Roberts, 1983; Kelliher *et al.*, 1997). Table 3.5 shows the annual transpiration from a range of different forests in Europe, averaging out over a six-month growing season at just under 2 mm d^{-1}. Kelliher *et al.* found a very similar figure for boreal coniferous forest. The reason for this similarity is not clear, but it implies that the different trees may be attempting to maintain a similar throughput of water.

Evaporation of intercepted water

Section 3.3 showed how a substantial proportion of incoming precipitation can be intercepted by vegetation and subsequently evaporated. Some is evaporated

Table 3.5 Annual forest transpiration (adapted from Roberts, 1983)

Species	Location	Annual transpiration (mm a^{-1})
Sitka spruce	Yorkshire, England	340
Norway spruce	Germany (site 1)	362
Norway spruce	Germany (site 2)	279
Norway spruce	Wales	320
Scots pine	Germany	324
Scots pine	Norfolk, England	353
Scots pine	Berkshire, England	427
Oak	Germany	327
Oak	UK	320
Beech	Belgium	344
Average		*340*

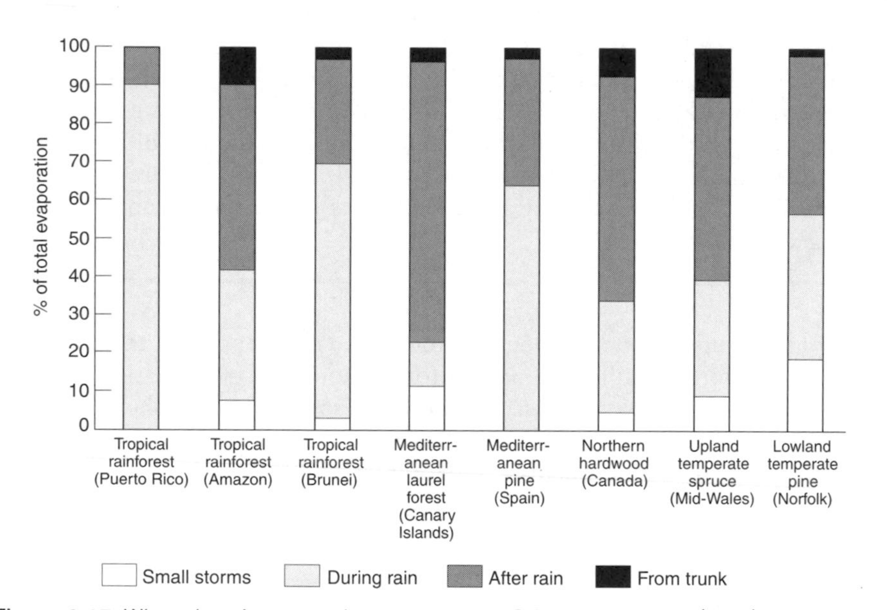

Figure 3.17 When does intercepted water evaporate? Data from Gash (1979), Schellekens *et al.* (1999), Dykes (1997), Llorens *et al.* (1997) and Aboal *et al.* (1999a)

during the precipitation event itself, and the remainder evaporated once the precipitation has finished. Figure 3.17 shows the modelled proportions (Gash, 1979) of intercepted water that is evaporated in small storms that do not saturate the canopy, whilst the canopy is wetting up, once it is saturated and precipitation is still falling, after the precipitation has finished, and from the trunks. Typically between 30 and 40% of the intercepted water is evaporated whilst it is raining, with a larger proportion evaporating after the rain has finished. The lowland maritime tropical forests studied by Schellekens *et al.* (1999) and Dykes (1997), however, lose a far higher proportion of the intercepted water during the rainfall event, as does the pine plantation in

Spain studied by Llorens *et al.* (1997). On the other hand, only a very small proportion of intercepted water was evaporated during rainfall in the laurel forests of the Canary Islands (Aboal *et al.*, 1999a), and this was attributed to the presence of thick cloud during rainfall. Klaassen *et al.* (1998) measured evaporation from intercepted water during rainfall, and found that only around 22% of the intercepted rain was evaporated whilst it was raining. Clearly, the timing of the evaporation of intercepted water varies in some way with climatic characteristics.

The rate of evaporation from intercepted water can be substantially higher than the rate of transpiration: Rutter (1967) estimated that the rate could be four times as high under the same environmental conditions. This is partly possible because the "canopy resistance" for the evaporation of intercepted water is essentially zero, and the plant therefore does not exert any control on the loss of intercepted water, but other factors are involved too. The evaporation rate of intercepted water has often been calculated (or inferred) to be greater than the amount of radiant energy available. At one extreme, Schellekens *et al.* (1999) estimated that the rate of evaporation of intercepted water in a Puerto Rican rainforest could be more than 10 times greater than the amount of radiant energy available. This means that some other source of energy must be available (Shuttleworth, 1989). One potential source is the advection of warm air from upwind (Shuttleworth & Calder, 1979; Schellekens *et al.*, 1999). Another is heat released upon the condensation of water vapour in the air above the forest (Schellekens *et al.*, 1999), and another potential source of energy is heat trapped within the forest cover. Turbulent mixing within the canopy maintains canopy temperature at approximately the same as air temperature, and canopy temperature is therefore lower than would be the case if there was no turbulence: the emission of long-wave radiation is therefore reduced. Shuttleworth (1989) also suggests that forest stands can capture more solar radiation than other vegetation types due to internal reflection within the stand. The availability of this "extra" energy means that intercepted water can be evaporated at night when there is no radiant energy and also when stomata are closed so there is no transpiration (Singh & Szeicz, 1979; Stewart, 1977).

Intercepted snow evaporates at a different rate from intercepted rain, because a snow-covered canopy is aerodynamically smoother than a wet canopy. For a given windspeed, there is therefore less turbulence above a snow-covered canopy, and the rate of evaporation of intercepted snow will be lower than that of intercepted rain. Calder (1990) calculated that the aerodynamic resistance of a snow-covered coniferous forest was an order of magnitude lower than that of a forest wet with rain, and Lundberg *et al.* (1998) showed that the rate of evaporation of intercepted snow from a Sitka spruce plantation near Aviemore in Scotland was just over a third that of the evaporation from intercepted rain under the same atmospheric conditions. However, more water can be stored on vegetation as snow than as liquid, so although the *rate* of evaporation of snow may be substantially lower than that of rain, over a season the difference in total evaporation may not be so large. The higher turbulence over a forest than over open ground means that evaporation of intercepted snow is generally larger than the evaporation of snow lying on the ground (Nakai *et al.*, 1999).

Evaporation from wetlands

Wetlands can broadly be defined as ecosystems which have a significant excess of water for a large proportion of the time, although there are many different types of wetland with different vegetation assemblages, areas of open water and duration of saturation. There have been very few detailed studies of the transpiration rates of wetland vegetation (Gilman, 1994), and available evidence suggests considerable variability depending on the wetland vegetation type, the area of open water and the heat storage capacity of the wetland. Burba *et al.* (1999), for example, studied evaporation rates in a prairie wetland in Nebraska, showing that the open water acted as a store of heat providing energy both to plants and the water body during the night. Although the daytime evaporation rates of the two plant species on the wetland were greater than daytime open water evaporation rates, over a 24-hour period more water was evaporated from the open water. In contrast, Parkhurst *et al.* (1998) found in a much smaller prairie wetland that less heat was stored during the day and evaporation followed the diurnal energy cycle. Allen *et al.* (1997) argued that the rate of evaporation from vegetation floating on water – in the specific case of the Florida Everglades – should be little different from, and probably slightly lower than, the rate of evaporation from the open water. Evaporation rates from bogs and fens, which tend to have little open water, are generally substantially lower than evaporation rates from open water. Campbell & Williamson (1997) estimated that evaporation from a raised bog in New Zealand was only 34% of the open water rate, because the plants living in the nutrient-poor bog environment had very high stomatal resistances. La Fleur & Roulet (1992) also measured evaporation rates from bogs close to Hudson Bay well below the open water rate.

Total evaporation from an area

The previous sections have considered separately evaporation from open water, evaporation from bare soil, transpiration and the evaporation of intercepted water. The total evaporation from a portion of the land surface (such as a catchment) is the sum of all four: what is their relative importance? There are two major factors determining the contribution of each of these sources of evaporation to total evaporation. The first is, rather obviously, the proportion of the land surface covered by open water, bare soil and different types of vegetation. The second is the climatic characteristics of the area.

Where the vegetation is sparse, evaporation from the soil may constitute a large part of the total, particularly when the soil is wet. Evaporation from wet bare soil is much greater than transpiration from plants under the same meteorological conditions. Gash *et al.* (1997), for example, found during the HAPEX-Sahel experiment that evaporation from bare soil could account for over 50% of the total areal evaporation immediately after rain, but less than 20% after just a couple of days: total plant transpiration varied little following rainfall. High spatial and temporal variability in rainfall therefore translates into very high spatial and temporal variability in evaporation. Gash *et al.* also

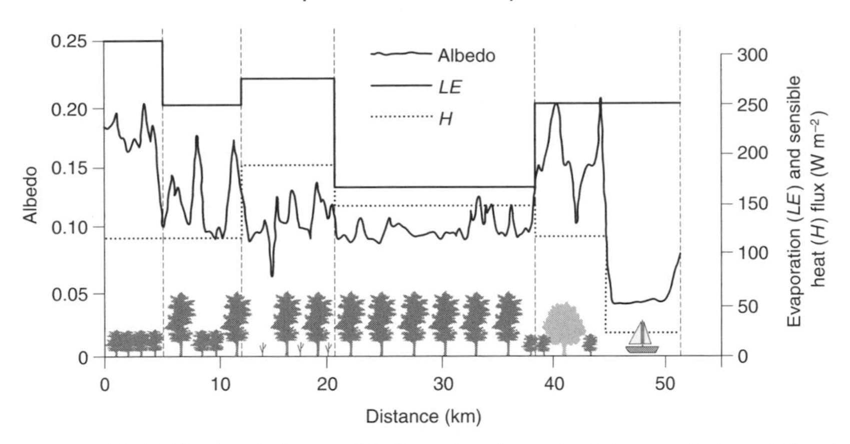

Figure 3.18 Variability in albedo, sensible heat flux, latent heat flux on 14 June 1994 along an agricultural–forest–agricultural–lake transect, eastern Sweden (Frech & Jochum, 1999)

Table 3.6 Annual precipitation (*P*), transpiration (*T*) and evaporation of intercepted water (*I*) (from data in Ward & Robinson, 2000; Grelle *et al.*, 1997; Murukami *et al.*, 2000; Granier *et al.*, 2000)

	P (mm)	*T* (mm)	*I* (mm)	*I*/*T*
UK conifers:				
Balquihidder (Scotland)	2500	280	710	2.5
Plynlimon (Wales)	1820	310	520	1.7
Stocks (NW England)	980	340	370	1.1
Thetford (E England)	600	350	210	0.6
Swedish conifers[1]		243	74	0.3
Japanese conifers				
Mature	1343	333	182	0.5
Young	1480	294	83	0.3
Beech woodland, France[1]	462	255	120	0.5
Heather (Scotland)	2500	170	350	2.1
Grass (Scotland)	2500	160	200	1.3
Rainforest	2640	990	330	0.3

[1] Summer season only

showed how different patches of vegetation translated different proportions of the incoming energy into evaporation: there was considerably more evaporation from the fallow grassland than the patterned woodland ("tiger bush"), and least from the cultivated millet. The mix of land covers within an area will therefore affect both total evaporation and also the distribution of evaporation over time. The effect of mixed land cover on evaporation is illustrated in Figure 3.18, which shows evaporation along a transect across agricultural land, forest and open water in eastern Sweden (Frech & Jochum, 1999).

Table 3.6 compares the total annual or seasonal amounts of transpiration and evaporation of intercepted water for a number of different vegetation types and climate zones. The difference between different vegetation types is clear. The Scottish grass and heather, for example, have similar transpiration totals, but there is less interception loss from grass; both, however, have far less interception than the forest. Within the UK coniferous forests shown in Table 3.6, the relationship between transpiration and interception varies with climate. Interception losses are considerably higher in the upland locations, where it rains more frequently. The tree canopy is therefore wetter for longer, and evaporation from the intercepted water can take place for longer. At Thetford, however, rain is less frequent and therefore stores are full less frequently. Here, transpiration is dominant: the maximum rate of transpiration is lower than the rate of evaporation of intercepted water, but transpiration can continue at this rate for much longer. Interception loss is small relative to transpiration in the Amazonian rain forest for the same reason. The balance between transpiration and interception may vary through the year, and it can be expected that interception will be relatively more important during winter. In the Swedish example (Grelle *et al.*, 1997) interception loss was a small proportion of total evaporation. The Japanese example (Murukami *et al.*, 2000) shows how the age of the forest vegetation affects the balance between transpiration and interception. The younger stand has a lower interception loss than the mature stand because the vegetation is less dense.

Many forests consist of a number of layers, all of which may contribute to total stand evaporation. Approximately 13% of the interception loss in the Scots pine woodland at Thetford (Table 3.6), for example, originated from the bracken understorey (Gash & Stewart, 1977). Evaporation from moss and small shrubs contributed 15% of the summer evaporation in a Swedish coniferous forest (Grelle *et al.*, 1997), and Schaap *et al.* (1997) estimated that forest floor evaporation contributed between 7 and 13% of the total annual forest evaporation for a Douglas fir forest in the Netherlands. Substantially higher proportions have been found in other forest environments, however, with approximately 54% of the total summer evaporation in a Siberian pine forest deriving from the 30 mm lichen cover on the forest floor (Kelliher *et al.*, 1998). This has been attributed to the low proportion of incoming radiation intercepted by the trees.

Estimating evaporation

As indicated above, it is difficult to measure evaporation directly. Consequently, a large number of equations have been produced to estimate evaporation – usually potential evaporation – from available meteorological data.

Many of these equations are empirical, often derived from pan evaporation data, and parameterise relationships between measurable data (usually temperature) and evaporation. Examples include the Thornthwaite formula (Thornthwaite, 1948: using temperature and day length), the Hargreaves equation (Shuttleworth, 1993: using temperature and incoming solar radiation), the Blaney–Criddle formula (Shuttleworth, 1993: using temperature, relative humidity, sunshine hours and windspeed), and the Hamon formula (Hamon, 1963).

Penman (1948) was the first to derive an equation combining the energy used to drive evaporation with a description of the turbulent diffusion of water vapour into the atmosphere (and hence it is often called a "combination equation": Box 3.6). The aerodynamic part (essentially equation 3.6) is described by an empirical function involving windspeed and vapour pressure deficit. Monteith (1965) refined the Penman equation to include an explicit representation of the turbulent exchange processes, to produce what has become known as the Penman–Monteith equation (Box 3.6). Although the Penman–Monteith equation is theoretically most accurate, because it represents the physical processes involved, it is quite difficult to apply in practice because it needs data on aerodynamic and surface resistances. The surface resistance value is most uncertain, especially how it varies with soil moisture stress. The Penman–Monteith equation is often applied "in reverse" with observations of evaporation E in order to estimate r_s. Nevertheless, the Penman–Monteith formulation is at the heart of climate model estimations of evaporation (Chapter 8), the MORECS operational potential evaporation estimation procedure used in the UK (Hough & Jones, 1997), and the evaporation estimation procedure recommended by the UN Food and Agriculture Organisation (FAO) (Allen *et al.*, 1998).

Box 3.6 The Penman and Penman–Monteith equations

The Penman equation is:

$$PE = \frac{\Delta R_n + \gamma E_{ap}}{\Delta + \gamma} \text{ mm d}^{-1}$$

where R_n is net radiation (mm d^{-1}), E_{ap} is the "aerodynamic" term, Δ is the slope of the relationship between saturation vapour pressure and temperature at temperature T (hPa °C^{-1}), and γ is the psychometric constant (=0.66 hPa °C^{-1}). The aerodynamic term is defined by an empirical function, and E_{ap} *in millimetres per day* is

$$E_{ap} = 0.26(e_s - e)(1 + 0.536u)$$

where u is windspeed in m s^{-1}, e is the vapour pressure and e_s is the saturation vapour pressure (equation 3.1), both in hPa. The slope of the saturation vapour pressure and temperature relationship at temperature T is the derivative of equation 3.1, which is

$$\Delta = \frac{4098\, e_s}{(237.3 + T)^2}$$

If net radiation is given in mm d^{-1}, then the Penman equation above can be used to calculate potential evaporation in mm d^{-1}.

(continued)

(continued)

The Penman–Monteith equation is:

$$E = \frac{1000}{\lambda\rho_w}\left[\frac{\Delta R_n + 86.4\,\rho_a c_p (e_s - e)/r_a}{\Delta + \gamma[1 + r_s/r_a]}\right] \text{mm d}^{-1}$$

where r_a is the aerodynamic resistance (in s m^{-1}: equation 3.7) and r_s is the surface resistance (typically the canopy resistance for a vegetated surface: also in s m^{-1}). ρ_a is the density of the air (kg m^{-3}), and c_p is the specific heat capacity of the air (1.013 kJ kg^{-1} °C^{-1}). λ is the latent heat of vaporisation (MJ kg^{-1}: equation 3.3), and ρ_w is the density of water (1000 kg m^{-3}). In the Penman–Monteith equation above, net radiation R_n is in MJ m^{-2} d^{-1} (11.574 MJ m^{-2} d^{-1} is equal to 1 W m^{-2}). Note that different textbooks express the Penman–Monteith equation in slightly different ways, depending on the units used. Because r_s can vary with soil moisture content and water availability generally, the E in the above equation is actual evaporation: if the maximum surface conductance is used, then the equation estimates *potential* evaporation.

The energy term of the Penman equation tends to be larger than the aerodynamic term by a factor of around 4, and a simplified version of the Penman equation is the Priestley–Taylor equation (Priestley & Taylor, 1972):

$$PE = \alpha \frac{\Delta}{\Delta + \gamma} R_n$$

where PE is in the units of R_n, and α is a factor, generally taken to be 1.26. The equation is widely used to infer potential evaporation directly from radiation and temperature, but is unreliable in very dry areas or where there is significant advection (Ward & Robinson, 2000).

There have been several comparisons of the different procedures to estimate evaporation from meteorological data (e.g. Jensen *et al.*, 1990; Fenessey & Vogel, 1996; Federer *et al.*, 1996; Vörösmarty *et al.*, 1998; Beyazgul *et al.*, 2000). Some studies have compared estimates with pan evaporation data, whilst others have compared estimated evaporation with precipitation minus runoff. All have indicated that the Penman–Monteith equation generally gives the best estimates.

Evaporation when water is limited

The previous sections have alluded to what happens to the rate of evaporation when water is limited: evaporation from bare soil declines once the top of the soil dries, plants reduce their transpiration rates by closing stomata if soil moisture falls, and may not be able to access water through their roots if the soil becomes too dry. Because of the lack of detailed understanding of the physical processes involved, numerous attempts have been made to define

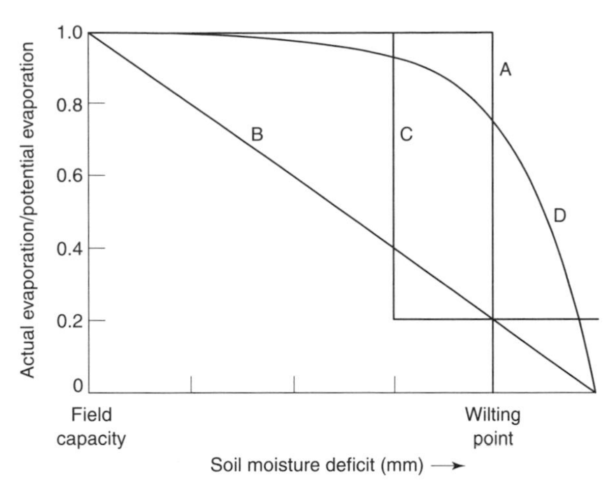

Figure 3.19 Reduction in the rate of evaporation as soil moisture content falls

empirical relationships showing how evaporation falls below the potential rate once soil moisture falls.

When water infiltrates into soil following rainfall, it fills the pore spaces. Water drains down through the soil, but some is held in the soil pores against gravitational forces (see Section 3.5). This amount of storage is termed the *field capacity*, and if soil moisture contents are at or above field capacity it is generally believed that evaporation (and transpiration) can proceed unchecked. Plants can continue to extract water from the soil as soil moisture content falls below field capacity, but plants cease to be able to access water once soil moisture content reaches *wilting point* (before the soil is completely dry). There are, however, several possible ways in which evaporation and transpiration can fall as soil moisture content varies between field capacity and wilting point. Figure 3.19 shows four possible pathways.

Pathway A assumes that evaporation proceeds at the potential rate until soil moisture content reaches the wilting point, and then suddenly ceases. This may occur when roots extend widely through the soil (Ward & Robinson, 2000). Pathway B assumes a linear decline, because soil moisture movement falls as the soil dries (Section 3.5). Pathway C assumes that plants can continue to evaporate at the potential rate up to a given soil moisture deficit, and then evaporation falls rapidly to a low figure. This pathway was proposed by Penman (1948), who defined the soil moisture deficit at which evaporation begins to decline rapidly as the *root constant*, dependent on vegetation type. Pathway D assumes a negative exponential decline, with the rate of reduction in evaporation falling more rapidly as the soil dries: Figure 3.13 gives empirical evidence for this pathway. In practice, it is possible that each of these pathways will exist in different situations. Functions such as those shown in Figure 3.19 are widely used to convert estimates of potential to actual evaporation both operationally and in hydrological models, and some procedures (such as MORECS: Hough & Jones, 1997) use more complicated two-layer representations.

Evaporation and transpiration: an overview

Evaporation is really at the heart of the hydrological cycle, and hydrologists over the past few years have placed an increasing emphasis on measuring, understanding and modelling evaporation and transpiration. Vegetation clearly plays a very important part in evaporation, both by intercepting water and by transpiring, and both vegetation composition and plant physiological properties are significant in determining the total evaporative flux from an area (and, indeed, therefore the partitioning of energy: Chapter 8). But so too are the climatic characteristics of a site, with the relative importance of transpiration and the evaporation of intercepted water for a given vegetation type depending on how frequently rain falls.

A number of different techniques for measuring evaporation have been developed over the years, with direct measurement becoming feasible (technically and financially) only during the 1990s. Kite & Droogers (2000) conducted a systematic review of several different techniques in a small area of western Turkey, involving direct measurement on the ground, the use of estimation equations, remote sensing and hydrological models. They found that, in practice, it was difficult to determine the "best" method, and concluded that different procedures were appropriate for different purposes.

Colour plate III shows an estimate of annual average potential evaporation over land, calculated using the Penman–Monteith formula and climate data from New *et al.*'s (1999) gridded data set, together with the ratio of actual to potential evaporation (the actual evaporation is calculated from a hydrological model applied in each grid square, and is as shown in Plate I). The mapped ratio of actual to potential evaporation reflects the availability of moisture: in temperate and tropical regions, annual actual evaporation is generally greater than 50% of the potential evaporation.

3.5 Soil moisture

The soil is the interface between the atmospheric inputs of precipitation and energy to the catchment system and the outputs of runoff and streamflow. The soil influences the amount of net precipitation (throughfall plus stemflow) that infiltrates into the ground, and water is both directly evaporated from the soil and extracted by plants through transpiration. The rate at which the remaining water reaches the river channel is a function of soil characteristics. This section therefore examines the role of the soil layer in the water balance, looking first at soil water contents and their variability over time, next at the movement of water within the soil, and finally at the process of infiltration – the means by which water enters the soil.

Soils are generally made up of different layers or horizons, each with different properties. The "soil profile" characterises the variation in soil properties with depth, and varies between soil types, depending on the characteristics of the bedrock, the time over which the soil has developed, and the processes operating on the soil profile (which are largely a function of climate). The soil

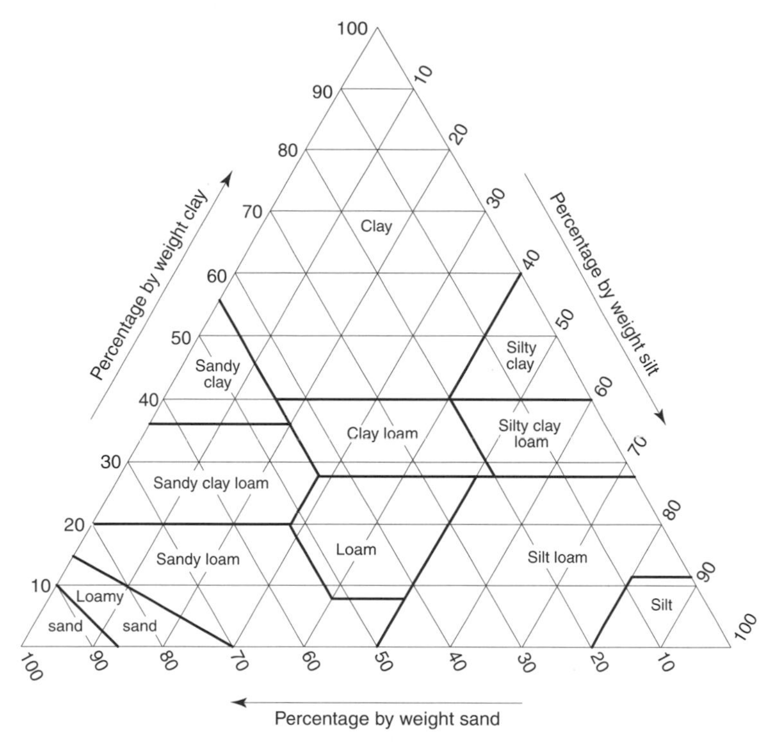

Figure 3.20 Soil-texture triangle, showing texture terms applied to soils with differing proportions of sand, silt and clay (Dingman, 1993)

system is made up of three *phases*: the solid phase, consisting of mineral and organic particles, the gaseous phase, consisting of the air stored within the soil, and the liquid phase, consisting of the water and the associated dissolved matter held within the soil. Water and air are held in pores in the soil, and the size and distribution of these pore spaces depend on soil *texture* and *structure*. Soil texture is a function of the size distribution of the mineral and organic particles in the soil. Figure 3.20 shows a classification of soil texture, based on the percentage of sand (0.05 to 2 mm diameter), silt (0.002 to 0.05 mm diameter) and clay (less than 0.002 mm diameter), and Figure 3.21 gives some estimates of the porosity of different soil types: the porosity is defined as the proportion of pore spaces in a unit volume of soil. Soil structure is determined by the way in which the soil particles are aggregated together into *peds*. Lines of weakness between peds can be important flow pathways, and soil structure is a function of soil formation, soil stability and, significantly, vegetation and disturbance. Structural voids can also store water.

Soil water content

The amount of water held in soil is typically expressed in terms of the proportion of the total soil volume that is made up of water:

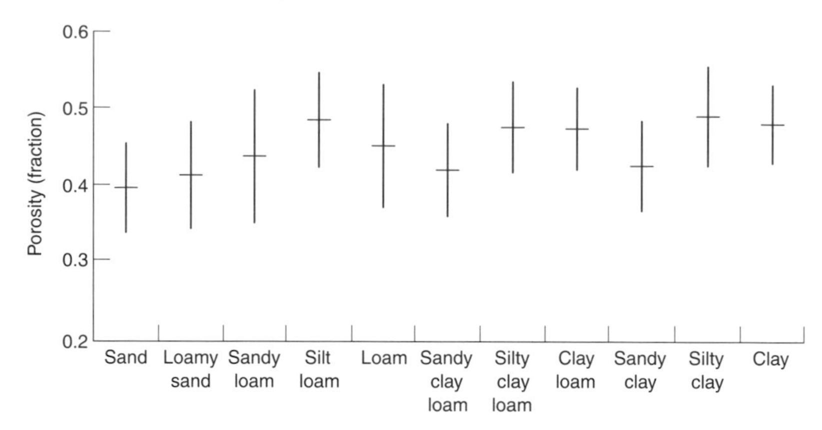

Figure 3.21 Porosity of different soil texture types (adapted from Clapp & Hornberger, 1978)

$$\theta = \frac{V_w}{V_t} \text{ cm}^3 \text{ cm}^{-3} \qquad \textbf{3.9}$$

where V_t is the volume of the soil column and V_w is the volume of water. Water is held in the soil by the balance of two opposing sets of forces. Gravitational forces or pressures pull the water down through the soil. They are opposed by a combination of *capillary*, *adsorptive* and *osmotic* forces which hold the water to the soil. Capillary forces arise from surface tension and bind water to the soil particles. A thin film of water can be adsorbed upon the soil particles by electrostatic forces, whereby water molecules are attracted to the charged faces of the soil particles. Adsorptive forces are usually small and difficult to distinguish from capillary forces, and the two together are frequently termed *matric forces*. Osmotic forces arise due to solutes in the soil water, and are generally negligible (Ward & Robinson, 2000).

The matric and osmotic forces together make up the total suction (denoted by the symbol ψ) holding water in the soil (see Box 3.7 for a note on terms). A force equal to or greater than this suction is needed to extract the water from the soil, and the suction that needs to be exerted varies with soil moisture content: the less the amount of water in the soil, the more tightly that water is held. The relationship between soil water content and suction is known as the *moisture characteristic curve* and varies with soil texture. The smaller the pores, the more tightly the water is held within the soil. Figure 3.22 shows some example soil moisture characteristic curves for different soil materials, showing the variation with soil moisture content and soil texture (note that there are several different sets of empirical functions relating suction to soil water content: see for example Clapp & Hornberger (1978); Brooks & Corey (1964); van Genuchten (1980); Saxton *et al.* (1986)). In practice, soil moisture curves are slightly different depending on whether the soil is wetting or drying: this is known as hysteresis, but is usually ignored.

Section 3.4 introduced, rather informally, two important soil concepts: field capacity and wilting point. Field capacity is defined as the water which can be

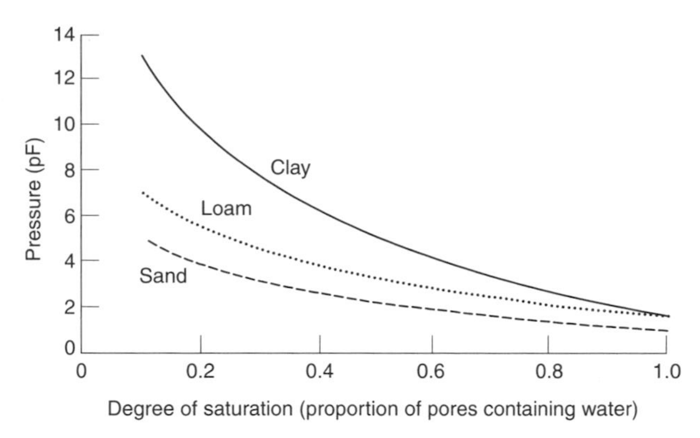

Figure 3.22 Soil moisture characteristic curves for different soil moisture contents and soil types (adapted from Clapp & Hornberger, 1978). The soil is saturated at 1 and completely dry at 0

Box 3.7 Soil moisture: terms and units

The terms suction, tension, pressure and potential are all used to describe the forces holding water in soil: they all mean the same. The forces are also measured in a range of units. Sometimes the forces are expressed in pressure units, either in bars or kPa (1 kPa equals 100 bar), but are more usually expressed in terms of the head of water (either in centimetres or as $pF = \log_{10}(-\psi)$, where ψ is in cm). A pressure of 0.1 bar corresponds to 10 kPa, −100 cm and a pF of 2. Note that the pressure head is negative, because it is acting against atmospheric pressure.

held in the soil against gravity, and this is usually taken to correspond to a pressure of around −333 cm (pF approximately 2.5). Wilting point is the point at which plants cannot extract water: this is conventionally taken to be −15 000 cm (pF approximately 4.2). The difference between field capacity and wilting point is the available water capacity. A *soil moisture deficit* exists if soil moisture is below field capacity.

Table 3.7 gives some values of the volumetric soil water content at field capacity, wilting point and when the soil is saturated (all the pores are filled) for three different soil textures. When saturated, a column of clay contains around 48% water, whilst a column of sand would only hold around 40%: the clay can hold more because although it has smaller pores, the total pore space is greater than in sand. When clay is at field capacity, the soil column is 40% water, but a sandy soil at field capacity would only be around 17% water. A "dry" clay soil (as far as the plants are concerned) would still comprise close to 29% water, whilst a dry sandy soil would only be around 7% water by weight. Water retention also increases as soil organic matter increases. Box 3.8 describes methods of measuring soil water content.

Table 3.7 Soil water content (θ), as proportion of total soil volume, at saturation, field capacity and wilting point, for three example soils (from Clapp & Hornberger's (1978) equations)

	Sand	Loam	Clay
Saturation	0.395	0.451	0.482
Field capacity	0.173	0.313	0.400
Wilting point	0.068	0.155	0.287

Box 3.8 Measuring soil water content

There are two groups of methods for estimating the amount of water held in soil. The first group aims to measure water content itself, using a variety of techniques (see Rawls *et al.*, 1993; Schmugge *et al.*, 1980; Boucher, 1997). The *gravimetric* method involves digging out a defined quantity of soil (typically 100 g), weighing it, drying the soil, then reweighing it and determining the amount of water originally held in the soil by the difference in weight. The method is accurate, but destructive: repeat samples from the same location are not possible. Two radiological methods are used to determine soil water content. One uses a neutron probe to measure the interaction between high-energy neutrons and the nuclei of hydrogen atoms in the soil, and the other measures the attenuation of gamma rays as they pass through the soil. Both need to be calibrated, usually using the gravimetric method, but are non-destructive and repeated measurements can be made at the same site. More recently, a number of procedures have been developed based on the measurement of the *dielectric constant* (a measure of ability to store, rather than conduct, energy) of the soil using an electromagnetic pulse between a pair of rods inserted into the soil. The dielectric constant of soil is a function of soil moisture content, and again measurements of dielectric constant need to be converted to soil moisture content using a calibrated relationship. A capacitance probe is a hand-held device for measuring dielectric constant (Robinson & Dean, 1993), whilst time domain reflectometry (TDR: Brisco *et al.*, 1992)) uses permanently-installed measuring rods.

Over the past two decades, considerable effort has been expended in attempting to measure soil moisture contents through remote sensing (Jackson *et al.*, 1996). Three sensors have been used: a passive microwave sensor, which measures microwave radiation emitted from the soil (Jackson, 1993; Jackson & Le Vine, 1996; Njoku & Entekhabi, 1996), an *active* microwave sensor (Blyth, 1997) and synthetic aperture radar (SAR), which determines soil moisture content by measuring soil dielectric properties. Ground penetrating radar (which is an instrument applied

(continued)

(continued)

at the ground surface) has also been used to estimate the soil dielectric constant (van Overmeeren *et al.*, 1997). In principle, remote sensing methods allow the estimation of the spatial pattern of soil moisture contents – which is extremely important – but in practice most remote sensing procedures developed so far are only capable of estimating moisture contents in the top few centimetres of the soil.

The second group of methods essentially measures matric potential and infers water content from the moisture characteristic curve. *Tensiometers* consist of a porous cup connected to a pressure-measuring device. The cup is buried in the soil and filled with water. Water flows between the soil and cup and the pressure potential inside the cup reaches equilibrium with that in the soil. As long as the cups are frequently refilled tensiometers can provide continuous measurements of matric potential if connected to a data logger, but they are not very accurate at low soil moisture contents (below around −800 cm: Hillel, 1980). *Piezometers* are small, narrow tubes with holes at the base sunk into the soil. Water enters and rises up the piezometer tube, reaching a level dependent on the pressure with which the water is held. *Electrical resistance* methods use electrical probes buried in a porous medium (such as gypsum) to measure the resistance across the medium, the water content of which reaches equilibrium with the soil water suction. The method is non-destructive, cheap – but not very reliable at high water contents. Again, measurements need to be calibrated.

Figure 3.23 summarises the distribution of water in a simple, idealised soil, and defines some terms. In the saturated zone all the pore spaces are filled with water, which can drain down no further. The water is at atmospheric pressure. The water table is the top of the saturated zone: the saturated zone may be "perched", or in other words resting on an impermeable layer, below which the soil is unsaturated. The unsaturated zone is also termed the vadose zone, and is frequently divided into the soil water zone – which contains plant roots – and the intermediate zone, below root depth. Shallow soils, of course, may not contain an intermediate zone. Immediately above the water table is the *capillary fringe*. Here, most or all of the pore spaces are filled, but the water is held by capillary forces against gravity, and pressure is less than atmospheric.

Soil water movement

Experiments by the French engineer Darcy in the mid-nineteenth century showed that the total discharge through a porous medium (in his case sand packed into iron pipes) between two locations was proportional to the difference in pressure head between the two points. The total pressure on water h in

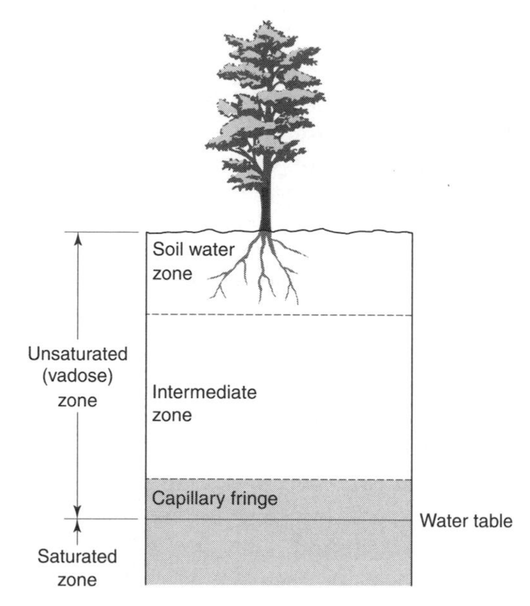

Figure 3.23 The distribution of water in a soil

the soil at a point is equal to the elevation head z – the height above some arbitrary datum – plus the matric potential ψ, which depends on soil water content θ:

$$h = z + \psi \tag{3.10}$$

Matric potential is negative, and so offsets the elevation head. Water within the soil will flow from high to low pressure. Darcy's Law calculates the water flow per unit area (q) between two points from:

$$q = -K\frac{dh}{dl} \tag{3.11}$$

where dh is the difference in total pressure between the two locations, dl is the distance, and K is termed the *hydraulic conductivity*. It is a measure of the ability of the soil to transmit water, and depends on soil properties (particularly pore size distribution) and, significantly, soil water content θ. The minus sign in Darcy's Law arises because the gradient dh/dl in the direction of flow is negative, whereas flow is considered to be positive. If the movement of water in the soil is assumed to be vertical, then equation 3.11 can be rephrased as:

$$q_z = -K\left(\frac{d(z+\psi)}{dz}\right) = -K\left(\frac{d\psi}{dz}+1\right) \tag{3.12}$$

where q_z is the rate of flow in the vertical direction. Both K and ψ depend on soil water content θ.

The hydraulic conductivity of the soil has a very significant influence on the rate of flow in soil. Figure 3.24 shows the *saturated* hydraulic conductivity

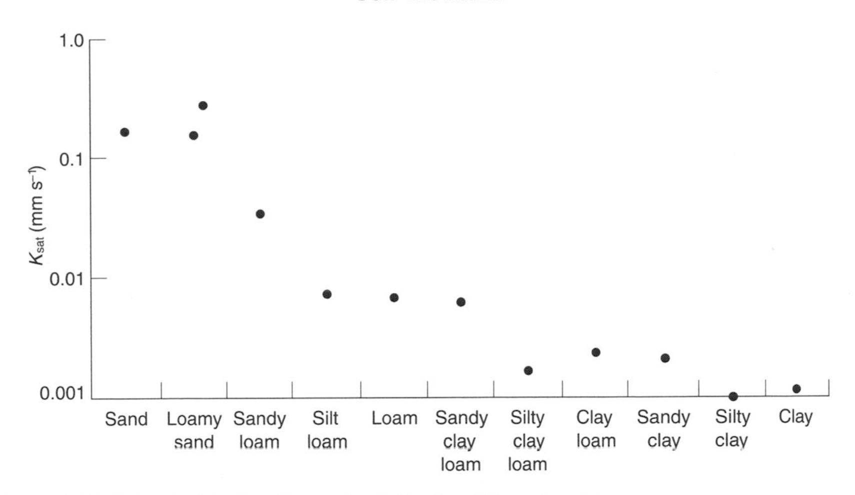

Figure 3.24 Saturated hydraulic conductivity for different soil types

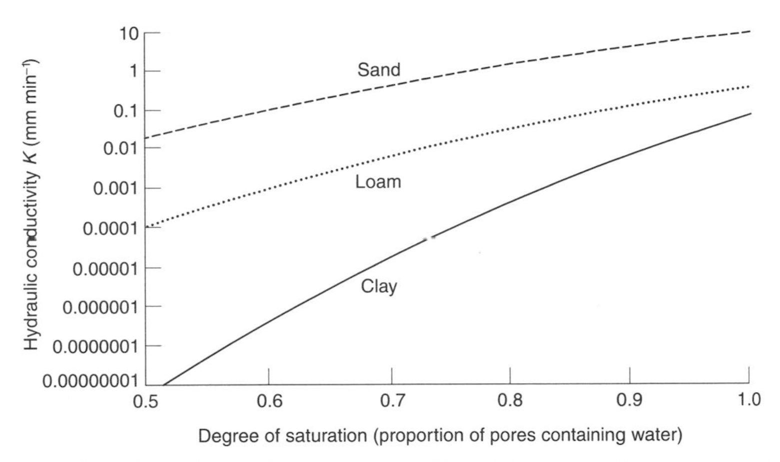

Figure 3.25 Variation of hydraulic conductivity with soil type and soil water content

for 11 soil types: this is the hydraulic conductivity when all the soil pores are full of water. There is clearly a very substantial difference between soil types, with the maximum rate of water movement in a clay soil nearly two orders of magnitude lower than that in a sandy soil where the pores are larger. Figure 3.25 shows the variation of hydraulic conductivity K with soil water content θ for three different soil types (based on Clapp & Hornberger's (1978) empirical equations). Note the logarithmic scale, and the very rapid decline in soil water movement as the soil dries. When the soil is "half dry" (50% of the pore space is filled with water) the flow rate in a sandy soil is a thousand times less than when the soil is saturated: in a clay soil the rate of decline is even more rapid.

Darcy's Law enables the calculation of a "snapshot" of the rate of flow between two points. The rate of change in the amount of water stored in a block of soil is equal to the difference between the inflow to that block and the outflow from it:

$$\frac{d\theta}{dt} = -\frac{dq_z}{dz} \qquad \mathbf{3.13}$$

Substituting equation 3.12 into equation 3.13 gives:

$$\frac{d\theta}{dt} = \frac{d\left[K\left(\frac{d\psi}{dz}+1\right)\right]}{dz} \qquad \mathbf{3.14}$$

which is known as the Richards equation: again, both K and ψ depend on θ.

Darcy's Law, and therefore the Richards equation, characterise the physical principles of flow through homogeneous soils. Soil profiles are not, of course, homogeneous, and in practice it is therefore necessary to apply the equations separately for each soil layer. More significantly, both the form of Darcy's Law and the calculation of hydraulic conductivity assume that water is moving in the soil slowly from pore to pore. In practice, many soils have larger void spaces known as *macropores* (Jones, 1997a), which may be formed by structural cracks within the soil, plant roots, worms and, at the most extreme, burrowing animals. Water will flow much more rapidly in macropores than through the soil matrix (perhaps up to around 3 mm s^{-1} – compare with Figure 3.25), although the effect of macropores depends on their continuity and interconnectivity (Beven & Germann, 1982). Flow in macropores will occur only where the soil is locally saturated, because otherwise any water will be drawn into the smallest pores first, where matric potential is highest. The main consequences of macropores are that water may "bypass" the soil matrix itself (and thus not be available for plants), recharge groundwater when the soil is dry (Section 3.6), generate rapid streamflow (Section 3.7), and affect significantly the transport of material to rivers and groundwater (Chapter 5).

Infiltration into the soil

Infiltration is the movement of rain and melting snow into soil. The maximum rate at which water can move into the soil is termed the *infiltration capacity.* If the surface of the soil is initially dry, then the infiltration capacity is high because the gradient in matric potential close to the surface is high. As water infiltrates, the surface of the soil will get wetter and the gradient will reduce, so infiltration capacity falls. As the surface layer of the soil becomes saturated, then the matric pressure gradient approaches zero and the infiltration capacity approaches the saturated hydraulic conductivity (equation 3.12 with $d\psi/dz$ equal to zero).

If the rainfall rate is less than the infiltration capacity, then water will infiltrate into the soil at the rainfall rate. If, however, the rainfall rate is greater than the infiltration capacity – perhaps because the surface layer of the soil has become saturated and the infiltration capacity has fallen – then water will

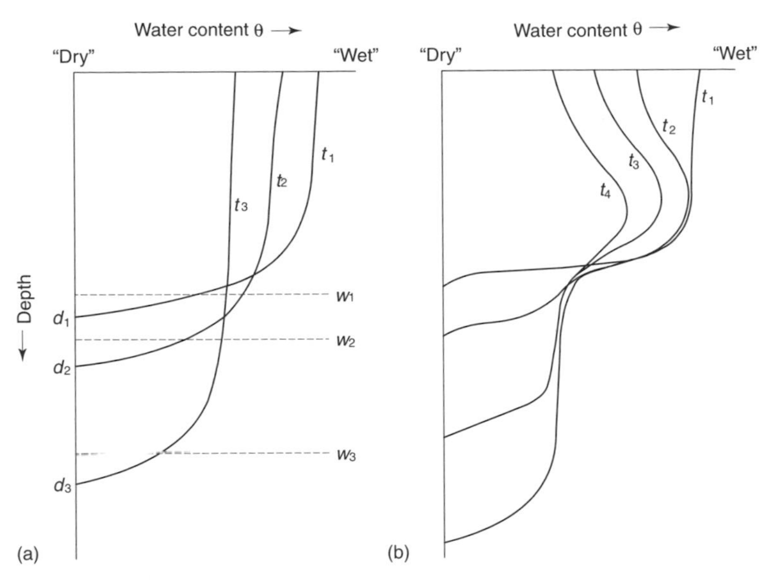

Figure 3.26 Variation in soil moisture content after infiltration with depth and time (modified from Dingman, 1993)

infiltrate at the infiltration capacity and excess water will be ponded on the surface or will flow across the surface. Water infiltrating into the soil will initially accumulate in the top few centimetres. Drainage of water from this surface layer depends on the hydraulic conductivity (Darcy's Law), and as long as the rate of drainage out of the layer is slower than the rate at which water is infiltrating into the soil, water will continue to accumulate. However, as the top layer becomes wetter, hydraulic conductivity increases, and a point may be reached when the rate of flow out of the top layer exceeds the rate of flow into that layer. Water will then begin to move down the soil profile, and the *wetting front* advances: the wetting front is the border between "wet" soil above and "dry" soil below. It is diffuse in fine soils, but can be very sharp in coarse-grained soils. Figure 3.26 shows typical patterns of variation in soil moisture content over depth and time following an input of rainfall (Dingman, 1993). At time t_1, the soil is wet at the surface and dry at depth d_1. The wetting front is at w_1. At time t_2, water has moved from the upper part of the soil, and the soil is now "wet" down to depth d_2; the wetting front is at w_2. In coarse soils or where the amount of water in the surface layer is greater than the field capacity, then a "bulge" in the soil water content profile develops (Figure 3.26b) because water drains rapidly downwards from the surface and fills available pore spaces. A similar pattern appears when evaporation depletes the upper soil layers.

The infiltration characteristics of a soil are extremely important in terms of getting water into the soil. Box 3.9 summarises the methods used to estimate infiltration capacity. The minimum infiltration capacity for a soil is, as indicated above, the saturated hydraulic capacity, which varies with soil texture.

Box 3.9 Measuring infiltration capacity

Infiltration rates in the field are measured using a *ring infiltrometer*. This is essentially a cylindrical ring pushed into the soil. Water is applied to the infiltrometer, and the rate of infiltration can be estimated by measuring the rate at which the level of ponded water decreases. Alternatively, the rate at which water needs to be added to maintain a constant level of ponding can be measured. Tension infiltrometers (Reynolds & Elrick, 1991) can supply water to the soil at different pressures, so that water can be prevented from infiltrating through pores of a certain size: they are used to determine the effect of pore size on infiltration, and also to determine macropore infiltration. Finally, infiltration pathways can be identified by spreading a tracer (such as a dye) on the surface and carefully removing layers of soil after having applied water.

However, the saturated hydraulic conductivity *at the surface* may be very different from that in the underlying undisturbed soil for several reasons. The presence of organic matter on the surface may substantially increase infiltration capacities – although large amounts of leaf litter can locally inhibit infiltration. Vegetation also alters the structure of the soil surface, and infiltration capacity tends to be highest where there is dense vegetation (Cerda (1997) found significant seasonal variations in infiltration capacity in a Mediterranean catchment due to differences in vegetation through the year: infiltration rates were much higher in summer). Some soils contain clay minerals which swell when wet, substantially reducing infiltration rates: these same soils crack when dry, and the cracks may allow rapid infiltration. Other types of macropore, created by worms or decayed vegetation, for example, may encourage infiltration (Heppell *et al.* (2000), for example, found that 86% of the infiltration into a clay soil near Oxford occurred through earthworm burrows). Where vegetation is sparse and overland flow occurs, fine sediment may be carried into the larger pores, effectively reducing infiltration capacity. Some soil types, particularly in semi-arid areas, form an impermeable crust when wet (Section 3.7): Peugeot *et al.* (1997) found that infiltration capacities of some Sahelian soils were reduced by a factor of between 3 and 6 when crusts formed. If a soil with high water content freezes, the frozen surface may become nearly impermeable, although on the other hand frost action can open up the soil structure. Finally, many human activities have very direct effects on infiltration capacity (Chapter 6).

Water will continue to move down the soil profile until either it meets an impermeable layer or the rate of flow slows down and finally ceases. This will occur some time after the supply of infiltrating water at the surface stops. With no continued replenishment, the water content in the soil above the wetting front will begin to decline, and hydraulic conductivity will therefore reduce, slowing the movement of water further down the soil profile. Also, the gradient of matric potential decreases as soil water content becomes more uniform.

Because infiltration is such an important hydrological process, many procedures have been developed to estimate the rate at which water will infiltrate into soil (see Rawls *et al.*, 1993). A simple empirical relationship was developed by Horton (1933):

$$f_t = f_c + (f_o - f_c)\hat{e}^{-\beta t} \qquad \textbf{3.15}$$

where f_t is the infiltration rate at time t, f_o and f_c are the maximum and minimum infiltration rates respectively, and β is a parameter. The Green–Ampt model (Green & Ampt, 1911; Rawls *et al.*, 1993) is based on Darcy's Law, and in general terms has the following form:

$$f_t = K_{ga}\left[1 + \frac{(\theta_s - \theta_i)\psi_f}{F_t}\right] \qquad \textbf{3.16}$$

K_{ga} is the effective hydraulic conductivity (related to, but not the same as, the hydraulic conductivity outlined above), θ_s and θ_i are the proportional soil moisture content at saturation (i.e. porosity) and at the start of the rain respectively, ψ_f is the effective matric potential at the wetting front, and F_t is the cumulative infiltration to time t. The effective matric potential at the wetting front can, like the effective hydraulic conductivity, be estimated from soil texture properties. Essentially, equation 3.16 shows that the infiltration rate at a point in time is a function of the hydraulic conductivity, initial soil moisture deficit and, importantly, time (as represented by the cumulative infiltration F_t). The integral of equation 3.16 relates cumulative infiltration to time:

$$t = \frac{F_t}{K_{ga}} - \frac{\psi_f}{K_{ga}}(\theta_s - \theta_i)\ln\left[1 + \frac{F_t}{(\theta_s - \theta_i)\psi_f}\right] \qquad \textbf{3.17}$$

The Green–Ampt model is usually applied by calculating t for a given F_t from equation 3.17 and using equation 3.16 to determine the infiltration rate at time t. The Green–Ampt model has been heavily amended to cope with differing soil layers and soil crusts (Rawls *et al.*, 1993).

Variation in soil moisture over space

The previous sections have been very one-dimensional, and have focused on soils at a point and the vertical movement of soil water. In practice, of course, soil moisture varies considerably over space, for a number of reasons.

The physical characteristics of soil vary over space, and this variation is usually more continuous than the discrete boundaries placed around different soil "types" implies. Small variations in soil texture can have very significant implications for hydraulic conductivity and the moisture characteristic curve (as implied in Figures 3.22 and 3.25). Superimposed onto this variability in soil properties is variability in vegetation, including amount of organic litter at the surface. As indicated above, the presence of organic litter can substantially increase infiltration rates. Several studies (for example Tricker (1981) and Wilcock & Essery (1984)) have shown how the spatial variability in infiltration capacity across a small catchment is largely a function of the amount of organic litter. Soil depth often varies with topography, and this obviously affects the absolute amount of water the soil can hold. Soil moisture content therefore varies across space with soil properties (and also the location of trees). This variation is greatest in dry conditions (e.g. Gottlein & Manderscheid, 1998). Beldring *et al.* (1999) showed how in a part of eastern Sweden the spatial variability in soil moisture content was controlled by the depth of the soil.

As indicated above, water tends to move vertically down the soil profile until either it stops because the rate of flow at the wetting front has slowed, or it meets a relatively impermeable barrier such as a different soil layer or the interface between soil and bedrock. If this impermeable barrier has a slope – as it probably will if it is on a hillslope – then water may move downslope along this barrier. This water will accumulate at the foot of the hillslope, or indeed at any topographic concavity. Water may also move downslope if, on infiltrating vertically downwards, it encounters a "preferential pathway", such as a macropore or a layer of soil with a higher hydraulic conductivity. If this preferential pathway has a gradient, water will flow downslope (Terajima *et al.* (2000) used a fibrescope to observe soil pipes, and found in their study site that they followed the gradient of the hillslope). The consequence of this *lateral flow* is that soil water is redistributed across space, and some parts of the catchment will be much more likely to experience saturated conditions than others. Anderson & Burt (1978) in a classic study mapped the spatial distribution of soil moisture contents across a hillslope, showing systematically higher soil moisture contents in hillslope concavities.

This spatial variability in soil moisture contents is important for many reasons: it influences the partitioning of energy into latent heat and sensible heat (see Chapter 8), and it affects the way streamflow is generated (Section 3.7).

Where does soil water go?

Water that infiltrates into the soil ultimately goes in one of three directions: evaporation, groundwater recharge and streamflow. The evaporation of water from bare soil was considered in Section 3.4. Evaporation creates a suction gradient at the soil surface, stimulating the upward movement of water. This water is supplied to the soil surface by the process of capillary rise, the rate of which is influenced by the depth to the water table and the soil texture: the closer the water table to the surface, the greater the rate at which water can be drawn up through the soil. In arid and semi-arid areas, capillary rise may extend over several metres – perhaps also bringing up minerals from deep in the soil profile to be left behind at the surface as salt when the water evaporates (Ward & Robinson, 2000). Roots also extract water from the soil, and as noted above, roots may also extend down for several metres.

Water that flows laterally through the soil is termed *throughflow*. Throughflow may move water from one part of a hillslope to another part, from which it can be evaporated, but it may deliver water to the river channel: this is examined in Section 3.7.

Water that continues to drain vertically through the soil will, if unimpeded by an impermeable layer, recharge underlying groundwater: this is examined in the next section.

3.6 Groundwater

Roughly 97% of all the freshwater in the world that is not held in ice sheets or glaciers is stored as groundwater (Table 2.1), around 60% of which is deeper than 1 km. Some of this groundwater is "fossil" groundwater, formed in earlier wetter periods (such as the large aquifers beneath parts of the Sahara desert), but most is actively being recharged at present. Groundwater "slows" the hydrological cycle by delaying the movement of water from precipitation to streamflow. The average residence time of groundwater is of the order of 300 years (Ward & Robinson, 2000), although there is very considerable variation around this figure: water may be stored in shallow surface aquifers for just a few days.

An *aquifer* is a layer of rock or unconsolidated deposits which can store and transmit water: an *aquitard* is a material which does not. The terms are, however, relative, and very few natural materials can store no water at all. The base of an aquifer may be formed by an aquitard, but at depths greater than around 10 km compression means that pore spaces are very small and virtually all rocks can be considered impermeable (Price, 1996). A distinction is often made between bedrock aquifers – which may be at some depth – and surface deposit aquifers, which are usually formed in relatively thin layers of unconsolidated deposits such as floodplain or glacial gravels. Bedrock aquifers are generally composed of sedimentary deposits.

Aquifer properties

Water is held in aquifers in pore spaces, cavities, caverns, joints and fractures, depending on the characteristics of the rock (but virtually all water in surface deposit aquifers is held in pore spaces). The *porosity* of an aquifer is defined as the percentage of the total volume of a rock which is represented by voids. Igneous and metamorphic rocks tend to have a porosity of less than 10%, although weathered granite can have a porosity up to 55% (Table 3.8). The porosity of gravel can vary between 25 and 40%, and chalk has a porosity varying between 5 and 30%. The ability of water to move through an aquifer is influenced also by the size of the voids and their degree of interconnectedness, and is characterised by its *permeability* or *hydraulic conductivity*. This varies considerably between different material types (as shown in Section 3.5 for soils): the hydraulic conductivity of gravel may range between 100 and 10 000 m d^{-1}, whilst in unfractured crystalline rocks it may lie between 10^{-9} and 10^{-3} m d^{-1}

Table 3.8 Typical aquifer properties (from several sources)

Material type	Porosity (%)	Hydraulic conductivity (m d^{-1})	Specific yield (%)
Clay	35–60	10^{-7}–10^{-3}	1–5
Sand	30–50	0.01–1000	10–30
Gravel	25–40	100–10 000	20–30
Chalk	5–30	1000–5500	1
Sandstone	5–30	10^{-5}–0.01	5–20
Weathered granite	35–55	10^{-5}–0.1	
Crystalline rocks	0–10	10^{-9}–10^{-5}	

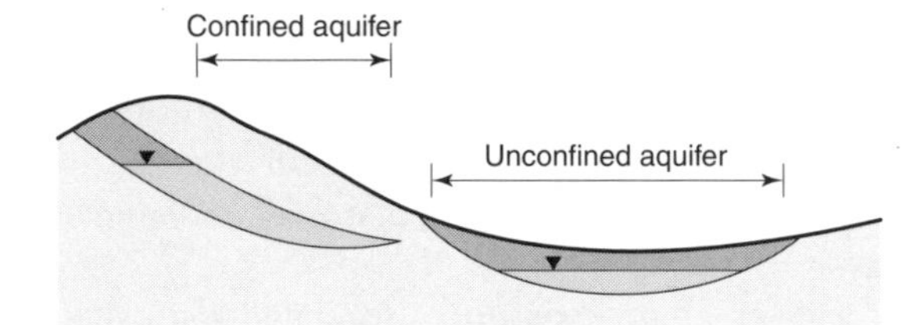

Figure 3.27 Confined and unconfined aquifers

(Table 3.8). Significantly, hydraulic conductivity varies through an aquifer, and also varies with direction. The mean vertical hydraulic conductivity of alluvial gravels, for example, is typically an order of magnitude less than the horizontal hydraulic conductivity, due to the manner in which the deposits are laid down (Beyerle *et al.*, 1999).

An important distinction is between *unconfined* and *confined* aquifers (Figure 3.27). An unconfined aquifer has no cap, and the water table is at the point where the porewater pressure is equal to the atmospheric pressure: the water table can rise as high as determined by the amount of water stored in the aquifer. Water in a well sunk into an unconfined aquifer would rise to the same level as the level of saturation in the aquifer. Above the water table is the capillary fringe (as in a soil), where most of the pores are full of water held by capillary forces. A confined aquifer, however, has an upper boundary, defined by an aquitard. Water in a confined aquifer is therefore held under pressure, and water in a well sunk into a confined aquifer would rise up above the top of the aquifer, with the height of the water above the top of the aquifer dependent on the pressure at which the water is held. The height to which water in such a well would rise is termed *the potentiometric surface*. If the pressure under which the water is held is such that the water would be forced to the surface, the well is termed an *artesian* well. In order to be recharged, of course, a confined aquifer must be unconfined somewhere. A perched aquifer is an aquifer lying on a discontinuous impermeable layer, perhaps some distance above the main body of groundwater.

In an unconfined aquifer, a change in the amount of water stored will produce a change in the water level. The *specific yield* is defined as the volume of water which will come out of a given volume of aquifer (precisely, the volume of water produced per unit area of aquifer for a unit fall in water table): it is usually expressed as a percentage of the total aquifer volume. A specific yield of 20%, for example, would mean that 0.2 m^3 of water would come out of a 1 m^3 block of aquifer. The specific yield is less than the porosity, because some water is held against drainage in the aquifer by capillary and other forces. The amount of water left behind after drainage is termed the *specific retention*, and is analogous to field capacity as used in soils. The porosity of an aquifer is equal to the specific yield plus specific retention. The relationship between specific yield and specific retention depends on the pore size distribution. Clay, for example, has a high porosity, but because the pores are small, specific retention is high and specific yield is very low (Table 3.8). Chalk also has high porosity, but low specific yield – typically 1% or lower. Gravel, on the other hand, has a lower porosity but because the pore sizes are larger, specific yield may be substantially larger than that from chalk.

In a confined aquifer, a change in the amount of water stored will not produce a change in the water level, but it will produce a change in the potentiometric head. The *storativity* of a confined aquifer is analogous to the specific yield of an unconfined aquifer, and is defined as the volume of water produced per unit area of aquifer for a unit fall in potentiometric head. Storativity is typically several orders of magnitude lower than specific yield for a given rock type.

Box 3.10 Measuring the water level

The water table or potentiometric surface is typically measured in boreholes or wells sunk down to the aquifer. A borehole has a smaller circumference than a well (typically less than 10 cm), and tends to be lined with a screen at the bottom allowing water to enter the borehole and rise to the level of the water table or potentiometric surface. Boreholes may be sunk either to monitor groundwater levels or for the extraction of groundwater, but wells are dug exclusively to extract water.

Recharge processes

Groundwater can be recharged in two main ways. *Direct recharge*, or percolation, arises when rain water (or melting snow) infiltrates into the soil and moves down through the soil profile. It therefore only tends to occur when the soil is saturated, and the rate of recharge will depend on soil texture characteristics.

It may take several years for water to percolate down to the water table. However, if the soil moisture content is high and the water table is shallow, then an input of rainfall may produce a very rapid rise in water table. This occurs because new water entering the top of the soil "pushes" water out of the bottom, into the aquifer. Hewlett & Hibbert (1967) termed this process *translatory flow*. Rapid recharge can also occur when water infiltrates through macropores and fissures (Bloomfield, 1996). Percolation may therefore be very variable over a short distance (e.g. Cook & Kilty, 1992). A certain amount of rainfall is generally necessary before direct recharge occurs. Taylor & Howard (1996) and Gburek & Folmar (1999) found in central Uganda and east-central Pennsylvania respectively that at least 10 mm of rainfall was needed in an event to initiate percolation.

Indirect recharge occurs when groundwater is recharged by surface water, either as overland flow or streamflow. The surface water may run across the land surface as overland flow, reach an area of soil with high infiltration capacity, and infiltrate down to the groundwater. This obviously only occurs where overland flow is important, as in arid and semi-arid areas (see Section 3.7), and where there is spatial variability in soil characteristics and hence infiltration capacity. Indirect recharge may occur through pools and ponds, fed by surface runoff and small streams. This is the dominant means of groundwater recharge in parts of the Sahel, for example (Leduc *et al.*, 1997; Desconnets *et al.*, 1997). Here, the land surface is very flat and the drainage network ill-defined: there are many small hollows into which water can drain. Desconnets *et al.* showed that the amount of recharge from a pool was a function of depth. If a pool was greater than 1 to 2 m deep, then most water would infiltrate, whereas if it was shallower most of the water would evaporate. This arises because the bottoms of the pools are clogged with clay, and a certain pressure is needed to force water into the pores of the clay. Different types of pools were identified, with most recharge deriving from pools in valley bottoms or large "sink" pools. As a consequence of the localised recharge, fluctuations in the water table were very variable. In semi-arid south east Australia, however, recharge from sinkholes and swamps only contributes around 10% of annual recharge (Herczeg *et al.*, 1997), largely because the topography is different from that in the Sahel: there are fewer opportunities for indirect recharge.

Another source of indirect recharge is river runoff. Aquifers in many arid and semi-arid areas are recharged by channel infiltration. As the river flows down the river channel water infiltrates downwards into the floodplain gravels: this is termed *transmission loss*, and is responsible for many rivers in dry areas failing to reach the sea. This infiltration is of course very irregular, and is often restricted to rare flood events. Ponce *et al.* (1999), for example, showed that infrequent events (occurring less than once per year) accounted for between 14 and 39% of the total recharge into a large aquifer in Baja California, Mexico, crossed by several ephemeral streams, and Shentsis *et al.* (1999) showed how recharge in a wadi in the Negev desert, Israel, was very small in most years but could exceed seven times the mean in some years. Such recharge is a function of runoff volume and duration, and also the infiltration capacity of the wadi

Box 3.11 Estimating groundwater recharge

There are several possible ways of measuring or inferring the rate of groundwater recharge. *Percolation gauges* are essentially blocks of soil from which soil drainage is collected and measured: they are very similar in concept to lysimeters (Box 3.4). In unconfined aquifers, the *change in water level* is a function of the change in storage, and it is possible to estimate the change in storage by multiplying the change in water level by the aquifer specific yield (e.g. Headworth, 1972; Gupta & Paudyal, 1988). In confined aquifers, the change in potentiometric head must be multiplied by the storativity. A third approach estimates total recharge by estimating the proportion of river flow that derives from groundwater (see Chapter 4; Wittenberg & Sivapalan, 1999). Recharge rates can also be inferred by using tracers to date groundwater at different levels in the aquifer. Tracers used include chemical tracers such as chloride (Sami & Hughes, 1996), stable isotopes (Taylor & Howard, 1996), and radioactive isotopes such as tritium (Solomon *et al.*, 1993). Finally, recharge is often inferred by using water balance models to estimate drainage out of the bottom of the soil (e.g. Finch, 1998).

bed. River flows can also recharge aquifers in wetter regions. Many of the floodplain aquifers in alpine and prealpine countries, for example, are replenished by river water either during floods, through the river bed or through the river banks (Beyerle *et al.*, 1999). In the temperate east of England, Bradbury & Rushton (1998) showed how the South Lincolnshire Limestone aquifer was partly recharged by infiltration from streams draining from more impermeable areas upstream. The Danube loses part of its flow to groundwater recharge when it crosses permeable limestone in central Bavaria. Methods of estimating the rate of groundwater recharge are described in Box 3.11.

Groundwater movement

Groundwater flows from one point to another in proportion to the hydraulic gradient (the difference in water table height or potentiometric surface). Where flow is essentially through pores and relatively small voids, the rate of the flow can be estimated from Darcy's Law (equation 3.11) and the hydraulic conductivity of the aquifer is therefore fundamentally important. The change in storage at a location can be estimated by combining Darcy's Law with the continuity equation (much as was done to produce the Richards equation), but the resulting equation is in practice much more difficult to apply because water can move in more than one direction. Darcy's Law does not apply, however, where flow is primarily through large fissures or conduits, as is often found in karstic limestone. The determination of flow directions, velocities and water table change is complicated and will not be discussed more here (see, for example, Ward & Robinson (2000), Dingman (1993) or Hornberger *et al.* (1998) for a description). The main point to note, however, is that the direction of groundwater flow may be different from that of flow at the surface: it is defined by hydraulic gradients not topographic gradients. The groundwater "catchment" may therefore be very different from the surface catchment as defined by topography.

Where does groundwater go?

Groundwater essentially goes in four possible directions. It can accumulate in storage for a very long time and, for practical purposes, therefore be "lost" to the surface hydrological system. It can drain to rivers, and this is discussed more in Section 3.7. Groundwater in coastal aquifers can drain directly to the sea. Finally, water can be lost from shallow groundwater through evaporation. In fact, many shallow floodplain aquifers in semi-arid regions are largely depleted by evaporation and transpiration. Phreatophytes are plants whose roots tap into groundwater, and are found in dry environments. Examples include cottonwood and greasewood. Nicholls (1993), for example, calculated the rate of depletion of a shallow aquifer in the Great Basin, Nevada, due to transpiration from greasewood plants.

3.7 Runoff generation and streamflow

Streamflow and runoff

Streamflow is defined as the amount of water in a river channel, typically expressed as the volume of flow passing a defined point over a specific time period: it is usually measured as $m^3 s^{-1}$ (*cumecs*) or, for smaller streams, in litres s^{-1}, and often termed *discharge*. Runoff is rather more loosely defined as the amount of precipitation that is not evaporated. It is usually expressed as a depth of water (in millimetres) across the area that the precipitation falls on (typically a catchment). The depth of runoff can be estimated by simply distributing the streamflow measured at the catchment outlet across the catchment area – but there are circumstances in which not all the runoff generated in a catchment reaches the catchment outlet. Some of the runoff generated at the top of a slope may infiltrate into soil further downslope, for example. Some of the river flow may percolate from the river bed into groundwater, or some may be evaporated from wetlands or lakes along the river course. Streamflow and runoff are therefore related – but not necessarily the same. They are usually equivalent in practice in temperate environments. Box 3.12 summarises the methods used to measure streamflow.

This section explores the processes of runoff or streamflow generation. In general terms, it is possible to draw a broad distinction between "quickflow", which is the short-term result of an input of precipitation to the catchment,

Box 3.12 Measuring streamflow

Streamflow is the volume of flow passing a given point over a defined time period. For very small streams, it is possible simply to collect all the water over a period of time (known as *volumetric gauging*), but in the vast majority of cases this will not be feasible. Streamflow is therefore determined by measuring stream velocity and multiplying by the river cross-section area. "Snapshot" estimates can be made by measuring velocity with a current meter or by dilution gauging, which measures the time taken for an input of some tracer (such as salt) to move between two points. Continuous streamflow estimation requires a continuous measurement of flow velocity, which has only become feasible since the early 1990s with the advent of "cheap" electromagnetic and ultrasonic flow gauges which measure velocity by the attenuation across the river of electromagnetic waves and ultrasound respectively. In the absence of continuous velocity measurements, hydrometric agencies have been forced to measure the depth of water (termed stage) and use stage-discharge ratings to estimate discharge from water depth. Water depth can be measured by a float and recorded on a continually-moving chart, although in many countries depth is now measured by pressure transducer and data recorded on a data logger. Central to this "indirect" approach is the nature of the relationship between water depth and discharge. In small streams it is possible to install a structure, such as a weir or flume, at which it is possible to develop from hydraulic principles a theoretical relationship between depth and flow. More frequently, an empirical stage-discharge rating is developed from snapshot gaugings. Unfortunately, such ratings can vary significantly over time, particularly where the river channel is very dynamic. Other problems arise when the river channel may be seasonally choked with vegetation or ice-bound. Also, it is generally very difficult to define the relationship between depth and discharge once the river floods onto the floodplain, so estimates of flood discharge can be very uncertain. However, despite all these limitations, streamflow is probably the most accurately estimated component of the catchment water balance.

and "baseflow", which is the slower drainage of stored water and which sustains river flows when there is no precipitation, although in practice there is a graduation between the two.

Runoff generation processes

There are four main ways in which precipitation (rainfall or snowmelt) becomes runoff and streamflow (Figure 3.28).

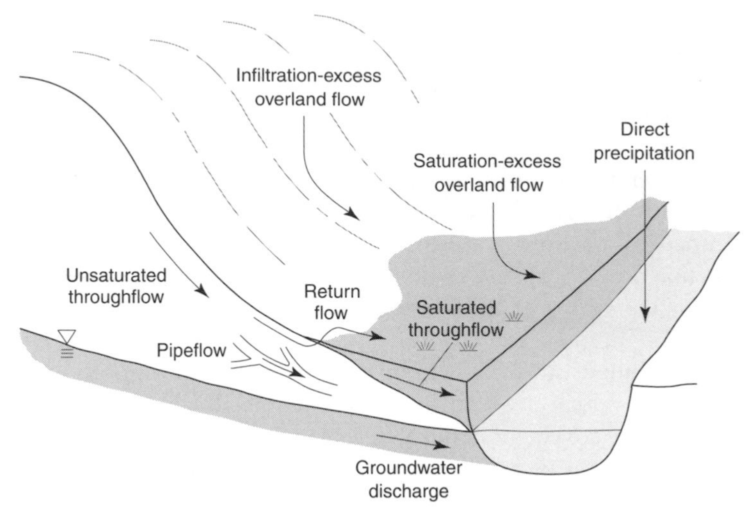

Figure 3.28 Runoff generation processes

Direct precipitation is precipitation falling directly into the open water of the river network, lakes and ponds. This tends to be very small, because only a small proportion of a catchment is typically covered with open water, but it does of course produce an immediate response to precipitation.

Overland flow is water that runs across the surface of the catchment towards the river channel. There are three types. *Infiltration-excess* overland flow (or *Hortonian* overland flow, named for the hydrologist who first described it) occurs when the rainfall rate is greater than the infiltration capacity. Water cannot infiltrate, so runs off the surface in a downslope direction. *Saturation-excess* overland flow occurs when precipitation cannot infiltrate into soil because the soil is saturated. Water moves fast as overland flow, so precipitation would soon reach a river channel. *Return flow* occurs when water that is flowing downslope through the soil reaches an impermeable barrier – such as an area of saturated soil – and emerges at the surface.

Throughflow is water that flows through the soil above the groundwater (Box 3.13 describes how it can be measured). Again, there are three broad types. *Unsaturated throughflow*, rather obviously, is lateral downslope flow in unsaturated soil. This is very slow because the hydraulic conductivity of unsaturated soil is low (Section 3.5). *Saturated throughflow* is lateral downslope flow in saturated soil. This is substantially faster than unsaturated throughflow. Both unsaturated and saturated throughflow tend to occur in defined parts of the soil, typically at the boundary between permeable and relatively impermeable soil layers, at the interface between the soil and the bedrock, and in layers with high permeability. *Pipeflow* is flow in macropores and soil pipes within the soil. This is faster than saturated throughflow, although the time it takes for water to reach a river depends on the connectivity of the macropore network. Rapid throughflow through pipes, macropores, highly-permeable layers or at the boundary between permeable and impermeable layers is frequently termed *preferential flow*.

Box 3.13 Measuring throughflow

Throughflow is typically measured by digging an intercepting ditch along a slope and collecting the water in troughs. This is generally accurate, and the major complication lies in scaling up from the sample site to the hillslope as a whole and indeed the catchment, because of the large spatial variability in soil properties and hence throughflow. Pipeflow is measured by collecting water discharging from a pipe. The routes taken by water through the soil can be determined by adding a dye to the surface of the catchment and either digging up the soil to see the pattern of staining or observing coloured discharges out of pipes.

Groundwater discharge is drainage from water stored as groundwater. This usually provides river baseflow, with the amount of discharge related to the volume of water stored in the aquifer (and therefore in unconfined aquifers the level of the water table). The relationship between storage and discharge is, however, not linear, with the rate of discharge falling more rapidly as the store reduces: a quadratic relationship is often assumed. Groundwater discharge tends to be slow, although it can be very rapid where the water table is shallow, as it usually is close to the river channel. Here, the groundwater may be very rapidly recharged, leading to a rapid rise in water table and the formation of a groundwater "mound" close to the river channel. The hydraulic gradient close to the river is increased, and water drains quickly into the river, resulting in rapid streamflow response to rainfall (Sklash & Farvolden, 1979).

The relative importance of these four different pathways varies between catchments, depending on catchment physical properties, climate and time. The next three sections examine how these pathways operate in three different types of environment, broadly distinguishing between "humid", "dry" and "cold" environments. Different mixes of processes operate under these different conditions, but it must be remembered that a given catchment may experience different conditions at different times in the year.

Humid environments

This section considers the response to rainfall in catchments that are "wet" for most of the time, looking first at temperate climates. Such catchments are characterised by relatively deep soils (compared to drier environments), and relatively low rainfall intensities (compared to tropical humid environments).

Infiltration-excess overland flow is very rare in humid temperate environments, except where the soil surface has been disturbed, because infiltration capacities are generally high. Virtually all rainfall therefore infiltrates. As it does so, the soil becomes wetter and the proportion of the catchment that is saturated increases. This part of the catchment is termed the *contributing area* (sometimes *source area*), because it is where most of the quickflow response to rainfall originates. Saturation-excess overland flow is generated from it, saturated throughflow is much quicker that throughflow in unsaturated soils, macropore flow may be triggered (flow only occurs in macropores once soil moisture reaches a certain value), and small streams can be supported in saturated contributing areas.

Contributing areas are dynamic (Hewlett & Hibbert, 1967; Dunne & Black, 1970). During a storm the contributing area expands as the saturated area increases, and so does the area generating quickflow. The rate and location of expansion are largely determined by topographic features such as zones of convergence, and several example maps of the expansion of the contributing area have been produced (e.g. Anderson & Burt, 1978; Dunne *et al.*, 1975; Gineste *et al.*, 1998). Variations in soil characteristics can also affect the rate of change in contributing areas, as found in the New Forest in southern England (Gurnell, 1978; Gurnell *et al.*, 1985: soil characteristics are indexed by vegetation type: Figure 3.29). Contributing areas within a catchment are not necessarily contiguous, but must be connected in some way if they are to create quickflow. In some cases this connection may be through overland flow (return flow from the upslope saturated contributing area plus saturation-excess overland flow), but in thicker soils the connections may be through soil pipes and macropores (as shown in many examples: e.g. Bonell *et al.*, 1984).

The variation in the contributing area through a storm may mean that the catchment responds in a highly non-linear way to rainfall. Sidle *et al.* (2000), for example, showed in a small, steeply incised catchment in Japan how once shallow groundwater developed in topographic hollows, subsurface flow rates increased substantially and saturation-excess overland flow could be generated. Similarly, Uchida *et al.* (1999) showed how different pipes contributed to streamflow as the catchment became more and more saturated.

The moisture status of the catchment also varies *between* storms, with the streamflow response of a given amount of rainfall depending on the antecedent conditions, and in particular the degree of saturation. The physical properties of aspects of the flow generation process may alter over time as well, and Uchida *et al.* showed how the ability of a pipe network in a small Japanese catchment to transmit water varied substantially over time as sediment moved through the pipes.

Early attempts to identify different sources of streamflow were based on "hydrograph separation" techniques, which are graphical procedures to separate quickflow from baseflow based on the rate of change of flow. However, all such techniques are empirical and involve subjective judgements. More recently, the contribution of different sources of water to the hydrograph has been assessed using the chemical properties of water: a characteristic used in this way is known as a tracer (Box 3.14). Tracer studies tend to show that most quickflow reaches the stream channel through subsurface routes and, significantly, that much of the water that forms the quickflow hydrograph did not fall in the rainfall event that triggered the hydrological response. Most quickflow comes from old "pre-event" water through a process known as piston flow. Water entering a saturated zone from above (by infiltration or lateral throughflow downslope, for example), displaces water out of the saturated

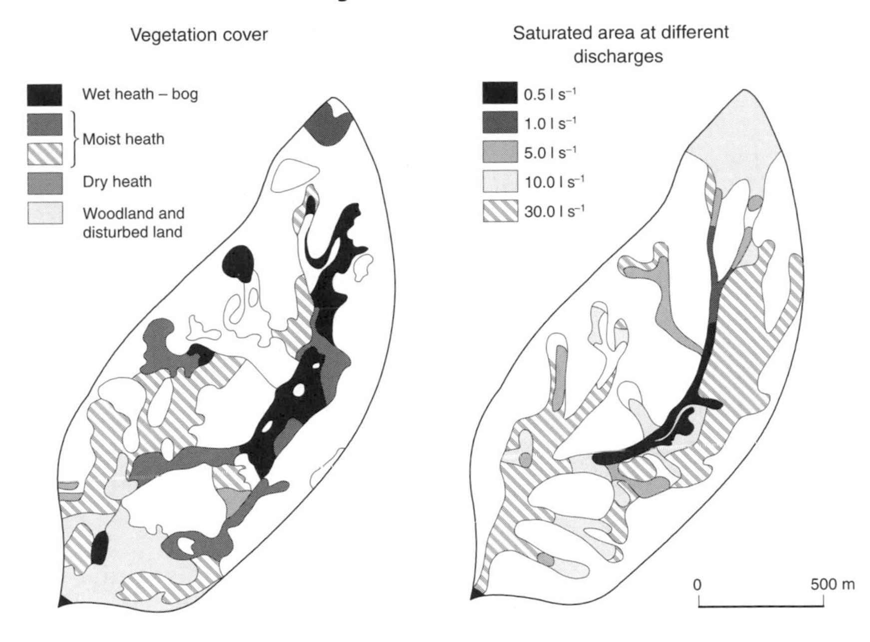

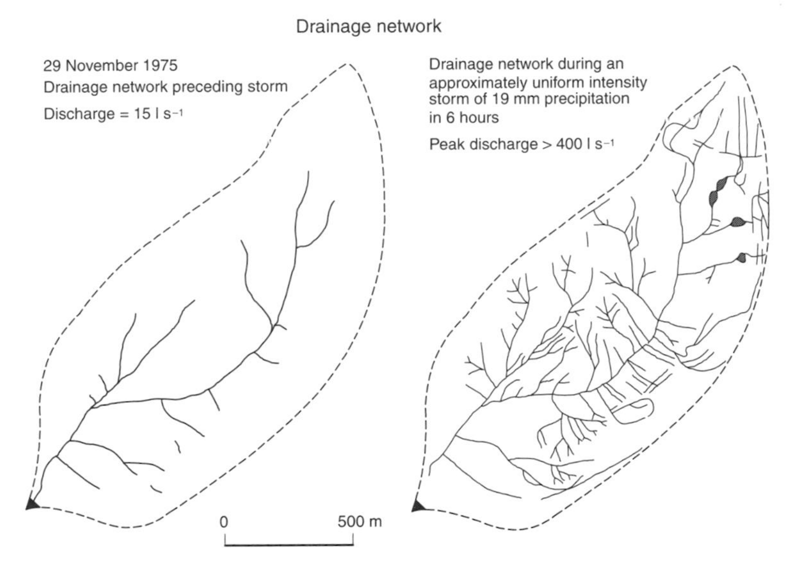

Figure 3.29 Variation in surface drainage network and area of saturation in a small heathland catchment in the New Forest, England (Gurnell, 1978; Gurnell *et al.*, 1985)

zone into the channel. Laudon & Slaymaker (1997), for example, found that between 60 and 90% of quickflow in a small subalpine basin in British Columbia, for example, comprised pre-event water, and Hoeg *et al.* (2000) found in a catchment in the Black Forest of Germany that event water made only a small contribution to quickflow. Event water tends to become more

Box 3.14 Using tracers to identify streamflow components

The chemical properties of water reflect its history. It is therefore possible to infer the pathways followed by water from its chemical characteristics, by looking at the concentrations of defined *tracers* (Buttle & Peters, 1997). Tracers that have been used include the concentrations of particular substances dissolved in the water (chloride is frequently used: e.g. Nyberg *et al.*, 1999), natural isotopes (the isotopes of oxygen ^{18}O (e.g. DeWalle *et al.*, 1998) and deuterium are often used: Laudon & Slaymaker, 1997), total dissolved ions (as measured by electrical conductivity: Chapter 5), acid-neutralising capacity (Neal *et al.*, 1997a) and temperature (Kobayashi *et al.*, 1999). A tracer must be conservative, or in other words it must not interact with other chemicals in the water. If there are two potential sources of water, the total load of the tracer is:

$$C_t Q_t = C_1 Q_1 + C_2 Q_2$$

where C represents concentration, Q is discharge, and the subscripts 1, 2 and t represent sources 1 and 2 and the total respectively. Q_t and C_t can be measured, as can C_1 and C_2 (the concentrations of the tracer in, for example, soil water and groundwater). The discharge from source 1 can be calculated by substituting $Q_t = Q_1 + Q_2$ into the above equation to yield:

$$Q_1 = \frac{Q_t (C_t - C_2)}{C_1 - C_2}$$

Most tracer studies assume two components: isolating three components is harder, but still possible (e.g. DeWalle *et al.*, 1988).

important during the falling limb of the hydrograph (as observed for example by Evans *et al.* (1999) in the Catskill Mountains in New York). Transit times of water through the catchment system can also be estimated using some tracers. Studies have shown, for example, that it can take several weeks for water to move through the shallow subsurface zone (e.g. DeWalle *et al.*, 1998; Nyberg *et al.*, 1999; Soulsby *et al.*, 2000), and rather longer through shallow and deep groundwater.

Humid tropical environments are, like humid temperate environments, characterised by precipitation falling as rain, although the intensities can be substantially higher. Soils also tend to be thinner than in temperate environments. Infiltration capacities, however, can be very high, largely because of the presence of significant amounts of organic matter and litter. Infiltration-excess overland flow is therefore rare, despite the high rainfall intensities, except where the soil is disturbed or where stemflow concentrates water (Herwitz, 1986). As in temperate environments, rainfall infiltrates and saturates the soil, resulting in saturation-excess overland flow (as observed for example in northern Queensland by Bonell *et al.*, 1983), return flow from perched water tables (Dykes & Thornes, 2000) and rapid movement through the thin soil layer. This rapid flow can occur through soil pipes (Elsenbeer & Lack, 1996). The main difference with temperate environments is that wet contributing areas are more widely distributed through a humid tropical catchment because conditions are generally wetter (Cassells *et al.*, 1985).

Dry environments

These environments are characterised, rather obviously, by a general lack of water, but the rainfall that does occur tends to be very localised and very intense. A broad distinction is drawn here between "semi-arid" and "arid". Semi-arid environments often have "wet" seasons, and get some rain each year. Consequently, there is at least a seasonal vegetation cover, and soils – albeit often shallow – develop. Arid environments, in contrast, may have no rain for several years and the land surface is covered with exposed bedrock and unsorted surface deposits.

The high intensity of rainfall and thin soils in *semi-arid areas* together tend to generate infiltration-excess overland flow. This is particularly exacerbated where the soil develops a crust on being wetted, as happens for example across much of semi-arid Africa (e.g. in the Sahel: Casenave & Valentin, 1992), in semi-arid China (Zhu *et al.*, 1997) and in India (Bajracharya & Lal, 1999). Soil crusts develop for several reasons, including raindrop impact, erosion, inwashing of fine sediment, swelling of clay particles and the growth of lichens and algae, and crusts typically have hydraulic conductivities and infiltration capacities between a third and a sixth of uncrusted soils (e.g. Peugeot *et al.*, 1997). The degree of crusting and resulting infiltration capacity depend on soil type (Casenave & Valentin, 1992), and the amount of overland flow generated from a rainstorm therefore depends very much on exactly where in the catchment the rain falls (as found by van de Giesen *et al.* (2000) in a semi-arid catchment in the Ivory Coast). Where the infiltration capacity is low, the volume of runoff generated depends largely on the intensity of rainfall. Where the infiltration capacity is higher, however, rainfall will infiltrate, saturate the soil and generate saturation-excess overland flow: the volume of runoff generated in these circumstances depends on the total volume of rainfall (Martinez-Mena *et al.*, 1998). In contrast to temperate catchments, the response to a given amount of rain therefore depends not on the moisture status of the catchment before the rainfall – because it is usually dry – but on where the rain falls in the catchment: the catchment *surface* properties are most important. In both environments only a portion of the catchment generates runoff, but for different reasons.

Not all the overland flow generated from the surface of semi-arid catchments reaches a river channel. A variable proportion of it infiltrates into soil and then either evaporates or recharges groundwater. This can happen in several ways. Water may, for example, collect in depressions in the land surface, forming short-lived pools. This is common across much of the Sahel of west Africa, where the flat topography produces a very "disorganised" drainage network consisting of a large number of small endorheic catchments. Desconnets *et al.* (1997) showed that in a part of Niger whether water collecting in these pools evaporated or recharged groundwater depended largely on the depth of the pool. In neither case does the water reach a river channel. Bromley *et al.* (1997) described a different situation in another part of Niger, where the gently sloping landscape is covered with a patterned vegetation type known as tiger bush. This vegetation type consists of alternating bands of shrub vegetation (10 to 30 m wide) and bare soil (50 to 100 m wide). When it rains, crusts develop on the bare soil and infiltration-excess overland flow is generated. This flows downslope towards the band of vegetation. Here the infiltration capacity is considerably higher and soils are deeper, so the overland flow infiltrates – and then goes no further downslope. The infiltrated water is then used by the plants and transpired. In fact, the vegetation could not be sustained without "harvesting" water from upslope. A similar phenomenon has been observed with linear acacia groves in central Australia (Pilgrim *et al.*, 1988).

Arid areas are exposed to similar intense rainfall events to semi-arid areas, albeit less frequently. The main distinguishing feature, however, is the nature of the catchment surface. Large parts of arid catchments are composed of bedrock or very coarse surface deposits such as talus slopes (Yair & Lavee, 1976). Unweathered bedrock has a very low infiltration capacity, and infiltration-excess overland flow is easily generated. Coarse surface deposits, consisting of large cobbles and stones, also – somewhat surprisingly – have a low infiltration capacity. This is because a large proportion of the surface is covered with hard stone, and this offsets the effect of the large spaces between the stones. The actual infiltration capacity of a stony surface – and thus generation of overland flow – depends on factors such as whether the stones are embedded and their spacing (Poesen *et al.*, 1990; Lavee & Poesen, 1991). As in semi-arid environments, the surface characteristics of the catchment are therefore very variable and determine how much runoff is generated from a given amount of rainfall. Again, little of the runoff generated within a catchment may reach a river, but the beds of gullies and the steep sides of wadis may themselves be impermeable and generate large volumes of runoff (Yair & Lavee, 1985). Much of the runoff that reaches the river channel (or wadi) infiltrates, as outlined in Section 3.6, and does not reach the basin outlet.

Cold environments

The distinguishing feature of "cold" environments is the presence of snow or ice. This section considers first streamflow generation processes when precipitation falls as snow, before looking at the particular features of rivers draining catchments with permafrost or glaciers.

The process of *snowmelt* was described in Section 3.2. As water melts, it either infiltrates into the soil or runs off as overland flow, depending on the catchment surface. Figure 3.30 (Dingman, 1993) shows three modes of

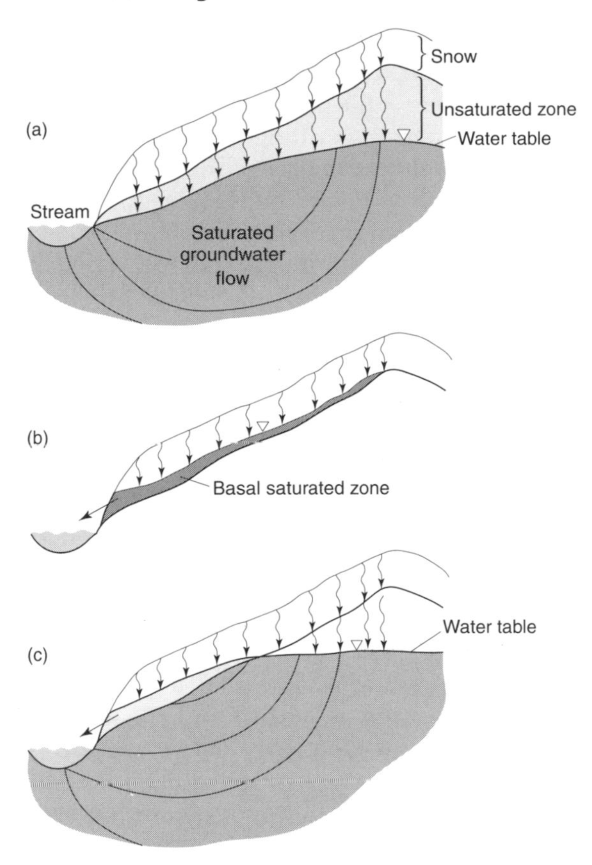

Figure 3.30 Three modes of snowmelt-runoff generation (Dingman, 1993)

snowmelt-runoff generation. In the first mode, melting water infiltrates into unsaturated soil and flows towards the stream as throughflow (or percolates to groundwater). In the second mode, the surface is impermeable and meltwater runs along the base of the snowpack downslope to the stream. The surface may be impermeable because it is bedrock, or perhaps because the soil is frozen. Soil frost depth depends on the depth of snow cover (thick snow acts as an insulator) and when the snow first fell. Snowfall early in winter insulates the soil and may prevent it freezing, but snow may not fall until the soil has been frozen by low air temperatures. The effect of soil freezing on the rate of snowmelt runoff is unclear. Shanley & Chalmers (1999) found in a catchment in north east Vermont that the depth of soil frost had an effect on rates of snowmelt runoff from small plots, but no clear effect at the catchment scale (in a 111 km^2 catchment). At this scale many other factors, such as the rate of snowmelt, the relative amounts of rainfall and snowmelt and the amount of snow under different vegetation types, had a greater effect on the volume of runoff. The third mode of snowmelt in Figure 3.30 occurs when the water table rises to the surface within the snowpack, resulting in saturation-excess overland flow occurring at the base of the snowpack. The areal extent of this saturated zone will of course vary through the period of snowmelt – and the area contributing rapid runoff will increase as melt continues.

The presence of *permafrost* influences streamflow generation in arctic and subarctic basins. Not only may the melting of the active layer during summer release water into the soil, but the presence of the permafrost layer itself affects flow pathways above. Carey & Woo (1999) compared north-facing and south-facing slopes in the Yukon, Canada. The south-facing slope had no permafrost. Snowmelt in spring infiltrated into the soil, recharging the soil moisture reservoir but producing little lateral flow through the soil to the stream: the soil moisture reservoir is depleted largely by summer evaporation. When the snow on the north-facing slope melts, however, it cannot infiltrate far into the soil because of the layer of permafrost. The top layer of the soil (actually the organic layer) therefore becomes saturated, and lateral subsurface flow occurs both through the matrix and through soil pipes. Saturation-excess overland flow and return flow also occur. The flow in the stream is therefore generated from the north-facing slope. Quinton & Marsh (1999), also in Canada, found that subsurface flow through the organic-rich upper layers of the soil and soil pipes was extremely fast, when the soil was saturated. When the water table falls below the conductive upper layer, then the rate of subsurface flow reduces rapidly. The area generating rapid runoff (in response to snowmelt or rainfall inputs) therefore depends on water table elevation, and varies over time.

Streamflow in *glacier-fed rivers* is determined not just by snowmelt and rainfall, but also by the melt of the glacier during summer. Glacier melt is driven by inputs of energy, and varies both through the melt season and through the day. There is therefore a diurnal cycle, with the greatest discharge in mid-afternoon, and the amplitude of this diurnal cycle is at a maximum in mid to late summer. Hodson *et al.* (1998) showed that the seasonal and diurnal cycles in glacier melt in a small high-Arctic basin in Svalbard varied not only with energy inputs, but also the rate of recession of the snowpack covering the glacier. The smaller the area of the glacier covered with snow, the greater the melt for a given amount of energy. Hannah *et al.* (1999) explored the variation in hydrograph shape through the melt season in a small glacierised basin in the Pyrenees, showing how the shape evolved as the meltwater routeing and storage conditions within the glacier changed through the season. Hydrograph shape and the magnitude of diurnal fluctuation varied from year to year, with greater diurnal variation occurring in a dry summer – an ablation-dominated melt season – than in a wet summer – when streamflows where dominated by precipitation falling on the glacier and surrounding catchment. Superimposed on the pattern of glacier melt are not only rainfall-induced peaks but also occasional outburst events. Through the summer season snow cover around the glacier reduces, and a greater proportion of the catchment is able to generate streamflow rapidly. Glacier outbursts may be triggered by heavy rainfall (as observed by Collins (1998) in the Findelengletscher in Switzerland in 1993), but may result from the release of ice-dammed lakes due to glacier melt. This may be infrequent or regular: the Gornersee, for example, is formed and released by the Gornergletscher in Switzerland each year (Collins, 1998).

Over the longer time scale, total streamflow volume from a glaciated basin may be influenced by glacier retreat or growth.

3.8 An overview

This has been a long chapter. It has described the basic hydrological processes, looking not only at streamflow but also at the processes of precipitation and evaporation. It is important to have covered these topics in some detail because they underpin understanding of the processes and consequences of global environmental change: these changes can only be interpreted in the context of the processes operating in the catchment. There are five main points to draw in conclusion to this chapter.

First, the role of vegetation cover has been highlighted. Vegetation intercepts water and strongly influences the partitioning of energy at the earth's surface. The total evaporation from an area (and hence both energy budgets and streamflow) is dependent on the mix of vegetation within that area. Vegetation strongly affects physical soil properties, influencing both the rate of infiltration and the movement of water through the soil. Stemflow concentrates water into certain parts of the forest floor.

Second, there is a very great diversity in hydrological processes at all spatial scales, from the hillslope to the globe. At the hillslope scale, diversity in processes is largely a function of different surface, soil and vegetation characteristics. *Hydrological response units* or *representative elementary areas* (REAs) can be identified (Wood *et al.*, 1990; Woods *et al.*, 1995; Blöschl *et al.*, 1995), within which hydrological behaviour (response to input) is assumed to be similar: these typically have a spatial scale of the order of 1 km, although there is no reason why they should be constant over time (Beven, 1995). At coarser scales, climatic variability begins to explain differences in processes. The implication of this diversity is that it may be difficult to generalise in quantitative terms.

Third, the pathways followed by water from sky to river or groundwater are varied, both over space and time: as will be shown in Chapter 5, these pathways are extremely important in controlling water quality.

Fourth, the understanding of hydrological processes has largely been based on field observations. Conventionally, these observations have been made by hydrologists concerned specifically with streamflow and, to a lesser extent, evaporation. More recently, however, field studies of hydrological processes have been undertaken to help in the understanding of chemical pathways (many of the tracer experiments which have helped in the understanding of water pathways through the soil have been conducted as part of investigations into the possible pathways taken by pollutants) and biogeochemical fluxes (many of the recent observations of transpiration have been made in parallel with measurements of carbon fluxes between vegetation and atmosphere).

Fifth, the chapter has hinted in a few places of the effects of human activity on hydrological processes: these are explored in depth in Chapter 6.

The next chapter explores different types of hydrological regime, and looks at the variation in hydrological behaviour over time.

Chapter 4

Patterns of hydrological behaviour

4.1 Introduction

The previous chapter reviewed the components of the water balance and the processes involved in converting "inputs" of precipitation and energy into "outputs" of evaporation, runoff and groundwater recharge. Different sets of processes operate in different places and at different times during the year, on inputs which also vary over space and time, in catchments which have variable physical characteristics. These layers of variability mean that patterns of evaporation, runoff and recharge also vary over space and time. This chapter explores these patterns, at the global scale, focusing particularly on runoff and recharge. First, however, it is important to review the many measures used to characterise hydrological behaviour. "Hydrological behaviour" is defined here to mean a pattern of variation both through the year and from year to year. The term "flow regime" is widely used in the hydrological literature, and tends to refer to patterns of variation through the year.

4.2 Indicators of hydrological behaviour

River flows

River flows vary over all time scales: minute to minute, day to day, month to month and year to year. Figure 4.1 summarises some of the key dimensions of this variability, used as indicators of hydrological behaviour. Some of these measures are *direct properties* of the hydrological data (such as a mean or standard deviation), but others are *derived statistics*, based on fitting some model to the data (such as a frequency curve).

Streamflow is usually measured in $m^3 s^{-1}$, and *specific discharge* is streamflow divided by basin area. This is often calculated for small streams in $l s^{-1} km^{-2}$, and is calculated in order to compare streamflow from streams with different basin areas. Over longer periods, such as a month or longer, the accumulation of water draining from a catchment (runoff) is expressed as a millimetre depth across the catchment (note that, unlike climatologists, hydrologists rarely express streamflow over durations of less than a day in terms of millimetres depth, because such numbers tend to be very small).

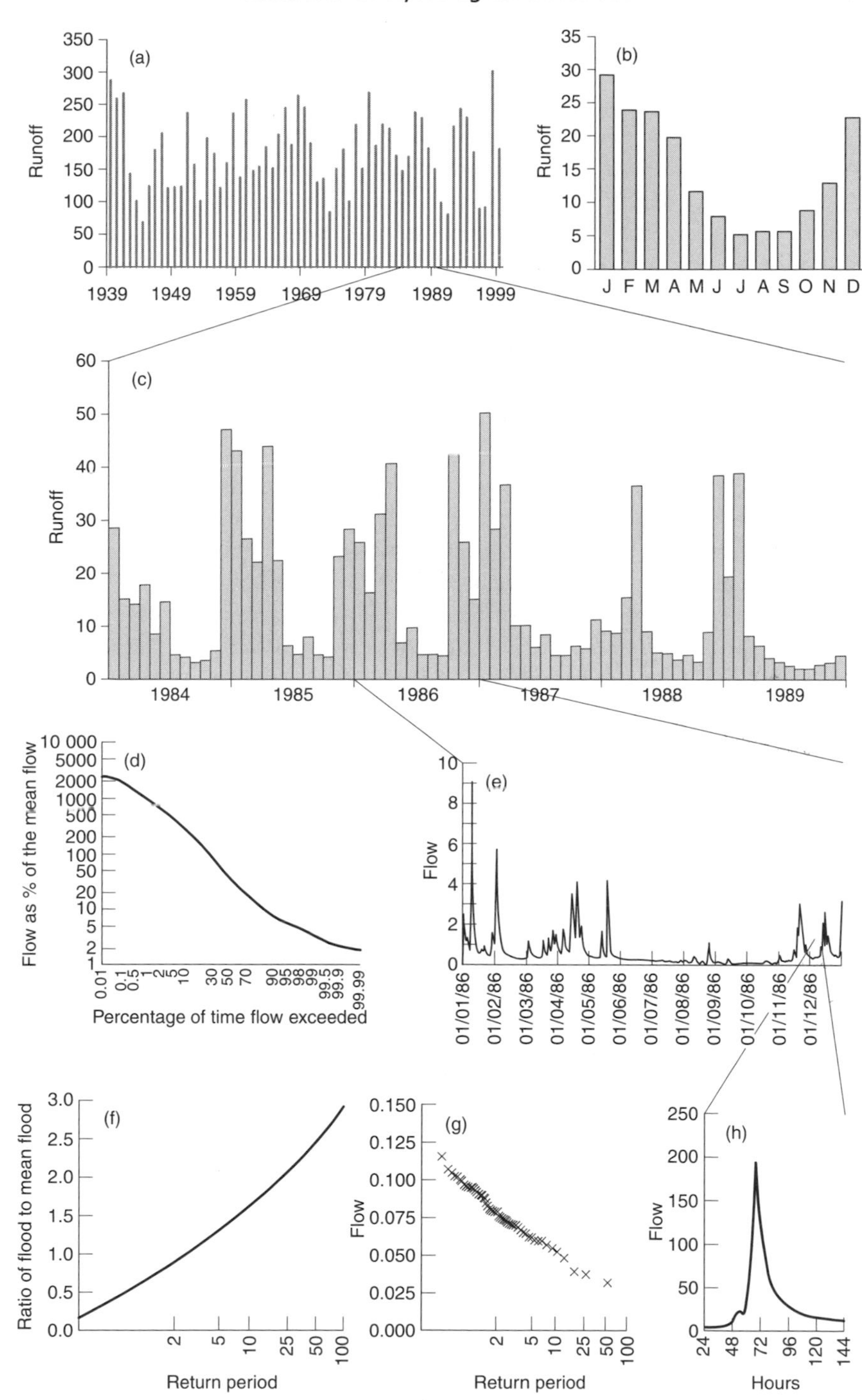

Figure 4.1 Dimensions of hydrological behaviour

The most basic hydrological indicator is the long-term (i.e. multi-year) average runoff, expressed either as an average streamflow or discharge, or as millimetres depth across the catchment. When average runoff is expressed in millimetres it is easy to calculate the ratio of average annual runoff to average annual rainfall, often termed the runoff ratio or the runoff coefficient. There is, however, no clear definition of "long-term", and currently no consensus amongst either hydrologists or hydrometric agencies collecting hydrological data to agree on standard reference periods (unlike climatologists, who routinely calculate averages over defined "normals", such as the 30-year period 1961 to 1990). This lack of consistency can make it difficult to compare "average" runoff from different catchments, as the data may span very different time periods.

Time series of annual runoff (Figure 4.1a) are produced by calculating the average streamflow or total runoff separately for each year. However, different hydrometric agencies use different definitions of a year, and frequently calculate annual runoff over the *water year* rather than the calendar year. A water year is broadly defined so that it begins at the end of the month with lowest flow, and is used so that a single wet season is not split between two calendar years. In the UK and the USA, for example, the water year starts on 1 October and finishes on 30 September, whereas in Germany the water year runs from November to October. In parts of west Africa, the water year begins in June. This variability in water year makes it difficult to compare time series produced in different countries.

Variation in annual runoff from year to year can be measured by calculating the standard deviation of the annual values, but this has the units of the runoff data. Relative variability – enabling comparisons between different catchments – is characterised by the *coefficient of variation* (CV), which is the standard deviation divided by the mean: the higher the CV, the greater the relative variability. The CV of annual runoff is an important hydrological indicator, as the volume of reservoir storage required to maintain a given supply is proportional to the square of the CV of inflows (McMahon *et al.*, 1992). The variability in annual runoff over time can also be characterised by constructing a relationship between magnitude and frequency of occurrence. This is done by fitting a frequency distribution to the data, as summarised in Box 4.1. From such a relationship it is possible to estimate the magnitude of runoff expected in any one year with a given probability or *return period* (Box 4.2), or vice versa.

River flow regimes are characterised by a graph of long-term mean monthly runoff (Figure 4.1b). The monthly values are sometimes expressed as a proportion of total annual runoff, and sometimes calculated as a ratio to the average monthly runoff.

The variation in flow from day to day (Figure 4.1e) is described by the *flow duration curve* (Figure 4.1d). This shows the proportion of time that flows are below a given flow value, and is produced by ranking the daily flows from smallest to largest and plotting them against the proportion of days with lower flows. The steeper the slope of the flow duration curve the greater the variability in flow from day to day. One index frequently abstracted from the flow

Box 4.1 Frequency distributions in hydrology

A frequency distribution defines the relationship between the magnitude of an event and the frequency with which that event is exceeded. An *empirical distribution function* plots magnitude of event against the proportion of events greater than or equal to that event, and from such a plot it is in principle possible to estimate the frequency of a given magnitude event. In practice, however, too few data are available and empirical distribution functions are therefore not very smooth – making it difficult to interpolate – and it is not possible to extrapolate to estimate the frequency of occurrence of events larger than the maximum recorded. The response is therefore to fit a *theoretical* frequency distribution to the sample of data, calculating the parameters of the distribution from the observed data. Hydrological data are generally skewed around the mean, so a symmetrical frequency distribution such as the normal distribution is not appropriate. One way round this is to transform the data, and the *log-normal* distribution is frequently used: this fits a normal distribution to the logarithms of the data. The other solution is to use a *skewed* distribution. Examples that are frequently used include the General Extreme Value (GEV) distribution, of which the Gumbel (EV1) and Weibull (EV3) distributions are special cases, the log-Pearson type III and the Generalised Pareto distributions. The more flexible the distribution, the greater the number of parameters, and the distributions used in hydrology tend to have two or three parameters (and rarely four). In mathematical terms, a frequency distribution defines the relationship between the magnitude of an event and the probability of occurrence of a *smaller* event. For example, the GEV distribution has the following form:

$$F(x) = F(x \leq X) = \exp\{-[1 - k(x - u)/a]^{1/k}\}$$

where x is magnitude, u, a and k are parameters and $F(x)$ is the probability of an event less than or equal to x occurring in any one year. Parameters can be estimated from sample data using a range of procedures, including the methods of moments, maximum-likelihood and l-moments.

There is a large literature on frequency analysis in hydrology, and the topic is covered in hydrology textbooks with an engineering emphasis (e.g. Stedinger *et al.*, 1993). The main points to emphasise, however, are that the credibility of an estimated frequency distribution depends on the degree to which the assumed distribution fits the data, the robustness of the procedure used to estimate parameters from the data, the extent to which the data conform to the assumption that they can be described by a single, smooth theoretical frequency distribution and, perhaps most importantly, the assumption that the nature of the relationship between magnitude and frequency does not change over time (see Section 4.4). As a general rule, most procedures give similar results within the range of the data, but can give very different magnitude/frequency relationships when extrapolating beyond the largest observed events.

Box 4.2 Return period and exceedance probability

Frequency distributions show the relationship between the magnitude of an event x and the probability of occurrence of smaller events $F(x)$ in any one year. This probability is known as the non-exceedance probability. Hydrologists are often interested in the *exceedance probability*, which is the probability of an event greater than or equal to x occurring in any one year. This is simply $1 - F(x)$. The *return period* is the average interval between events greater than or equal to x:

$$T = \frac{1}{1 - F(x)}$$

To illustrate: an event with a non-exceedance probability of 0.99 has an exceedance probability of 0.01 (or a 1% chance of occurring in any year), and a return period of 100 years. This does not mean of course that there is a 100-year interval between one event and another: it means that *on average over the long-term* these events occur once every 100 years. It is possible – and indeed has frequently happened – that one 100-year return period event is followed by another in the next year. The use of the concept of exceedance probability and return period assumes that the risk remains constant from one year to the next: this assumption is explored in Section 4.4.

duration curve is the flow exceeded 95% of the time (*Q*95). This flow is exceeded on all but 18 days in a year on average, although the number of days per year falling below *Q*95 usually varies considerably from year to year. Flow duration curves and *Q*95 are often expressed as a percentage of the mean daily flow, allowing comparison between catchments. The flow duration curve says nothing about the sequencing of high and low flow periods, and it is conceptually possible for similar flow duration curves to be produced from quite different patterns of flow variability from day to day.

"Drought" is conventionally interpreted to be a shortage of water, but this definition is difficult to quantify in practice. The extent of a "drought" is also dependent not only on the absolute lack of water, but more precisely on the lack of water *relative to demand for water*. Hydrologists, as opposed to water managers, therefore tend to focus on "low flows" rather than "droughts". Indicators of low flow include the lowest n-day average flow (where n is usually 7) in each year (Figure 4.1g), the number of days in each year below a defined threshold (such as *Q*95), the duration that flows are continuously below the threahold, and the volume of flow "missing" below the threshold (the flow deficit volume: Tallaksen *et al.*, 1997).

Indicators of high flows include the peak magnitude and the volume of flow above some defined threshold. Again, time series can be produced, summary statistics calculated and frequency distributions fitted (Figure 4.1f).

Most studies of flood frequency use annual maximum data, extracting only the largest flood in each year (water or calendar). However, some years contain more than one flood, and in some cases the second or third largest flood in a year may be bigger than the largest floods in other years. Another way of extracting flood data is therefore to select all the peaks over a defined threshold (defined such that there are, for example, on average three floods per year). This tends to produce a better indication of the frequency of occurrence of relatively small floods (with a return period of 10 years or less), but has little effect on the estimated magnitudes of rarer floods.

At the scale of the individual hydrological event (a response to an input of rainfall or snowmelt) there are four important indicators (Figure 4.1h). The first is the magnitude of the peak response, and the second is the volume of "quickflow". In practice it is very difficult to identify this second indicator (see Chapter 3), and most estimates use some form of graphical streamflow separation. The total volume of quickflow, or storm response, is often expressed as a percentage of the input precipitation. The third indicator is the rate of rise of the hydrograph, often measured by the time to peak (measured either as the time from the centroid of the input precipitation event to peak discharge, or the time from immediately before the rise to the peak). The final indicator is the rate of recession from the peak. As a general rule, streamflow declines from the hydrograph peak in an exponential manner, and is often described by the following function:

$$Q_t = Q_0 e^{-bt} \qquad \textbf{4.1}$$

where t is time from the beginning of the recession, Q_0 is discharge at the start of the recession, and b is a parameter, known as the recession coefficient. The "half-life" is defined as the time taken for flow to reach half its initial value (i.e. $Q_t = 0.5Q_0$), and can be calculated from:

$$t_{0.5} = \frac{\ln 2}{b} \qquad \textbf{4.2}$$

Of course, b can be calculated from the measured half-life. Other functions are sometimes used to characterise recessions (see Tallaksen, 1995), and recessions are sometimes divided into several different periods.

All these measures of hydrograph shape give values which vary from event to event (although for reasons outlined in Section 4.3 it can be expected that the time to peak and rate of recession are reasonably constant for a given catchment). One way of producing "characteristic" measures of hydrograph shape, and particularly peak flow and flow volume, uses the concept of the *unit hydrograph*. This concept assumes that, in a given catchment, a given input of rainfall over a given time period will always produce the same hydrograph response. It is therefore possible to produce a characteristic hydrograph by rescaling hydrographs produced from lots of different events. In actuality, of course, the response of a catchment to rainfall depends on antecedent conditions (Chapter 3), and the unit hydrograph concept is therefore rather unrealistic.

Nevertheless, several studies looking at environmental change or comparing different catchments have attempted to define characteristic unit hydrographs.

Groundwater recharge and levels

Many of the indicators of river runoff can also be applied to groundwater. Average annual recharge can be calculated, for example, as can time series of recharge (again, calculated over a defined water year). In principle, frequency analysis can be applied to time series of recharge, although in practice this is rarely done. The rate of rise of groundwater level can be calculated, but it is more frequent to calculate the rate of recession, using similar recession concepts as applied to streamflows.

Evaporation

The most important indicators of evaporation are the annual total, seasonal and monthly totals, and at the finest scale the variation through the day. Time series of annual, seasonal or monthly evaporation can be produced and in principle frequency analyses applied, but in practice, as with groundwater recharge, this is rarely done.

4.3 Variations in hydrological behaviour over space

Hydrological behaviour varies from one catchment to another, and at a larger scale from one region to another. This variability is a function of the variation in climatic regime over space at one level, with variations in catchment physical characteristics adding finer-scale variability. This section looks at this spatial variability in hydrological behaviour: the next looks at variations in this behaviour at a site over time.

Annual, monthly and daily flow regimes

The term "flow regime" describes the variation in river flow through the year, typically as represented by monthly runoff. Since the mid-twentieth century there have been many attempts to describe and classify river flow regime types, although the first attempt dates back to the 1880s when Voeikov differentiated between nine types of flow regimes, based primarily on climatic features (Arnell *et al.*, 1993). Lvovich (1938) developed this classification, identifying four sources of streamflow (snow, rain, glaciers and groundwater) and four dominant seasons (winter, spring, summer and autumn). Each of the four classes was divided into three gradations, resulting in a total of 144 possible regime types, of which only 38 actually exist. Perhaps the best known classification, however, is that of Pardé (1955), which too is based on the source of flow and the distribution of flow through the year. This classification recognises three main flow regime types: simple, original complex and changing complex. "Simple" flow regimes have one high flow season, one low flow season and one dominant

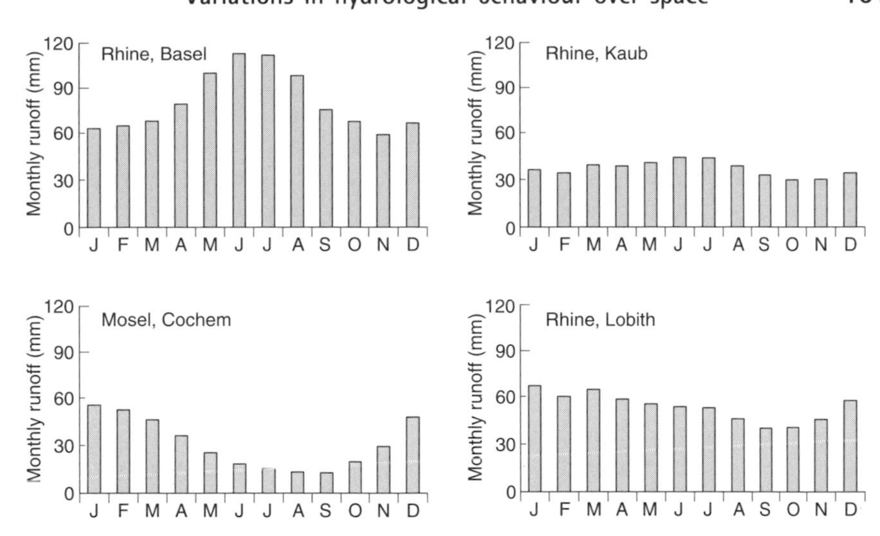

Figure 4.2 River flow regimes in the Rhine catchment: data from the Unesco data set "Discharge of Selected Rivers of the World", available from the University of New Hampshire on http://www.grdc.sr.unh.edu

source of streamflow. "Original complex" regimes have more than one high or low flow season and possibly several sources of streamflow dominating at different times of the year. Examples include regimes with a spring maximum due to snowmelt and an autumn maximum following rainfall, or regimes with two wet seasons producing two high flow seasons. "Changing complex" regimes arise when different parts of a catchment are exposed to different climatic controls. The classic example is the River Rhine (Figure 4.2). The headwaters of the catchment are fed by snow and glacier melt in spring and summer (as shown at Basel), and the summer peak can be seen in the Rhine as far downstream as Kaub (between Mainz and Koblenz). The next major tributary (the Mosel) has a winter maximum, and the flow regime on the Rhine downstream (as shown at Lobith on the Dutch border) reflects the different processes operating in different parts of the catchment. In his book, Pardé illustrated different regime types with examples from all over the world. Beckinsale (1969) used a similar classification procedure to construct a map showing the distribution of river flow regime types across the world: the map was largely based on climate zones.

Flow regime types in the Nordic region were classified by Gottschalk *et al.* (1979) based on the timing of high and low flows, and a variant on this classification was used by Krasovskaia *et al.* (1994) to describe flow regimes across Europe. A broad distinction is drawn between *Atlantic*, *Mediterranean* and *continental* regimes. The Atlantic pattern occurs across most of western Europe. It has highest runoff during winter and early spring, with lowest flows during late summer. Rainfall is generally similar throughout the year, and this pattern reflects the varying balance between rainfall and evaporation through the year: flows are low in summer not because of a lack of rain, but because high evaporation – and high soil moisture deficits – means summer rain does not generate much streamflow. The Mediterranean pattern looks similar, but here the range in flows through the year is greater because most rainfall is concentrated in winter. Continental regimes occur in the Nordic region, Europe east of central Germany, and mountain Europe. These continental regimes are all characterised by snowmelt during spring, but different regime types vary according to the magnitude of rain-generated peak flows and the timing of the low flow season (during winter or summer). The boundaries between these regime types is not sharp, and there are obviously gradations. Stanescu & Ungureanu (1997) explored regime types in southern and central Europe in more detail, identifying more sub-types on the basis largely of altitude and the relative importance of rain-generated and snowmelt-generated flows.

Figure 4.3 gives some examples of monthly river flow regimes in a number of catchments (all with an area less than 5000 km^2) across the world. The River Severn, draining east Wales and central England, has a typical "Atlantic" flow regime, with highest flows in winter. Rainfall is relatively constant through the year, but during winter evaporation is considerably lower. The River Narew in Poland, in contrast, has a significant snowmelt peak in spring, with some rainfall-generated flows in late autumn and winter: the Schuylkill in the eastern United States has a similar pattern. Further west in North America, the Little Falls River in Minnesota shows a classic mid-continental regime, with a very dominant snowmelt peak in spring and very low flows during winter. Note that it too has a secondary peak in autumn. The two African examples show regimes dominated by single rainy seasons, as indeed does the regime from the Nam Yong in tropical Thailand. The Zhuya in northern Siberia is an even more extreme version of the continental regime in the Minnesota example. Finally, the regime in the Quijos River in Brazil reflects wet conditions throughout the year but a peak in (Southern Hemisphere) winter, and the regime on the Iguaicu is characteristic of that in most of the tributaries of the Amazon.

At this scale of analysis, the role of climate is dominant. However, even with monthly flow regimes it is possible to identify catchments with large volumes of storage – either in a lake or groundwater – as the variation in flow through the year will be considerably smaller than in nearby catchments with less storage.

At the daily scale the effects of catchment geology are more apparent. Figure 4.4 shows some daily flow hydrographs for "typical" years for a number of catchments in Europe and North America. The two British catchments are close to each other (and therefore have the same climatic regime) and are approximately the same size, but have very different hydrological regimes, due to differences in geology: one is underlain by chalk, and most streamflow derives from groundwater discharge, and the other has a largely clay catchment (note that catchments containing both types of geology will show one regime type superimposed on the other). The streamflow regimes in the other catchments largely show the effect of different climate regimes, but at a finer scale than the monthly plots. The Raritan River in New Jersey has a temperate regime

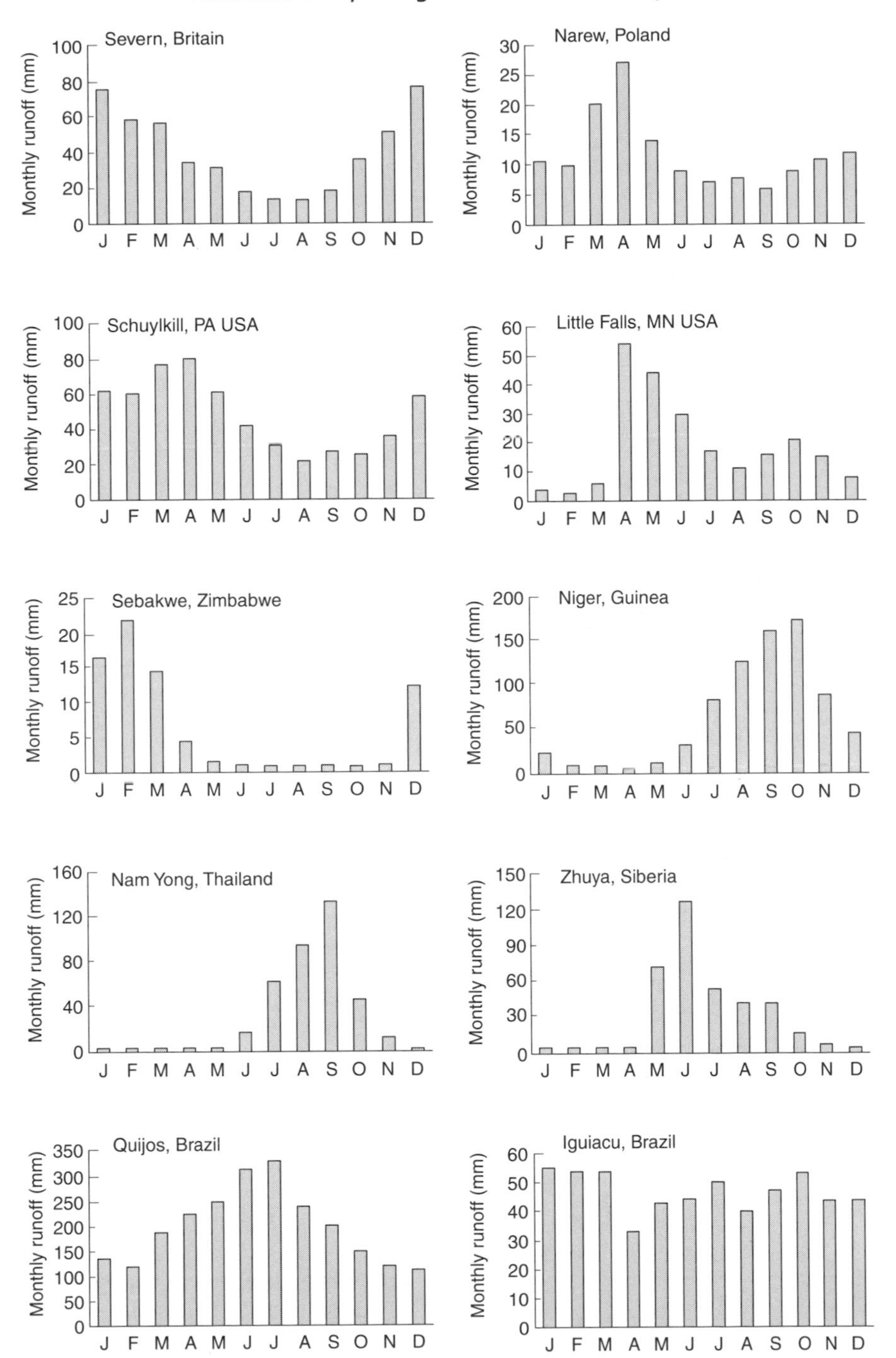

Figure 4.3 Monthly river flow regimes: data from the Unesco data set "Discharge of Selected Rivers of the World", available from the University of New Hampshire on http://www.grdc.sr.unh.edu

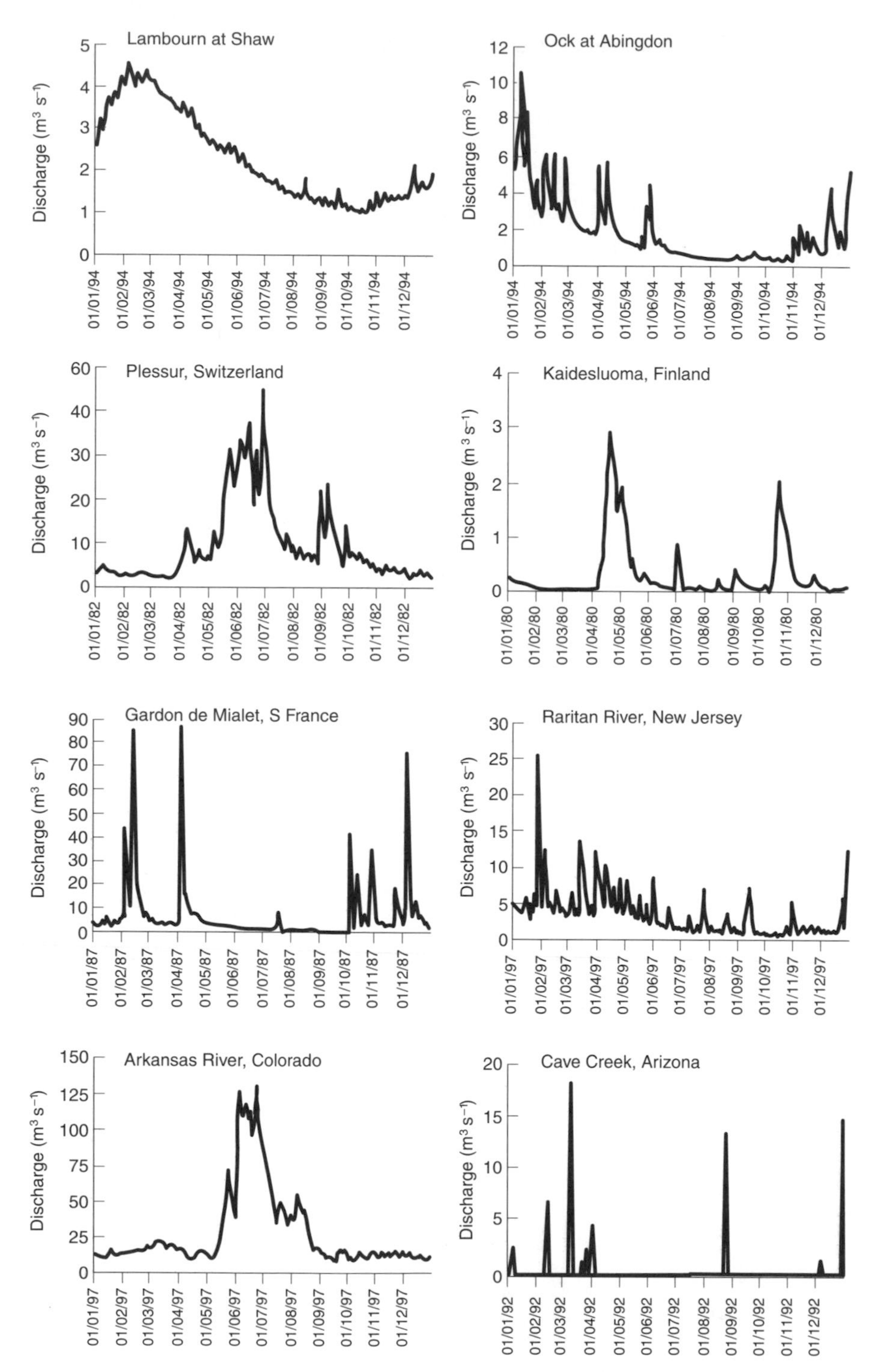

Figure 4.4 Daily flow regimes for eight example catchments. The data come from the UK National Water Archive and the European Water Archive, both held at the Centre for Ecology and Hydrology, Wallingford, and the US National Water Information System (http://water.usgs.gov/nwis)

similar to those across most of western Europe. The Kaidesluoma catchment in south east Finland has a typical continental regime, dominated by the snowmelt peak in later spring. The flow regime in the Plessur in Switzerland shows rain-generated peaks superimposed on snow melting in the Alps during later spring and early summer: a similar pattern is seen in the headwaters of the Arkansas River draining the Rocky Mountains in Colorado. The data for the Gardon de Mialet show a typical Mediterranean-type flow regime, characterised by rapid fluctuations in response to intense rainfall and higher flows during winter; Cave Creek in Arizona is a classic example of an arid catchment, and is dry for much of the year.

For a given catchment geology and climate regime, the variation of flows from day to day around the seasonal trend reduces as catchment size increases. This is partly because the larger the catchment the smaller the proportion being rained on at any one time, but mostly because the larger catchment is, rather obviously, integrating runoff over a larger area. A very small first-order catchment generates a very rapid response – all the quickflow leaves the catchment in a short period – which is routed along the river network. As this "pulse" of flow moves along the channel it is attenuated, and joined by similar pulses from other catchments. The time it takes for the quickflow generated by an individual event to pass a point therefore increases with distance downstream. The degree of attenuation, and thus the degree of "smoothing" of the daily flows depends on the shape of the catchment, the density of the river network and the slope of the river channels: this is explored in more detail below.

The proportions of total annual streamflow that derive from "quickflow" and "baseflow" vary between catchments, largely with catchment geological conditions. High levels of baseflow come from catchments with significant groundwater storage, and also from catchments with large lakes. Indices of baseflow (such as the Base Flow Index, or BFI) are determined graphically from plots of daily streamflow, by some form of smoothing and elimination of peaks. In the UK, chalk catchments typically have a BFI of over 0.9 (implying that more than 90% of their streamflow derives from groundwater discharge), whilst impermeable catchments have a BFI closer to 0.2 (Gustard *et al.*, 1992). In southern Australia, Lacey & Grayson (1998) calculated baseflow indices between 0.2 and 0.9, and showed a very strong relationship with combined geology/soil/vegetation classes.

Year-to-year variations in annual and seasonal runoff

An important characteristic of a hydrological regime is the variation in flow from year to year. This is generally indexed by the coefficient of variation of the annual flows. McMahon *et al.* (1992) analysed data from 974 catchments distributed across the world, calculating an average CV of 0.43. Figure 4.5 shows the variation in average CV between regions. The highest values are in the driest regions, where the year-to-year variability in rainfall is greatest. For

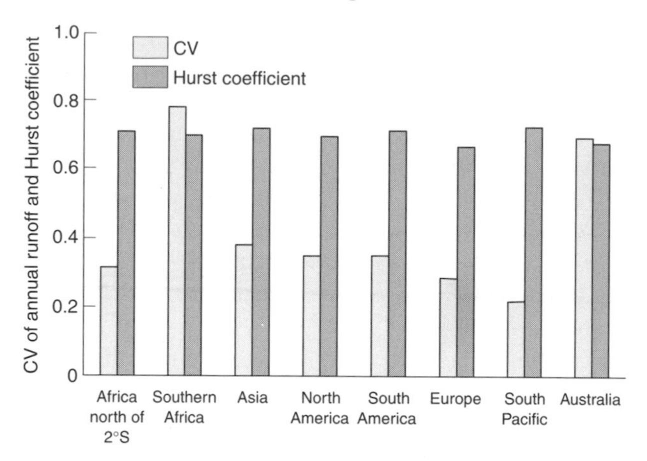

Figure 4.5 Variation in the regional average CV and Hurst coefficient (persistence) of annual flows (adapted from McMahon *et al.*, 1992)

a given climate type, the CV of annual runoff generally declines as catchment area increases, as larger catchments "smooth" out geographical variabilities. However, in some regions – such as the western USA (Vogel & Sankarasubramanian, 2000) – CV varies in a more complicated way with area, as the largest catchments include areas with very diverse climate regimes.

As river flows vary from year to year, so may regime type. In Europe, "simple" regime types (with a winter rain-generated peak or a single dominant snowmelt-generated peak) tend to be very stable, but mixed regimes vary from year to year (Krasovskaia, 1995). In one year the snowmelt peak may be the largest, whilst in another a summer rain-generated peak may be greater.

A time series of annual river flows may also show correlation from year to year, or "persistence": what happens in one year may be influenced by what happened the previous year, for two possible reasons. Either the catchment has a "memory" – where the volume of water in a store influences flows over several years, for example – or the atmosphere has some sort of persistent pattern (explored more in Section 4.4). One measure of persistence is the Hurst coefficient (Mandelbrot & Wallis, 1969), defined below:

$$\frac{\text{Range of cumulative departures from mean}}{\text{Standard deviation of original series}} = (0.5n)^h \qquad \textbf{4.3}$$

where n is the sample length and h is the Hurst coefficient. For a random time series h should be equal to 0.5, but time series of annual runoff tend to have values around 0.7: McMahon *et al.* (1992) calculated a global average of 0.69, together with the regional average values shown in Figure 4.5. There is little variation between regions. Persistence is generally not important in water resources terms where variability from year to year is low or the size of reservoir storage is small, but is significant where either variability is high or the reservoir is large relative to inflows.

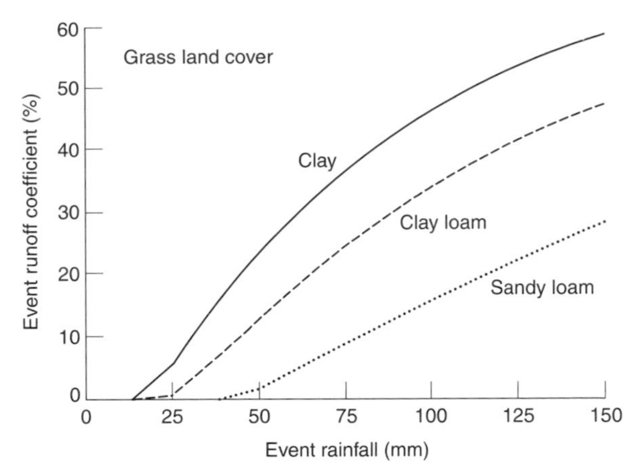

Figure 4.6 Variation in average event runoff coefficient with event precipitation and soil type. The graph is based on the US Soil Conservation Service runoff estimation method (see Dingman, 1993), assuming an "averagely-wet" catchment, and is indicative only

Variations in flood characteristics

In the humid eastern USA, the average ratio of quickflow to event rainfall (sometimes termed the *runoff coefficient,* which is not to be confused with the annual runoff coefficient) is less than 30% in around 90% of small catchments (Woodruff & Hewlett, 1970), and virtually always less than 50%, at least for rain-generated floods: the rest of the rainfall is either evaporated or replenishes soil and groundwater storages and reaches the stream later as baseflow. These figures are broadly characteristic of temperate humid environments. A rather higher proportion of *snowmelt* goes to quickflow, because evaporative losses are less and there is often less infiltration. In semi-arid areas a very high proportion of event rainfall may runoff from parts of the catchment (up to 90%: Peugeot *et al.*, 1997), but only a small proportion may actually become streamflow in a river channel because the overland flow infiltrates downslope and water in the channel infiltrates through the river bed.

The event runoff coefficient varies with event magnitude – the greater the total input, the greater the proportion going to quickflow – but in an individual catchment there may be very considerable scatter from event to event, largely due to the degree of saturation of the catchment. The greater the soil moisture deficit before the rainfall, the lower the proportion of rainfall that goes to quickflow.

Between catchments, the mean event runoff coefficient varies with soil type (Figure 4.6) and land cover. In a study site in Niger, for example, the maximum event runoff coefficient ranged between 10% and 90%, depending on the type and amount of vegetation (Peugeot *et al.*, 1997). The vegetation cover affected not only the interception and evaporation, but also the structure of the soil and its infiltration capacity: the consequences of this will be explored in more detail in Chapter 6.

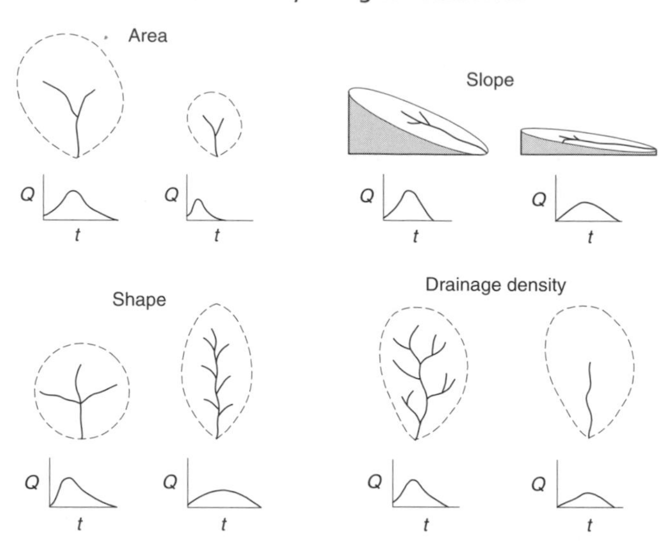

Figure 4.7 The effects of catchment physical characteristics on storm hydrographs (Jones, 1997b)

The time to peak of a hydrograph is much more consistent on a catchment from event to event, but varies considerably *between* catchments. In general terms, the shorter the main stream channel, the greater the density of the drainage network and the steeper the slope of the main channel, the shorter the time between rainfall or snowmelt input and the peak of the hydrograph (Figure 4.7). The shape of the catchment can be important too, with long thin catchments producing slower-peaked hydrographs. Lag-times are also shorter in catchments where quickflow is generated rapidly, such as those with contributing areas which can expand rapidly, although in large events the contributing area is usually at its maximum and the time to peak is largely a function of the drainage network. Empirical equations to predict the time to peak (e.g. Burn & Boorman, 1993; Sefton & Howarth, 1998) usually employ measures of catchment size, topography and channel network. The equation developed in the UK Flood Estimation Handbook (Institute of Hydrology, 1999), for example, has the following form:

$$T_p = {}^{4.27}\ DPSBAR^{-0.35} PROPWET^{-0.8} DPLBAR^{0.54} (1+URBEXT)^{-5.77} \qquad \textbf{4.4}$$

where *DPSBAR* is a measure of channel slope, *DPLBAR* is a measure of catchment area taking into account the length of flow paths, *PROPWET* is the proportion of the time the soils in the catchment are wet, and *URBEXT* is the fraction of the catchment covered by urban development. The equation indicates that the time to peak reduces as slope increases, catchment size decreases, the tendency towards dryness decreases, and urban area increases (see Chapter 6).

The magnitude of the *peak* of the flood hydrograph reflects both the total volume of event quickflow and its distribution over time. For a given volume of quickflow, the shorter the lag time the greater the peak flow. The variation in the translation of event rainfall into volume of quickflow from event to event means that it is not possible to infer the return period of peak streamflow directly from the return period of the event rainfall: the 10-year return period one-day rainfall does not necessarily produce the 10-year return period peak flow.

An important indicator of peak flood magnitude is the average annual flood, defined as the mean (or the median) of the time series of annual maximum instantaneous flood peaks. The magnitude of the average flood is a function of basin area and the magnitude of flood-generating events (rainfall or snowmelt), but is also influenced by slope, drainage density and soil properties. These catchment properties affect the proportion of the flood-generating event that becomes quickflow, and the routing of that quickflow to the catchment outlet.

The specific average annual flood is simply the average annual flood divided by catchment area. In the UK, this varies from well below 0.25 $m^3 s^{-1} km^{-2}$ in the south and east to over 1.5 $m^3 s^{-1} km^{-2}$ in the wet north west (NERC, 1975), largely reflecting variations in the magnitude of flood-producing events. The specific average annual flood in the UK declines slightly as catchment area increases, but in other parts of the world specific average annual flood declines much more rapidly with catchment area, particularly where rainfall is very localised and floods may therefore be generated only from parts of a basin. In southern Africa, for example, the specific mean annual flood ranges between around 0.06 $m^3 s^{-1} km^{-2}$ and 1.7 $m^3 s^{-1} km^{-2}$ for catchments with an area of 50 km^2, but is between 0.04 $m^3 s^{-1} km^{-2}$ and 0.3 $m^3 s^{-1} km^{-2}$ for 1000 km^2 catchments (Mkhandi & Kachroo, 1997).

Many attempts have been made by engineering hydrologists to relate the average annual flood to catchment properties, in order to be able to estimate the flood risk at sites with no gauged river flow data. These relationships typically have the form

$$\bar{Q} = cA^u D^v G^w R^x S^y \qquad \textbf{4.5}$$

where *A* is a measure of basin size (such as area or stream length), *D* is a measure of drainage density, *G* is a measure of channel slope, *R* is an index of "typical" flood-producing rainfall, and *S* is a measure of soil properties. Such empirical equations have been produced for many regions, including the UK (Acreman, 1985; NERC, 1975; Institute of Hydrology, 1999), large parts of continental Europe (Gustard *et al.*, 1989), and the USA (undertaken state by state by the US Geological Survey).

The variation in flood peak magnitudes from event to event in a catchment is characterised by the coefficient of variation (CV) of the annual maximum peaks or the slope of the flood frequency curve (which is a function essentially of CV and the skewness of the data). At the largest scale, the differences in CV and flood frequency curve between catchments is related to differences in the year-to-year variability in flood-producing rainfall or snowmelt, which is a characteristic of climate. Bates *et al.* (1998) found in south eastern Australia, for example, that the highest CVs were found in the catchments with the lowest

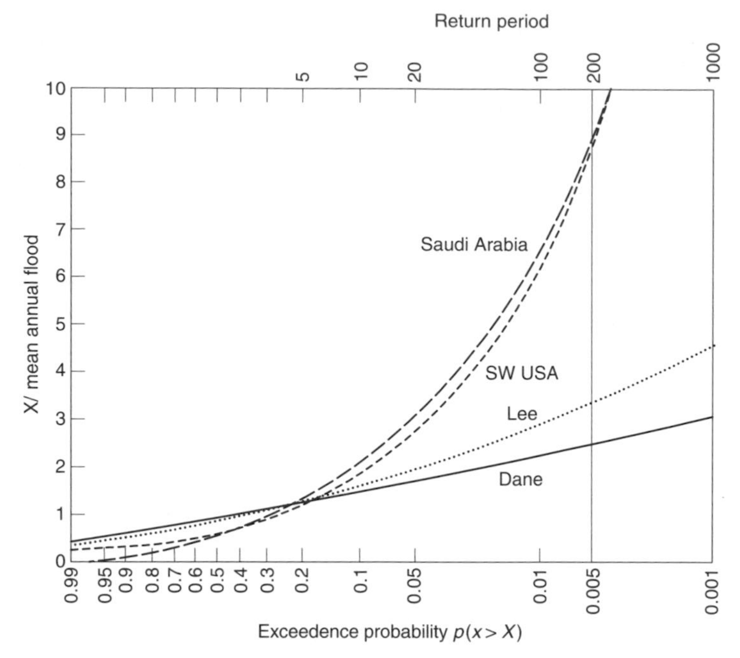

Figure 4.8 Some example flood frequency curves: the Lee and Dane are in south east and north west England, respectively

rainfall and the highest variability in rainfall from year to year. Flood frequency curves in semi-arid catchments are generally much steeper than those in temperate regions, because storm rainfall totals are much more variable: the 100-year flood can be 5 to 8 times as large as the average annual flood (Farquharson *et al.*, 1992), compared with between 2 and 3 times as large. Figure 4.8 shows some typical frequency curves from different environments, standardised by the average annual flood.

However, for a given climate regime the CV of annual maximum peaks varies with catchment properties, and particularly catchment area. Dawdy (1961) noted that CV tended to reduce as basin area increases, particularly where floods are generated by rainfall. This has been attributed to spatial variability in flood-producing rainfall across the catchment: the larger the catchment, the smaller the proportion of the catchment that is producing a flood at a given time, and the greater the likelihood that somewhere in the catchment will be producing a large flood in each year. However, detailed analysis of data from a number of areas, including Appalachia (Smith, 1992) and Austria (Blöschl & Sivapalan, 1997), has revealed a much more complicated pattern. In some circumstances, it appears that CV may *increase* up to a catchment size of around 100 km^2, and decrease thereafter (Robinson & Sivapalan, 1997; Blöschl & Sivapalan,1997), reflecting the complex interplay of a number of influences. These include the relationship between the duration of flood-producing rainfall and the time for the entire catchment to respond, non-linearities in the flood-generation process, and the magnitude of baseflow before the rainfall event (which will be influenced by hydrological regime and catchment storage). In semi-arid and arid areas, transmission losses along the river channel also tend to become more important as basin area increases, lessening the differences between different floods (Goodrich *et al.*, 1997).

As a general rule, flood magnitudes in catchments which have large amounts of storage are more dependent on the amount of precipitation accumulated over a long period than on the magnitude of short-duration storm rainfall. On the other hand, floods in a highly-responsive catchment with little storage capacity will be very dependent on the magnitude of the storm rainfall. The flood frequency curves in similarly-sized catchments exposed to the same climate will therefore vary with catchment storage.

At the extreme, floods downstream of catchments with large lakes or significant groundwater stores show very little variation from year to year as outflows are dependent on the total volume of water in store, which may change little over time. In some cases, however, such catchments may be prone to unusually large floods as some important threshold is crossed. Many of the largest floods from the highly permeable chalk catchments of southern England, for example, have come from heavy rain falling on top of an accumulation of snow: the precipitation falling as snow is stored on the catchment instead of infiltrating, and is "released" by the rainfall to flow directly to the channel.

Such flood events which depart from a smooth magnitude/frequency relationship are known as *outliers*, and reflect the operation of different flood-generating mechanisms in different years. In some environments, different flood-generating mechanisms operate at different times of the year. Most obviously, some catchments are exposed to snowmelt floods in spring and rain-generated floods in summer and autumn. These two different mechanisms can be expected to have different magnitude/frequency relationships, with in general less variability in the snowmelt floods from year to year. The maximum flood in any one year can in these catchments come always from snowmelt, always from rainfall – or from either. The resulting frequency distribution of annual maximum floods can therefore be a "mixed" distribution (e.g. Woo & Waylen, 1984): this is common in large parts of North America and northern Europe (incidentally, making it difficult to describe the annual maximum flood series by a single frequency distribution as described in Box 4.1). In other parts of the world different floods may be generated at different times by other processes. In southern Italy, for example, floods can be generated in winter by heavy frontal rainfall, but in summer by short-duration, very intense, convective rainstorms (Rossi *et al.*, 1984). Again, a mixture distribution may be appropriate here to describe the relationship between flood magnitude and frequency, and the peak flood in any one year can occur in any of the flood "seasons".

It is important to note here that not all "floods" are caused by an excess of rainfall or snowmelt: Box 4.3 summarises the different types of flood.

Box 4.3 Types of flood

If a flood is defined as an excess of water in a place that is normally dry, then there are several different types of flood. The most common is where a river overflows its banks due to a large input of rainfall or snowmelt. These are the types of flood that can be predicted and explained in terms of catchment physical characteristics and climatic inputs.

River overflow floods can also come from the release of ice-dammed lakes. Some massive landscape-forming floods were caused in North America and the Himalaya in this way during deglaciation, and much smaller floods occur still from the periodic release of water stored behind or within glaciers. The Gornersee, for example, builds up each year from water melting from the Gorner Glacier in Switzerland, and is released in late summer. Landslides too can create temporary dams which produce floods when breached. Floods are also generated when man-made dams fail, which may or may not occur during heavy rainfall.

Not all floods, however, are caused by water overflowing a river channel. Flooding can be caused by an unusually high rise in groundwater levels, such that the water table reaches the surface. Similarly, a rise in lake levels can lead to inundation of the surrounding land. Both these types of floods are generated by prolonged heavy rainfall or snowfall. Flooding can also be caused by surface runoff, particularly after intense short-duration rainfall and where the land surface has been disturbed (Chapter 6).

Variations in low flow characteristics

The rate at which flows fall from the peak of a hydrograph, in the absence of further precipitation, is largely a function of catchment soil type, geology and other sources of storage. Table 4.1 gives the recession constant and "half-life" for a number of European catchments with different geological and soil characteristics (Gustard *et al.*, 1989). The greater the amount of storage, the longer the half-life and the slower the rate of recession. Note that the half-life of the Kenwyn catchment is 177 days: in practice, the recession is usually interrupted before flows fall to half the starting value.

The flow duration curve characterises the variation of flows from day to day. Figure 4.9 shows flow duration curves for two British catchments. The curve for the Greta catchment, underlain by impermeable rock, is much steeper than that for the chalk-dominated Lambourn catchment, indicating that flows are much more stable and consistent in the chalk catchment. The flow exceeded 95% of the time (*Q*95) is 60% of the average flow in the chalk catchment, but only 10% of the mean in the other. Catchment geology and, more generally, the availability of storage within the catchment, control the slope of the flow duration curve and the proportion of days with flows below a defined

Table 4.1 Recession constants and half-lives for European catchments (Gustard *et al.*, 1989)

Catchment	Characteristics	Recession constant	Half-life (days)
Teeressuonoja, Finland	Impermeable bedrock, with moraine soil	0.0278	24.9
Hupselse Beek, Netherlands	Impermeable bedrock, with sandy soil	0.1048	6.9
Kenwyn, Cornwall, UK	Grits and shales, with brown earth soil	0.0039	177.2
Dowles Brook, Midlands, UK	Sandstone, with brown earth soil	0.0236	29.3

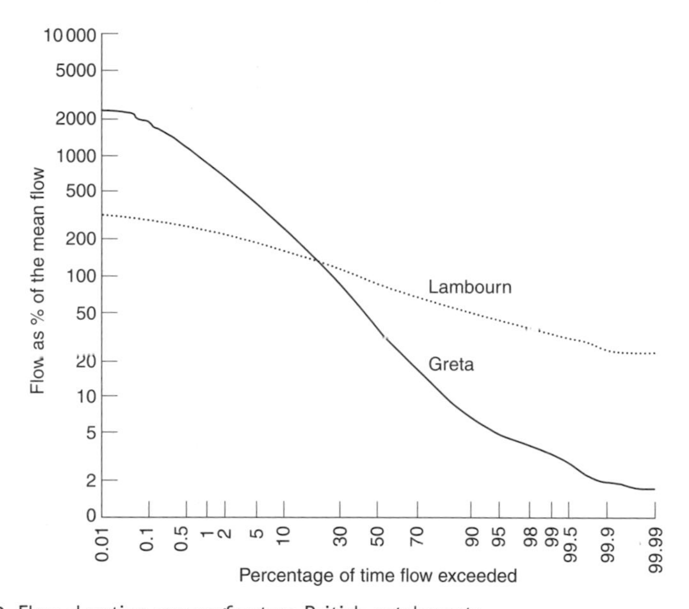

Figure 4.9 Flow duration curves for two British catchments

threshold. Typical *Q*95 values for catchments in the UK with different underlying geology are shown in Figure 4.10 (Young *et al.*, 2000). A map of *Q*95 across the UK, as a percentage of mean flow, is very similar to a map of geology.

Similarly, the mean annual minimum flow, averaged over different durations, is largely determined by catchment geology and storage (as shown in the UK for example by Gustard *et al.* (1992), in south west Germany by Schreiber & Demuth (1997) and in Massachusetts by Vogel & Kroll (1992)). Drought duration and deficit volume is a function of hydrological regime at one scale, and catchment geology at a finer scale. Tallaksen & Hisdal (1997), for example, identified areas in the Nordic countries prone to summer or winter droughts, largely on the basis of flow regime, and Demuth & Heinrich (1997)

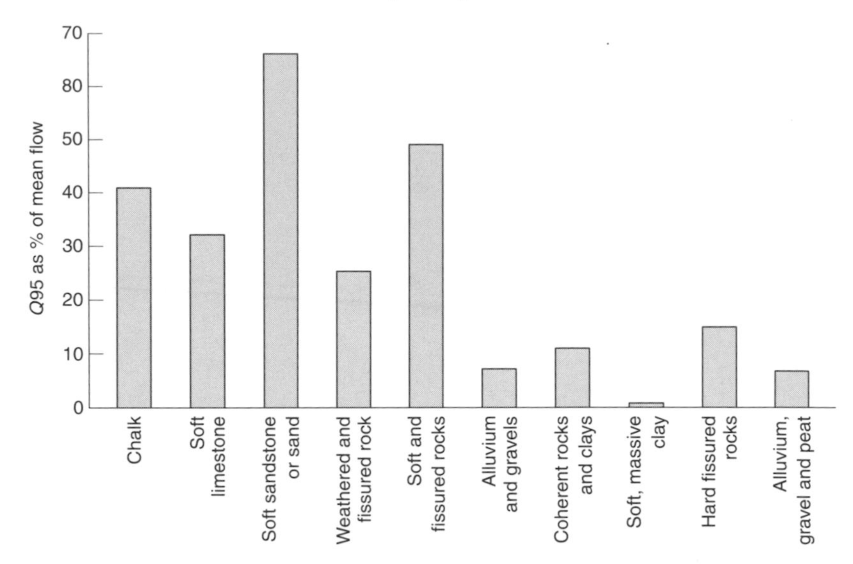

Figure 4.10 Variation in *Q*95 with geology: UK catchments (adapted from Young *et al.*, 2000)

showed how the variation in the variability in deficit durations across south Germany was largely determined by regime type. In both regions, areas with significant snowmelt were much more prone to winter droughts – caused by low temperatures – than summer droughts caused by high temperatures.

In some semi-arid and all arid catchments the lowest flow in a year is zero, and the duration of time when there is no flow is a more interesting indicator than the mean annual minimum flow or *Q*95. This duration is largely climatically driven (it depends on the duration of the wet and dry seasons), and as indicated in Chapter 3, there is generally very little storage in the soil in semi-arid or arid areas. Flows after the wet season has finished are therefore maintained either by groundwater drainage or by drainage from wetlands and lakes within the catchment. If there is neither, flow will soon cease.

Groundwater levels and recharge

The groundwater recharge season at a site depends partly on climatic characteristics and partly on the type of aquifer. Aquifers recharged by infiltration from rivers or pools, for example, are obviously only recharged when the rivers are flowing or ponds full. Direct percolation to aquifers tends only to occur when soil moisture deficits are reduced to zero. In tropical areas this will be during the wet season, but in temperate areas where there is less variation in rainfall through the year direct percolation occurs during winter when there is less evaporation. In these environments, rainfall falling during summer generally has little effect on groundwater recharge, except where the aquifer is very shallow and can be recharged when soil moisture deficits are briefly eliminated.

Groundwater levels therefore tend to rise after the recharge season progresses, reaching a peak sometime shortly after the recharge season ends, and falling to

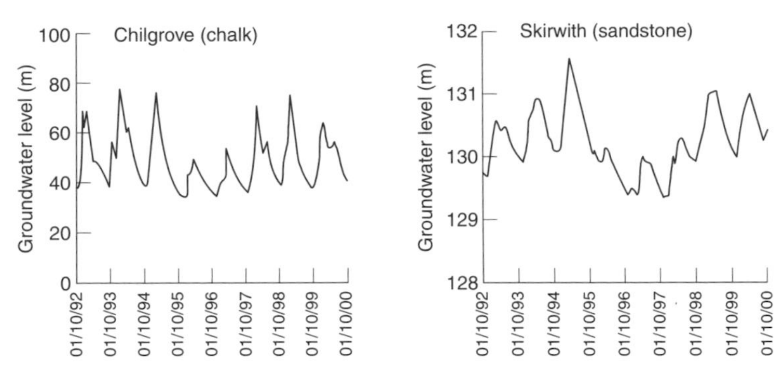

Figure 4.11 Groundwater well hydrographs for two English aquifers

a minimum just after the recharge season recommences. The lag between recharge season at the surface and aquifer water level response depends on the time taken for the water level to respond to inputs at the surface, but is usually just a few days. The amount of rise and fall in water table varies between aquifers with the aquifer specific yield. The lower the specific yield, the greater the rise in water table for a given input of water. Figure 4.11 shows the change in water level for eight years in two aquifers in southern England. The range in levels through the year in the chalk aquifer, with the lowest specific yield, is around 20 m (and in other chalk wells it can be close to 40 m), whilst in the Permo-Triassic sandstone aquifer the range is less than 1 m.

Variations over space: an overview

This section has explored how hydrological behaviour varies from catchment to catchment. At the broadest scale, this variability is determined by climate regime, which determines the distribution of rainfall through the year and from year to year, the volume of snowfall and the magnitude and timing of evaporation. At finer scales, catchment physical characteristics add detail to the climatically-driven regime types. The amount of flood quickflow generated by an input of water is largely determined by catchment soil moisture storage properties, but the rate of rise of the flood hydrograph is primarily a function of catchment morphology, and in particular the extent and configuration of the drainage network. Catchment geological and soil conditions affect the rate of decline from the hydrograph peak, and also catchment low flow characteristics. Recent research (e.g. Vogel & Sankarasubramanian, 2000; Robinson & Sivapalan, 1997; Blöschl & Sivapalan, 1997) has showed that the relationship between hydrological behaviour – particularly variation from year to year – and catchment area is not as simple as has generally been assumed.

4.4 Variations in hydrological behaviour over time

The previous section focused on variations in "average" hydrological behaviour over space, and showed how variation over time – the CV of annual runoff or floods, and persistence, for example – was an important dimension of behaviour in a catchment. Implicitly, it was assumed that this hydrological behaviour was "fixed", and did not vary over time. This of course is not the case, and the characteristics of a hydrological regime – in terms of both volume and timing of flow – may change over time. There are three possible explanations for such a change: the characteristics of the catchment and use of water within the catchment may be being altered by various human activities, there may be variations over time in the catchment's climate due to natural climatic variability, or changes in the catchment and climate due to global warming. The first of these potential effects is explored in detail in Chapter 6, and the last is examined in Chapter 7. The remainder of this section looks at the effects of natural climatic variability on hydrological characteristics.

The term "climate" is usually interpreted as meaning the aggregate or composite of weather over a number of years, and involves both extremes and variability, as well as the average. *Climatic variability* refers to the variation in climate from one averaging-period to another. Climatic variability occurs at the inter-seasonal scale (from winter to spring, for example), at the inter-annual scale (from one year to the next), at the decadal scale (from one decade to the next), through multi-decadal and centennial to millennial. These scales of variability are nested. One decade, for example, may have a particular pattern of inter-annual variability (characterised for example by the CV of annual rainfall), different from another decade. Similarly, one century may have a particular pattern of inter-decadal variability (characterised by the CV of decadal rainfall) different from another.

Through most of the twentieth century it was conventionally assumed that climate (at scales longer than a decade and shorter than glacial-interglacial cycles) was effectively constant (Hulme & Barrow, 1997): indeed, a 30-year period over which climate characteristics are calculated is termed a climatic "normal". However, during the last quarter of the twentieth century this rather static view of climate came to be increasingly replaced by a dynamic view, holding that climate varies on decadal time scales, and essentially challenging the conventional view that any one year could be seen as a random occurrence from a constant population with statistical properties adequately determined from a sample of around 30 years. Several developments led to this shift in perspective, including increased understanding of the nature of past climatic variability over the last few hundred or thousand years, the availability of longer records of observed climate data (showing considerable variation from decade to decade), the occurrence of a number of well-publicised regional-scale climatic anomalies (particularly the El Niño–Southern Oscillation event of 1982–83), and the possibility that human activities were leading to global warming.

This climatic variability will of course impact upon hydrological systems. This section first looks at variations in hydrological regimes over the last few thousand years, before examining in more detail variations in hydrological behaviour over the last century. The final part of this section relates variability in hydrological behaviour to climatic variability.

The "long term": variations in hydrological regime over the last few thousand years

There is abundant physical evidence that river flow and groundwater regimes have varied considerably over the last few thousand years. This evidence includes relict flood deposits, "underfit" streams (rivers flowing in valleys that are far too large for them), the dry valleys common in the chalk of southern and eastern England and northern France, and reserves of "fossil" groundwater found in the Sahara and some other dry environments.

The Quaternary Period, spanning the last two to three million years, comprises the Pleistocene and Holocene Epochs. Figure 4.12a summarises global

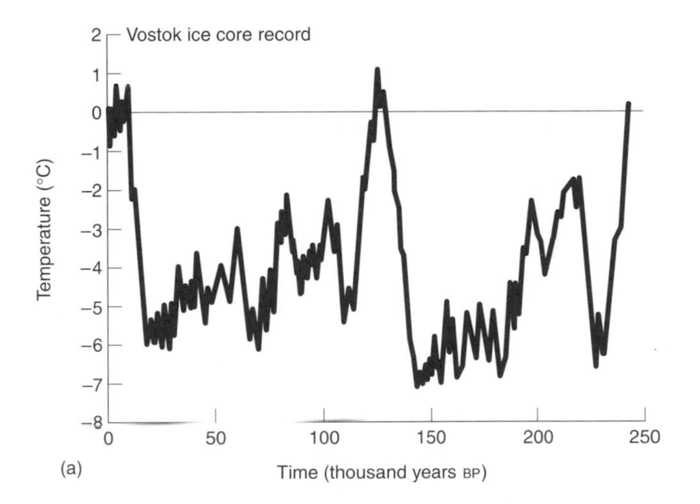

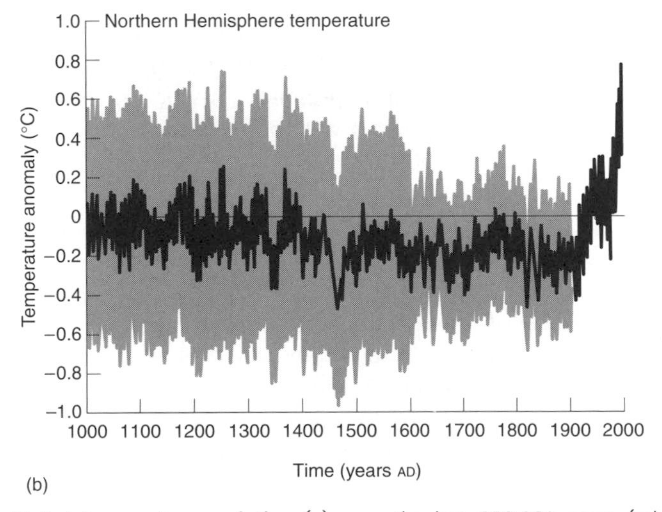

Figure 4.12 Global temperature variation (a) over the last 250 000 years (adapted from Jouzel *et al.*, 1987) and (b) over the last millennium (adapted from Mann *et al.*, 1999): the grey area indicates the possible error range in the reconstructed temperatures

average temperature change over the last 250 000 years of this period, as estimated from the Vostok ice core taken from Antarctica (Jouzel *et al.*, 1987). The Pleistocene Epoch included 20 or more cold episodes and warm phases, during which ice sheets advanced and receded. The most recent glacial period (known in Britain as the Devensian, and more generally as the Weichselian) occurred between roughly 117 000 and 11 000 years before present (BP: years before 1950), with the maximum ice sheet extent at around 21 000 years ago. In Britain, ice sheets reached as far south as a line from the Bristol Channel north east towards the Humber Estuary, and periglacial conditions prevailed everywhere south and east of the line. The Laurentide ice sheet reached 40°N in eastern North America. Global average temperatures were between 4 and 6 °C cooler than at present, and sea level was up to 150 m lower (Slaymaker & Spencer, 1998).

As the ice sheets decayed for the last (so far) time, climate became unstable and very large temperature changes occurred. Between around 13 500 and 10 000 BP temperatures rose to those similar to the present, but this increase was not regular. The increase was at first rapid, and by around 12 700 BP average temperature was generally close to or above the present value. Over the next 2500 years, however, temperatures fell, with an abrupt fall at around 11 300 BP. Temperatures remained low, almost at glacial levels, for around 1000 years, before rising rapidly again to 10 000 BP. This cold period is known as the Younger Dryas, and is particularly evident in northern and western Europe. Ice sheets and glaciers advanced again (and in Britain new glaciers formed after an ice-free period), and periglacial conditions returned.

By around 10 000 BP global temperatures had risen again and stabilised, and this is taken to represent the start of the Holocene Epoch. By this time, the continental ice sheets had largely disappeared. Climate has not, however, been constant through the Holocene, although fluctuations are small compared to those in previous periods. Widespread warmth and dryness, for example, occurred for a few centuries around 4000 BP, when temperatures may have been between 1 and 2 °C higher than at present. This period (sometimes termed the "Post-glacial climatic optimum") was followed by cool and wet conditions in the few centuries before 2000 BP.

The most recent two millennia (between 2000 BP and the present) have also seen climatic fluctuations (Figure 4.12b shows reconstructed temperatures for the Northern Hemisphere from AD 1000 to 2000: Mann *et al.*, 1999). The Medieval Warm Period between around AD 900 and 1300 again had temperatures slightly in excess of those of the late twentieth century, at least in Europe and North America, and was followed by a period of instability, replaced in the 1600s by the relative climatic stability of the Little Ice Age, which lasted in some places to the end of the nineteenth century. Since then temperatures have risen, with the additional influence of global warming appearing during the late twentieth century.

The causes of these climatic fluctuations have been the subject of much research, and there is still no complete explanation. Table 4.2 summarises the major natural causes of climatic change, indicating the time scales over which

Table 4.2 Major causes of natural climatic change (adapted from Harvey, 2000)

Cause	Time scale (years)	Global mean magnitude, lower and upper limits
Changes in land–sea geography	10^7–10^8	up to 5 °C
Changes in the earth's orbit	10^4–10^5	4–6 °C
Changes in greenhouse gas concentrations	10^2–10^9	2–3 °C
Changes in solar luminosity	10^1–10^9	0.1 °C
Transitions to new climate states	10^1–10^2	2–3 °C
Internal variability	10^1–10^3	0.2–0.4 °C
Volcanic activity	1–4	0.4 °C

they operate and the potential global effect of these perturbations (Harvey, 2000). Changes in land–sea geography reflect the effects of plate tectonics moving continents around the earth. Shifting a continent from low to high latitude, for example, results in a net increase in the absorption of solar radiation and a rise in temperature; closing a gap between continents alters ocean circulation patterns. Over a shorter time scale, changes in the earth's orbit due to the gravitational effects of other planets in the solar system affect the amount of energy received. There are three important cycles, with periods of around 94 000 years, 41 000 years and 25 920 years, and these cycles (known as Milankovitch cycles after the Yugoslavian mathematician who discovered them) provide the basis for the explanation of glacial and interglacial cycles. However, the orbital cycles do not provide the entire explanation, and other feedbacks are in force. Perhaps the most important is the effect of changes in ocean currents due to changes in radiation receipts, which alters the strength of the biological pump exchanging carbon dioxide with the atmosphere – thus affecting the concentration of carbon dioxide in the atmosphere. Changes in solar luminosity are small over the time scale of decades, centuries and millennia, but over the last few billion years solar luminosity has declined substantially. Abrupt climate changes can occur when the climate regime in a region shifts rapidly to a new state. One such shift that has occurred in the past is a weakening of the Atlantic Conveyor, the ocean current bringing heat towards western Europe from the tropical Atlantic: when this weakens, temperatures across western Europe fall. A weakening of the Atlantic Conveyor can be triggered by a large input of freshwater into the north Atlantic, and this is hypothesised to be the explanation of the sudden return to near-glacial conditions in the Younger Dryas (Broecker, 1987). The freshwater came from the sudden drainage of an ice-dammed lake, released down the St Lawrence Seaway as the Laurentide Ice Sheet decayed. Internal climatic variability may occur in the absence of any changing external forcing, through interactions between ocean, atmosphere and ice sheets. These fluctuations tend to be regional, rather than global, so their regional impact may be substantially greater than the global temperature change effect shown in Table 4.2. The El Niño–Southern Oscillation (ENSO) is an example of such a form of internal climatic variability. Finally, volcanic eruptions can eject large amounts of material into the atmosphere, reducing the amount of solar radiation reached at the earth's surface. These effects, however, generally only last for a few years.

Hydrological regimes can be expected to vary over the long term in response to these climatic fluctuations, and the effects of these fluctuations are apparent in the form of the river channel and floodplain sedimentology. Past hydrological regimes can therefore be inferred from deposits and landforms, and the sub-discipline of *palaeohydrology* has developed. This is concerned with the "reconstruction" of past hydrological regimes for several reasons, including a desire to understand landform development and the characteristics of change, reconstruct past climates, and provide data to enhance estimates of risk. Box 4.4

Box 4.4 Reconstructing hydrological regimes through palaeohydrological methods

Palaeohydrology essentially attempts to infer past process from various indicators of measured form. Indicators include (Baker, 1987; 1998) slackwater deposits, scour marks, palaeochannel dimensions, flood-scarred trees and the magnitudes of flood-transported boulders. Process – typically flood magnitude – is inferred using either empirical relationships between channel form and peak flow (see Reinfelds & Bishop, 1998) or the flow velocities required to entrain and transport sediments (Maizels, 1983). Studies in vertically-stable bedrock channels use slackwater deposits to construct flood chronologies, but in alluvial channels – predominant across most temperate regions – periods of enhanced flooding tend to be inferred from evidence of channel movement or incision, as continuous sedimentary sequences are rare. All these approaches, however, require accurate dating of the features used to reconstruct flows or levels, and the assumption that present relationships between form and process operated in the study area in the past. An understanding of the geomorphological linkages between channel form and process is therefore essential. One additional complication with inferring process from form is that river channels may change not only in response to climatic changes or human intervention, but also following tectonic changes, changes in sea level or the "capture" of a previously separate river system.

Past lake levels, particularly for lakes in closed basins, can be estimated by dating sedimentary deposits, and give an indication of the balance between precipitation and evaporation (Fontes & Gasse, 1991; Harrison & Digerfeldt, 1993).

River flows over the past 1000 years have been estimated from tree rings (e.g. Earle, 1993; Loaiciga *et al.*, 1993), but this relies on intermediate relationships with precipitation.

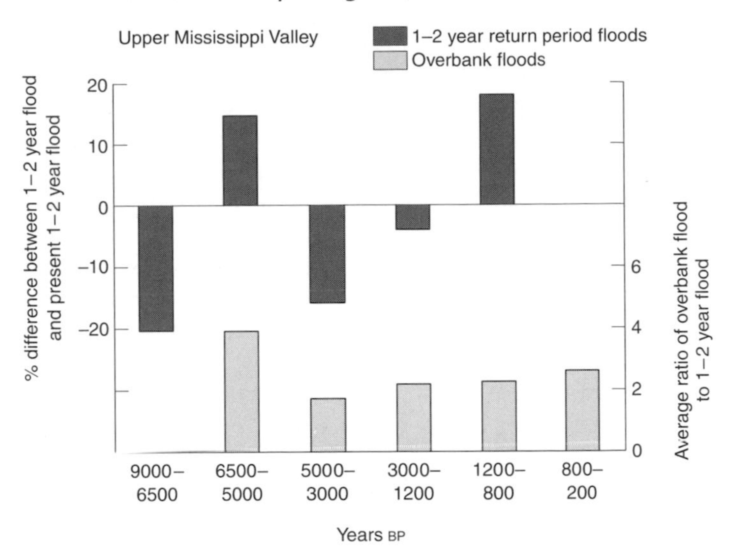

Figure 4.13 Flood magnitudes during the Holocene in the Upper Mississippi Valley (adapted from Knox, 1993)

summarises the main techniques used by palaeohydrologists to infer past flow regimes, reconstruct "palaeodischarges", and construct flood chronologies.

There have been many palaeohydrological reconstructions of river channel evolution in a large number of environments (see Starkel *et al.*, 1991 and Gregory *et al.* 1995 for overviews). One of the longest duration reconstructions is that of Fuller *et al.* (1998), which evaluated river channel change – in terms of aggradation and incision – in a catchment in north east Spain over 200 000 years. This period spans the entire last glacial cycle, and Fuller *et al.* showed how incision tended to occur during "warm" spells whilst aggradational episodes coincided with "cold" periods. They inferred from these results that river flood frequency varied substantially between the different periods. Computer simulation studies, rather than palaeohydrological reconstructions, indicate that the total freshwater runoff into the Atlantic Ocean from North America was up to three times as great over the period from 60 000 to 10 000 years ago compared to present (Marshall & Clarke, 1999). This increase in runoff was despite a decrease in continental precipitation, and was attributed to both very low evaporation and high levels of meltwater production.

Most studies, however, have focused on the Holocene (i.e. after deglaciation), and particularly on the last 5000 years. Knox (1993) constructed a flood chronology for the Upper Mississippi Valley extending throughout the Holocene, using palaeochannel dimensions to estimate the magnitudes of floods which filled the channel and estimates of the velocity needed to entrain sediment for larger floods. He showed (Figure 4.13) that the period between 6500 and 5000 years BP experienced a dramatic increase in the magnitude of channel-filling floods (more than 20% greater than present), and found evidence of a small number of very large floods. Between 5000 and 3000 BP, around the post-glacial climatic optimum, both channel-filling floods and overbank events were smaller than at present. Another abrupt change occurred around 3000 BP following a cooling, when flood magnitudes increased again. Another increase in flood magnitudes began around 1200 BP (around AD 800). Knox drew the conclusion that relatively small changes in climate during the Holocene – fluctuations of average temperature of less than 1.5 °C – could produce very substantial changes in flood frequency.

Climatic fluctuations during the Holocene, however, did not produce the same hydrological response everywhere. Ely (1997), for example, found an increase in flood frequency in the south western United States between 5000 and 3600 BP, when flood magnitudes were low in the Upper Mississippi, although other flood-rich periods were coincident. Five of the six largest floods during the Holocene in the Yellow River, China, occurred between 8000 and 6000 BP (Yang *et al.*, 2000).

There is rather more, higher-resolution, information available on river flow variations over the last 2000 years. Brown (1998) reviewed the fluvial evidence for hydrological changes in Europe during the Medieval Warm Period (MWP) and the subsequent deterioration in climate to the Little Ice Age. During the MWP, temperatures in northern Europe were approximately 1 °C higher than present, with drier summers (but probably with more intense summer convective storms) and little difference in winter rainfall. South of the Alps, rainfall during summer was higher than present. Increased flood frequency has been inferred at the end of the MWP in the early fourteenth century in upland Britain, Belarus, northern Germany, the upper Danube and eastern Spain (Brown, 1998; Barriendos Vallve & Martin-Vide, 1998). Over this period, documentary evidence can be used to supplement geological or geomorphological reconstructions, at least in Europe (and also in parts of South America; Prieto *et al.*, 1999), and there is considerable evidence of increased flooding during the Little Ice Age between AD 1600 and 1900 (Pfister & Brazdil, 1999). Figure 4.14 shows a thousand-year long record of summer flows for the White River in Arkansas (Cleaveland, 2000), reconstructed from tree rings, and flows from 1640 in the Burdekin River in north Queensland, Australia (Isdale *et al.*, 1998), estimated from fluorescent bands in corals of the Great Barrier Reef. Both these records show substantial persistent variations in flow over time with the strongest patterns in the Australian continent. River levels on the Nile at Cairo have been recorded for close to 3000 years, and Table 4.3 summarises the frequency of Nile floods from AD 622 to 1973 (Roberts, 1994): there are clear maxima during cool periods.

The instrumental period: the mid-1800s to the present

Although there is a long record from the Nile, systematic measurement of river and groundwater levels only began in the mid-nineteenth century, and records are very sparse until the middle of the twentieth century (Box 4.5 summarises some of the problems with constructing long flow records). Figure 4.15 shows

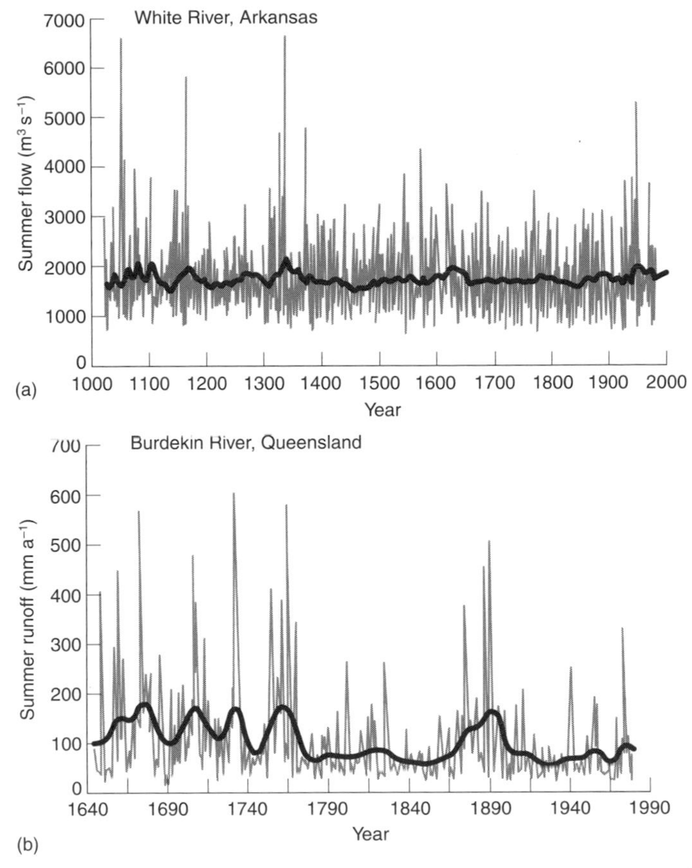

Figure 4.14 Variations in floods, summer and annual runoff: USA and Australia (Cleaveland, 2000; Isdale *et al.*, 1998). The thick line is a 30 year running mean

Table 4.3 Frequency of occurrence of Nile River floods (adapted from Roberts, 1994)

Period (AD)	Description	Number of floods	Floods per year	Number of occasions with four or more successive floods
622–999	Cool period	105	0.28	6
1000–1290	Medieval Warm Period	23	0.08	0
1291–1522	Deterioration	50	0.22	0
1523–1693	no data	no data	no data	no data
1694–1899	Little Ice Age	72	0.29	4
1900–1973	Recent period	23	0.31	0

Box 4.5 Constructing long records of river flows

There are in general terms three potential problems with long river flow records. First, the data may have been collected in different ways at different times. Before the advent of automatic recording devices, river levels were typically only measured once or twice a day, for example. The location of the measurements may have changed over time, and if measurements are made at a structure, such as a weir or lock, that structure may have changed. It may be difficult or impossible to correct for past measurement errors, which were not realised at the time. Second, the catchment or river channel may have changed significantly due to human interventions. Unfortunately, most long records tend to be from rivers which have had a long history of human use. Third, there may be breaks in the records which can be difficult to fill.

annual total runoff from six catchments, spanning approximately the period 1900 to 1990: summary statistics for each are given in Table 4.4. Annual runoff varies from year to year in each catchment, with by far the greatest relative variation in the Australian example. Annual runoff in the Niger River shows some very persistent departures from the mean lasting for several years. Flows were well above average between the mid-1920s and the mid-1930s, and well below between the 1980s and to at least 1990. The Australian example also shows evidence of decadal-scale variability. Long groundwater records provide indications of variations in recharge from year to year and decade to decade (Figure 4.16a), as do the few long records of evaporation. Figure 4.16b shows annual potential evaporation at Oxford between 1815 and 1996, as estimated from temperature (Burt & Shahgedanova, 1998).

Figures 4.15 and 4.16 show examples of hydrological time series at individual locations or catchments. There is in fact a great deal of geographical consistency in behaviour across space, and it is possible to identify regions where flows or recharge vary in a similar way from year to year. One way of doing this is to cross-correlate flows in one catchment to those in another, and identify areas with high correlation. This was done by Shorthouse & Arnell (1998) in Europe, who showed that catchments in northern Europe tended to vary over time in the same way, but in opposition to catchments in southern Europe. In other words, when annual or seasonal runoff was high in northern Europe, it tended to be low in southern Europe. Such an approach, however, involves the calculation of large numbers of cross-correlations. Another approach is therefore to undertake a *principal components analysis*, which seeks to identify consistent patterns in multiple data series. This has been done in North America (Bartlein, 1982; Lins, 1985; Guetter & Georgakakos, 1993; Lins, 1997), and most recently Lins (1997) identified 11 distinct and independent regional streamflow regimes across the United States, as shown in Figure 4.17.

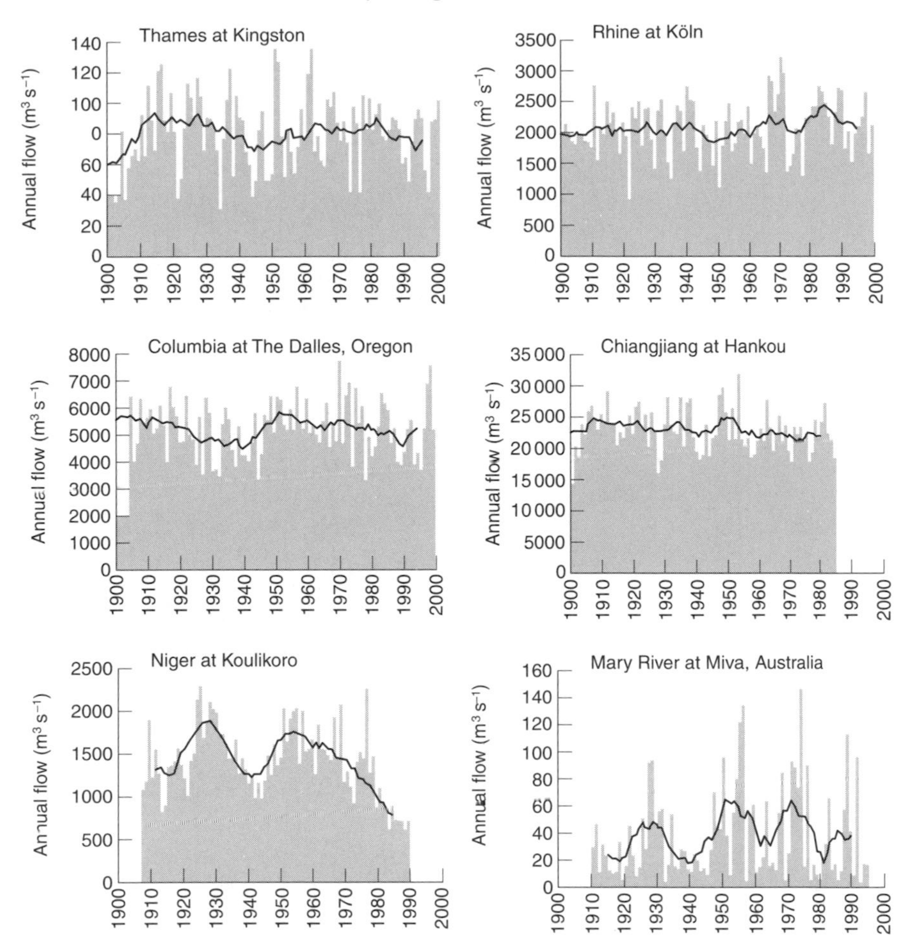

Figure 4.15 Variation in annual total runoff through the twentieth century in six large catchments

Table 4.4 Summary statistics for the river flow records in Figure 4.15

	Coefficient of variation of annual runoff	Hurst coefficient
Thames at Kingston, UK	0.266	0.603
Rhine at Köln, Germany	0.214	0.639
Columbia at The Dalles, Oregon, USA	0.182	0.682
Chiangjiang at Hankou, China	0.137	0.639
Niger at Koulikouro, Niger	0.282	0.859
Mary River at Miva, Australia	0.871	0.744

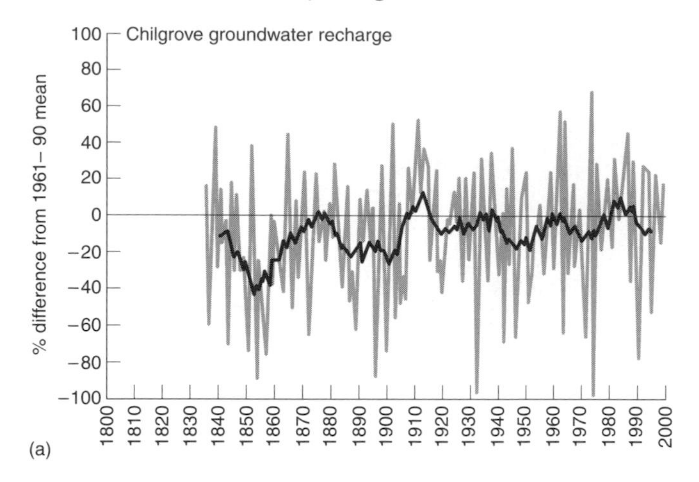

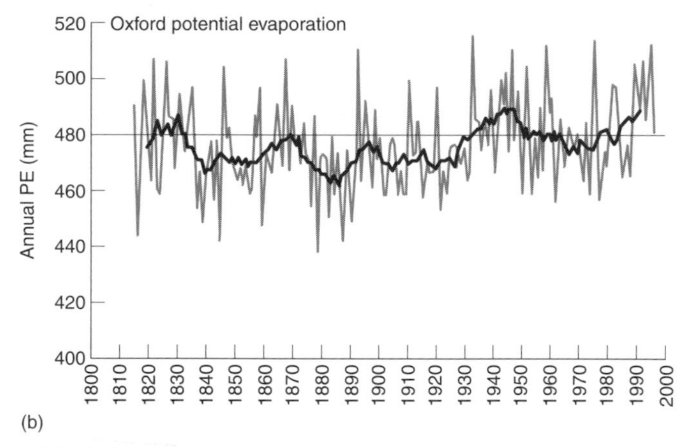

Figure 4.16 (a) Variation over time in estimated annual recharge at Chilgrove, West Sussex, southern England. Recharge is approximated by the range between the end of summer minimum level and the maximum level over the following few months. (b) Variation over time in annual potential evaporation at Oxford (adapted from Burt & Shahgedenova, 1998)

Within each of these regions streamflows vary broadly consistently from year to year, and the regions are largely independent of each other. The extent of each region varies through the year. One of the regions – in the western USA – is different from the others in the sense that it has two centres (one in the Pacific North West and the other in southern California, Arizona and New Mexico) which behave in a consistent way, but are negatively correlated: when flows are low in the Pacific North West, they are high in the other centre.

Figure 4.18 shows the variation over time in regional average streamflows for two "homogeneous" flow regions in the United States (Guetter & Georgakakos, 1993). In the north east USA flows were low during summer for several years in the mid-1960s, for example, but above average in the early

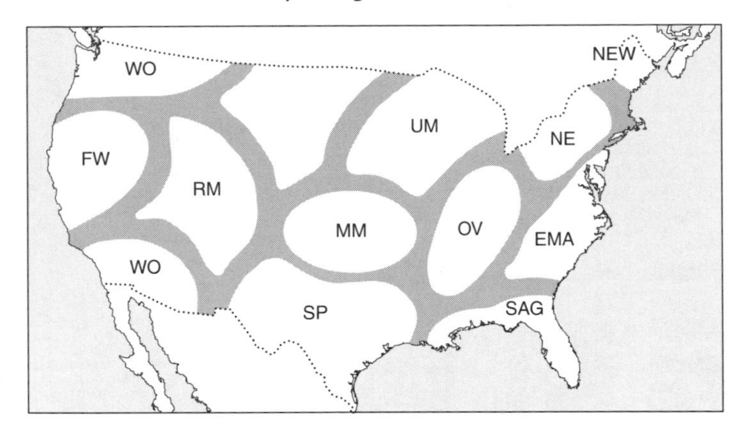

Figure 4.17 Hydrological regions in the continental USA with consistent patterns of variation in flow over time, based on Lins (1997): the boundaries are not precise, and vary through the year

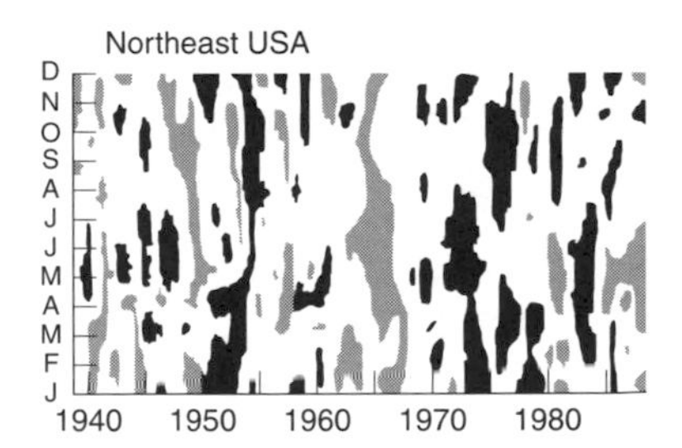

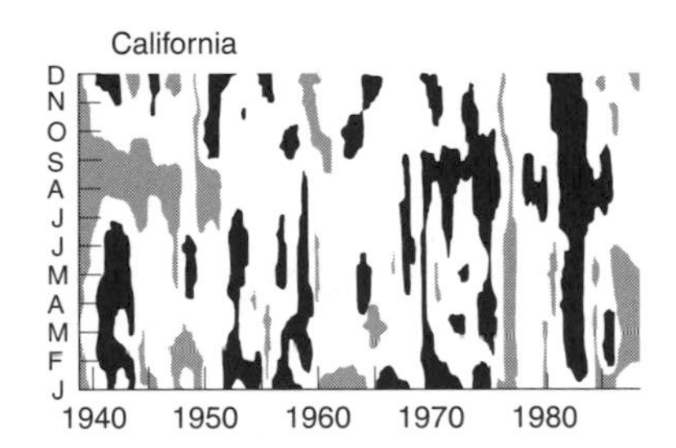

Figure 4.18 Variability of outflow from north east USA and California between 1939 and 1988 (Guetter & Georgakakos, 1993). Dark shading indicates a value in the top 25% of the monthly values, and light shading indicates a value in the bottom 25%

1970s. The plot for California shows more persistent patterns, with lower than average summer flows throughout the 1940s, for example, and consistently high flows throughout the year for the period around 1982–83.

Across Europe there is also evidence of above or below average streamflows persisting for several years. Figure 4.19 shows the variation in winter (December to February) runoff for 41 regions in western and northern Europe (Arnell, 1994). The late 1960s, for example, saw high winter runoff across most of western Europe, as did the early 1980s: winter flows were below average in the mid-1960s and the early 1970s. Figure 4.19 illustrates not only the temporal clustering of anomalous years, but also the strongly consistent pattern of variation across much of western Europe.

At the global scale, Probst & Tardy (1987) identified patterns in hydrological behaviour across 50 major world basins, using data up to the early 1970s. In North America, for example, runoff was generally above average around 1950 and the late 1960s: the reverse was the case in western and central Europe.

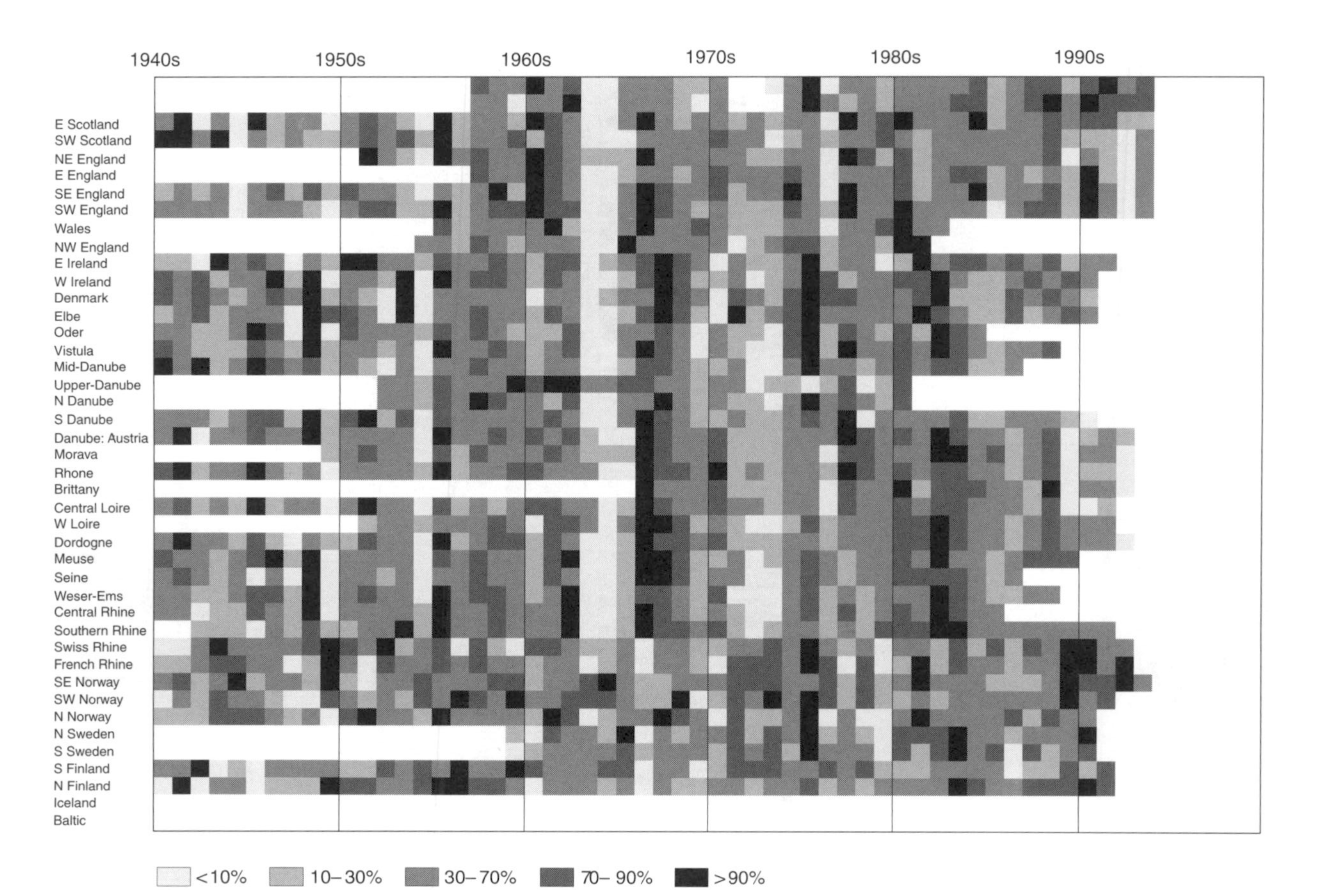

Figure 4.19 Variation over time in winter runoff across northern and western Europe (Arnell, 1994)

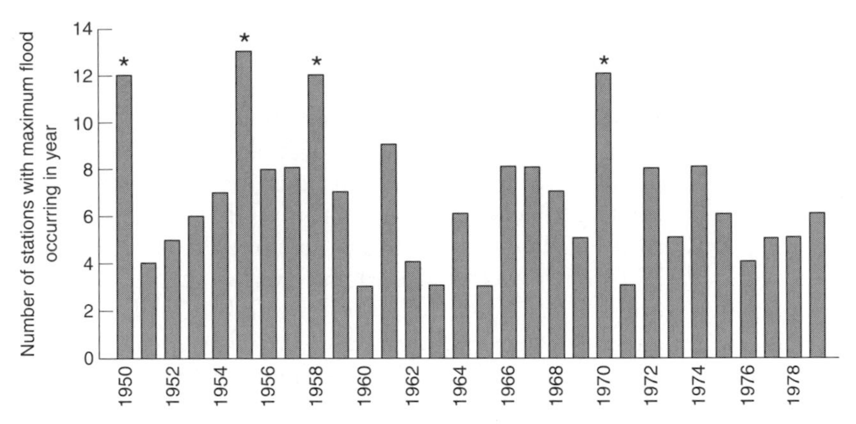

Figure 4.20 The number of stations with the largest flood over the period 1950 to 1979 occurring in each year (adapted from Burn & Arnell, 1993). The highlighted years have more maximum floods than would be expected in any one year, at a 5% level of signifiance

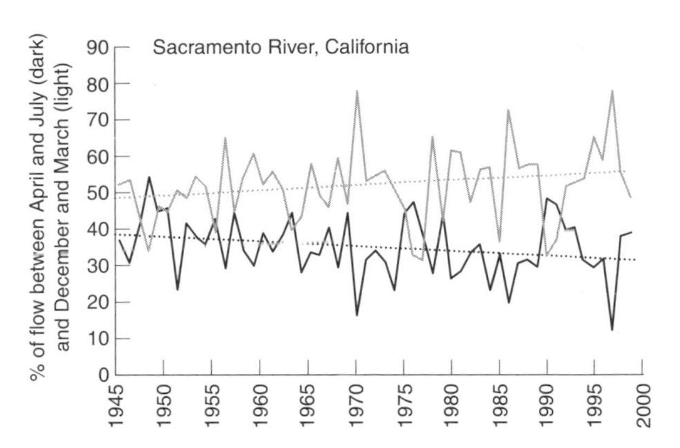

Figure 4.21 The percentage of annual streamflow in the Sacramento River basin occurring between April and July, and between December and March

Probst & Tardy's analysis showed both spatial patterns in variation over time, and evidence of clustering of anomalous years.

Other important indicators of hydrological behaviour which are known to show decadal-scale variability include the frequency of floods and low flows, and the timing of peak streamflow. Burn & Arnell (1993) grouped annual maximum flood data from 200 catchments into 44 regions worldwide, showing not only non-random variations in the number of floods from year to year (Figure 4.20) but also strong correlations in flood occurrence between different parts of the world. Aguado *et al.* (1992) calculated the percentage of flow occurring during spring in rivers fed by snowmelt in California (Figure 4.21), showing substantial fluctuations on a decadal timescale.

Understanding variability over time in hydrological behaviour

The clear patterns in the variation in hydrological behaviour over time introduced in the previous section are of course closely related to climatic variability. As a general rule, the variation in the volume of streamflow (or indeed recharge) is largely associated with the variation in precipitation from year to year. Because of the integrating effect of a catchment, in both space and time, the relationship between streamflow and catchment precipitation is not necessarily simple, but rather obviously wetter years tend to have higher streamflow and bigger floods whilst drier years have lower flows and larger droughts. The translation of precipitation change into streamflow (or recharge) change depends broadly on the runoff coefficient. The lower the runoff coefficient, the greater the percentage change in streamflow for a given percentage change in precipitation. Algebraically, this can be expressed as (Wigley & Jones, 1985):

$$\Delta R = \frac{\Delta P - (1 - \alpha)\Delta E}{\alpha} \qquad \textbf{4.6}$$

where ΔP is the fractional change in precipitation, ΔE is the fractional change in actual evaporation, ΔR is the fractional change in streamflow, and α is the runoff coefficient. With no change in actual evaporation, a 10% increase in precipitation would result in 50% more streamflow in a catchment with a runoff coefficient of 0.1, but only 16.7% extra in a catchment with a runoff coefficient of 0.6. Equation 4.6 only holds precisely, however, if changes in actual evaporation and precipitation are independent, and this is not necessarily the case: even with no change in potential evaporation, a reduction in precipitation may lead to a reduction in *actual* evaporation. In practice this means that the relationship between variability in precipitation and streamflow can be non-linear (as shown in California by Risbey & Entekhabi, 1996). For example, a 20% increase in precipitation may lead to more than twice the increase in streamflow than a 10% increase in precipitation.

Potential evaporation varies much less from year to year than rainfall (the standard deviation of annual potential evaporation at Oxford is just 15 mm, compared to a mean of 475 mm, for example: Burt & Shahgednova, 1998), so changes from year to year or decade to decade in streamflow are much more sensitive to variability in precipitation than evaporation. Figure 4.22 shows some example "sensitivity surfaces", plotting the modelled change in annual streamflow in two British catchments to changes in precipitation and potential evaporation applied equally throughout the year. The slope of the contours emphasises the greater sensitivity to precipitation variability than to variability in potential evaporation, and the spacing between them indicates the greater sensitivity in the drier Harper's Brook catchment than in the more humid Greta (note that the equal spacing of the contours implies that in these catchments the relationship between precipitation change and streamflow change is approximately linear).

Although variability in precipitation from year to year or decade to decade is the primary driver of variability in hydrological behaviour, in snow-dominated regimes the timing of peak streamflow is largely determined by temperature. The variation over time in the proportion of streamflow occurring

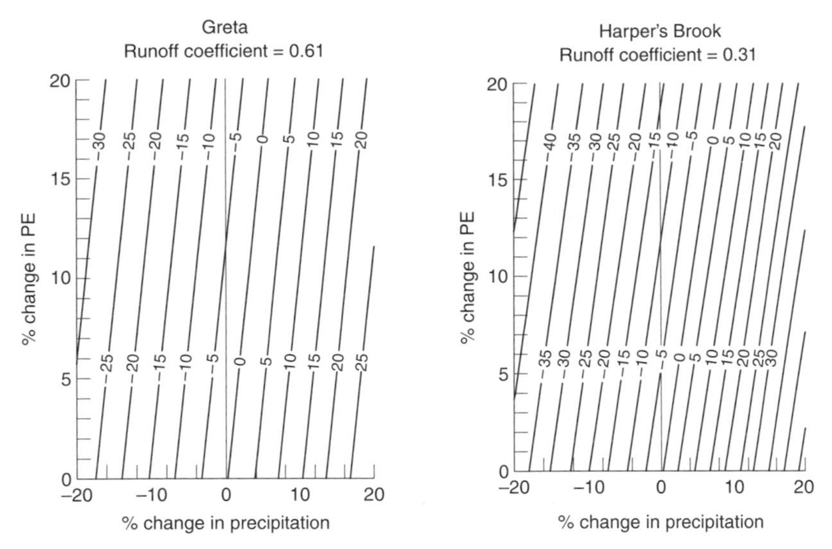

Figure 4.22 Sensitivity of average annual runoff to changes in precipitation and potential evaporation in two British catchments (Arnell, 1996)

during spring shown in Figure 4.21, for example, is largely explained by variations in winter temperature, and thus the proportion of precipitation that falls as snow (Aguado *et al.*, 1992).

Understanding the links between climatic variability and hydrological variability is, however, just one aspect of understanding the nature of variation in hydrological behaviour from year to year and decade to decade. The last two decades of the twentieth century saw major advances in the understanding of the dynamics of climate, and in the last decade hydrologists began to interpret patterns of hydrological variability both in a catchment and across a region in terms of climate *anomalies* (defined as departures from average).

It has long been known, of course, that the world can broadly be divided into different climate zones, with different average climates. However, within a particular area, climate in any one season, year or decade may be substantially different from the average. Indeed, it is possible to recognise certain "modes of variability" (also known as teleconnection patterns) operating in particular parts of the world (Barnston & Livesey, 1987): in some years the local and regional climate has one set of characteristics, and in others a different set. Each set of climate states is characterised by the recurrence of different patterns of precipitation, temperature and other aspects of atmospheric circulation such as pressure. By far the best known mode of variability is the El Niño–Southern Oscillation (ENSO: Box 4.6), which is focused around the south Pacific but has potentially global-scale implications. Barnston & Livesey identified 11 modes of variability in the Northern Hemisphere, most of them apparent only during certain seasons. The most prominent of these are the North Atlantic Oscillation (NAO: Box 4.7), affecting primarily the north Atlantic and western Europe, and the Pacific Decadal Oscillation (PDO: Box 4.8), affecting the north Pacific and western North America. An oscillation in the

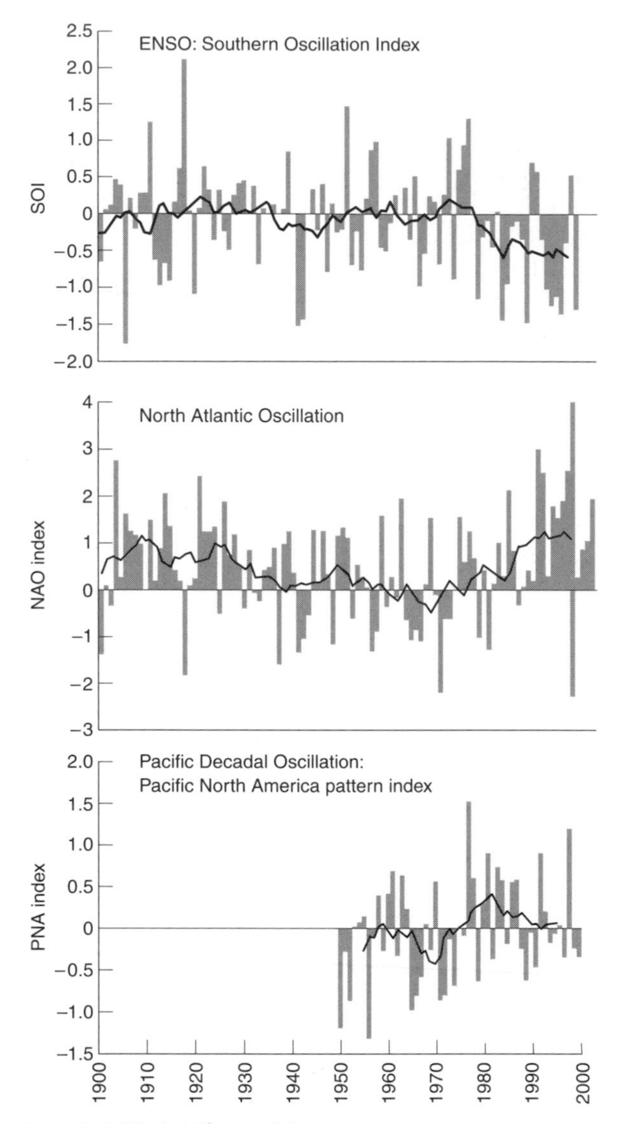

Figure 4.23 Indices of ENSO, NAO and PDO

tropical Atlantic Ocean (Box 4.9) influences precipitation in eastern South America and west Africa. Although the characteristics of these modes of variability and their geographical extent are increasingly well understood, the mechanisms which drive them remain unclear. In the most general terms, slowly varying surface conditions (sea surface temperatures, land surface moisture and vegetation (see Chapter 8), and extent of snow and ice cover) act as boundary conditions that significantly influence the climate of the atmosphere, but these surface boundary conditions are themselves influenced by the atmosphere. Figure 4.23

Box 4.6 El Niño–Southern Oscillation (ENSO)

ENSO is a complicated phenomenon involving both anomalous sea surface temperatures in the Pacific Ocean and anomalous atmospheric circulation across the Pacific (see Glantz (1996) for a review). In most years, cold water upwells off the coast of Peru in the eastern Pacific. This nutrient-rich water sustains a rich fishery. In some years, however, this cold water is replaced by warm water, and fish stocks fall substantially. This phenomenon tends to occur around Christmas, and was termed El Niño (the Christ Child) by Peruvian fishermen. Meanwhile, during the early part of the twentieth century it was noticed that the pressure gradient across the central Pacific occasionally changed direction: this was termed the Southern Oscillation. In the mid-1950s these two phenomena were linked and termed ENSO. In "normal" conditions, cold water upwells off the coast of Peru, and there is a pool of warm water in the western equatorial Pacific. This warm water causes air to rise and move eastwards at high altitude across the Pacific towards Peru, where it descends and then moves westwards at low levels. Under these conditions the western Pacific is "wet" and the eastern Pacific "dry". During an El Niño event (also known as a warm ENSO event), however, this situation essentially reverses. The westward-blowing surface winds weaken, the warm pool of water in the west spreads east. Rainfall is reduced in the west Pacific, but increased in the east Pacific. After between 12 and 18 months the surface winds strengthen again, reconstructing the warm water in the west Pacific. The precise details of the formation of El Niños are not known, and it has become clear that different events are created in slightly different ways and consequently have different patterns of impact. The effects of El Niño are felt over a much larger area than the Pacific Ocean: Figure 4.24 shows the typical rainfall and temperature patterns associated with El Niño during the Northern Hemisphere winter season (Ropelewski & Halpert, 1987; Ropelewski, 1992).

In some years, sea surface temperatures are *colder* than average off the coast of Peru, and winds blow more vigorously. The western Pacific is wetter than usual, and the eastern Pacific drier. Such occurrences are known as La Niña or cold ENSO events. La Niña events are less frequent than El Niño events.

Although the most obvious effects of warm and cold ENSO events are around the Pacific Rim, in southern Africa and southern Asia, there is evidence of an ENSO signal in European climate. There is a tendency for an increased frequency of depressions in the winter of a warm ENSO event and a more northerly depression track (Fraedrich, 1994), bringing higher than average temperatures and precipitation to northern Europe and the UK (Wilby, 1993).

(continued)

(continued)

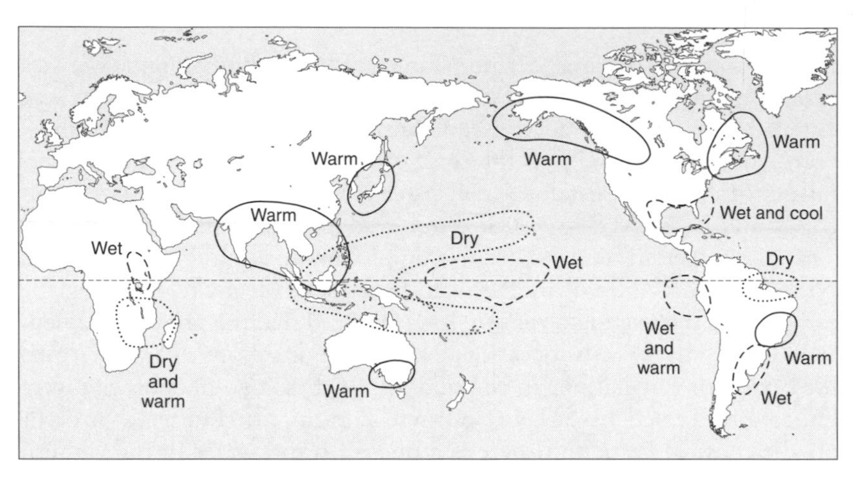

Figure 4.24 Typical rainfall and temperature patterns associated with ENSO for the Northern Hemisphere winter season (Ropelewski & Halpert, 1987)

Box 4.7 The North Atlantic Oscillation (NAO)

As far back as the mid-eighteenth century, it was observed that cold winters in Greenland tended to occur at the same time as warm winters in Denmark. In the 1930s, it was noticed that when pressure was lower than average over Iceland it tended to be higher than average over the Azores, and this was termed the North Atlantic Oscillation. In general terms, the greater the pressure gradient between the Azores and Iceland, the more vigorous the depressions moving across the Atlantic, particularly in winter. Northern and western Europe are therefore warmer and wetter, whilst southern Europe is drier (Hurrell, 1995). Meanwhile, temperatures are lower across eastern Canada and Greenland. The reverse applies when the pressure gradient is low.

There is a strong association between sea surface temperatures in the Atlantic and the NAO, but the nature of the link is not clear. Climate model simulations indicate, however, that the NAO is forced by sea surface temperature anomalies (Rodwell *et al.*, 1999), which themselves may be generated by slowly-varying processes within the Atlantic Ocean.

shows time series of indices of ENSO, NAO and the PDO. The main feature to note in the three indices is that there is evidence of non-stationary behaviour: each series is characterised by prolonged periods with above or below average values of the index. The NAO index, for example, shows a prolonged period

Box 4.8 The Pacific Decadal Oscillation (PDO)

An abrupt change in the winter circulation patterns over the north Pacific occurred during the 1970s: the Aleutian low pressure zone, which influences the intensity and location of depressions moving from the Pacific across North America, moved southwards and intensified, and sea surface temperatures rose in the equatorial and north Pacific Ocean. Temperature rose over the Pacific north west region of the United States, and precipitation reduced. A similar shift occurred between 1925 and 1946. This decadal-scale variability in sea surface temperatures and atmospheric circulation is termed the Pacific Decadal Oscillation (PDO: Mantua *et al.*, 1997). The most prominent atmospheric circulation anomaly associated with the PDO is termed the Pacific North America (PNA) pattern. In this pattern, higher than normal pressure occurs over the Aleutians, the Pacific north west, and in the south east USA, and low pressure anomalies occur over Hawaii and the northern Great Plains. Unlike ENSO, the PDO anomaly pattern persists for several years, and has been characterised as representing a shift in mean conditions (Trenberth & Hurrell, 1994). Although they operate on different time scales, ENSO and the PDO are related. The most "extreme" PNA patterns (warmest and driest in the Pacific north west and the south east USA) during a warm PDO phase tend to occur during El Niño events.

Box 4.9 Variability in the tropical Atlantic

Sea surface temperature anomalies in the tropical Atlantic Ocean are weaker than those in the Pacific associated with El Niño, but are linked to variations in the Inter-Tropical Convergence Zone (ITCZ) which in turn influences precipitation in north east Brazil and west Africa. When the tropical Atlantic Ocean north of the equator is warmer than average, the ITCZ is shifted to the north, causing severe drought in north east Brazil and high rain in the Sahel region of west Africa. Conversely, when the Atlantic Ocean south of the equator is warm, rainfall is low in the Sahel but high in north east Brazil. It has been suggested that this equatorial Atlantic oscillation is analogous to, but weaker than, ENSO, and is strongly influenced by ENSO (Latif & Grötzner, 2000).

in the 1960s and 1970s with a low pressure gradient, followed by a larger gradient in the 1980s and, particularly, 1990s. The PDO index shows a prolonged period with below average values between 1977 and 1988.

Since the early 1990s there have been many studies attempting to link hydrological variability with modes of atmospheric variability, focusing

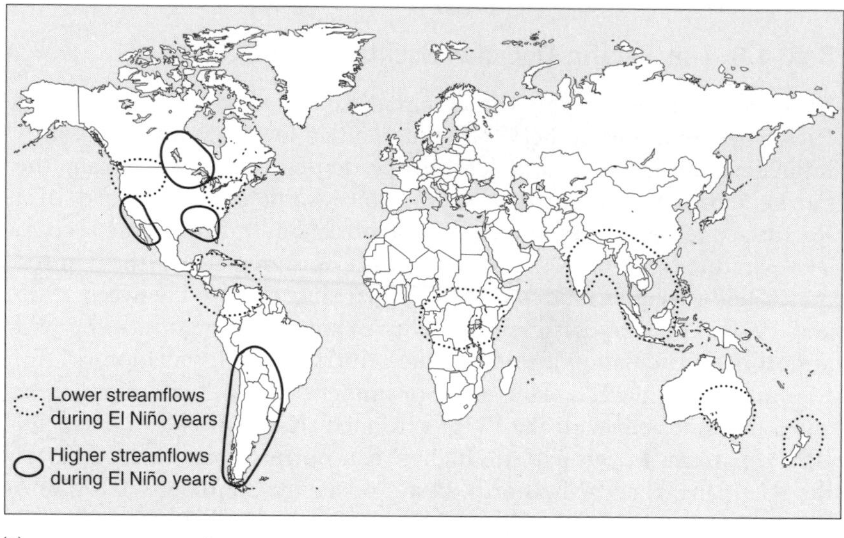

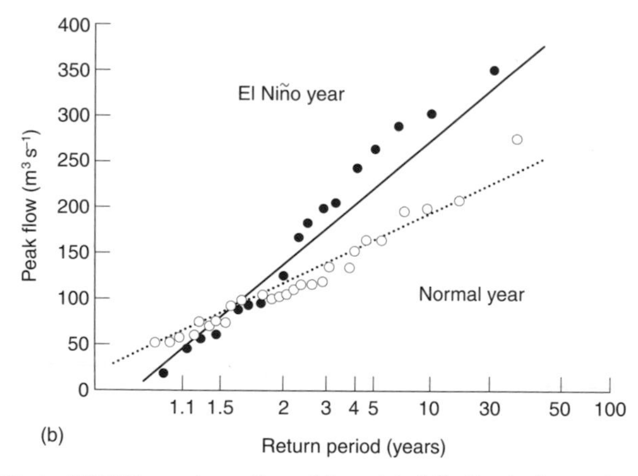

Figure 4.25 Effect of ENSO on streamflow: (a) a global (indicative) overview of streamflow anomalies during warm ENSO (El Niño) events; (b) flood frequency curves for a river in Chile (adapted from Waylen & Caviedes, 1990)

almost exclusively on links between ENSO and streamflow. Figure 4.25 and Table 4.5 summarise published associations between ENSO and streamflow. During El Niño events, streamflow tends to be *higher* than average in western South America (as indicated by the flood frequency curves in Figure 4.25b), south east South America, the Pacific south west of the USA, north central USA and the Gulf Coast of the USA. Streamflow tends to be *lower* than average in Australasia, central and northern South America (including the Amazon), the Pacific north west of North America, the north east of North America, and equatorial and southern Africa. Anomalies during La Niña events

Table 4.5 Associations between El Niño and La Niña and streamflow

Region	El Niño response	La Niña response	References
South America			
West coast (Chile, Peru)	Higher streamflows and floods		Waylen & Caviedes (1990)
South east (Parana, Uruguay basins)	Higher streamflows and floods		Mechoso & Irribarren (1992); Burn & Arnell (1993); Robertson & Mechoso (1998); Compagnucci & Vargas (1998); Berri & Flamenco (1999)
Central America and northern South America	Lower streamflow	Higher streamflow	Restrepo & Kjerfve (2000); Poveda & Mesa (1997); George *et al.* (1998); Amarasekera *et al.* (1997)
North America			
Pacific north west	Lower streamflow	Higher streamflow	Redmond & Koch (1991); Piechota & Dracup (1996); Leung *et al.* (1999); Piechota *et al.* (1997); Cayan *et al.* (1999); Kahya & Dracup (1993)
Pacific south west	Higher streamflow and floods	Lower streamflow	Kahya & Dracup (1994); Dracup & Kahya (1994); Ely *et al.* (1994)
North central USA, Canada	Higher streamflow	Lower streamflow	Guetter & Georgakakos (1996); Kahya & Dracup (1993); Burn & Arnell (1993)
North east USA	Lower streamflow	Higher streamflow	Kahya & Dracup (1993)
Gulf coast and Florida	Higher streamflow	Lower streamflow	Sun & Furbish (1997); Kahya & Dracup (1993)
Australasia			
South east Australia	Lower streamflow	Higher streamflow	Kuhnel *et al.* (1990); Chiew *et al.* (1998); Piechota *et al.* (1998); Simpson *et al.* (1993)
New Zealand	Lower streamflow (except far south)	Higher streamflow	Moss *et al.* (1994); Mosley (2000)
South Asia			
India, south east Asia, Philippines	Smaller floods	Larger floods	Burn & Arnell (1993)
Africa			
Nile basin (equatorial east Africa)	Lower streamflow		Eltahir (1996); Amarasekera *et al.* (1997)
Zaire River	Lower streamflow	Higher streamflow	Amarasekera *et al.* (1997)

are generally – but not always – the opposite of those during El Niño events. The strength of the relationship between ENSO and streamflow varies from location to location, but it is difficult to compare different studies because each uses slightly different techniques and indices of both ENSO and stream response. However, in a comparative study Amaraskera *et al.* (1997) showed that flows in the Nile and Parana rivers were more strongly correlated with ENSO than those in the Amazon or Zaire rivers.

El Niño and La Niña events last for several months, and the hydrological response in an area may be very delayed. This delay can arise for two reasons. First, the climate response in an area may lag the El Niño or La Niña event, as recorded in the Pacific Ocean, by several months. Second, the hydrological response to a climate anomaly may be delayed. This is particularly evident where winter precipitation is stored as snow and released during spring and summer by snowmelt: a winter climate anomaly will therefore manifest itself in spring and summer streamflows. Several studies have therefore looked at the sequencing of hydrological response to El Niño and La Niña events during and after the event. Figure 4.26a shows composite flow regimes for four regions of the USA for the years before, during and after El Niño and La Niña events (Dracup & Kahya, 1994). Each graph shows the long-term average monthly flow (solid line), together with the average sequence of flows during El Niño events (dashed line) and La Niña events (dotted line). In general, the response of streamflow in each region appears during the second half of the event year or during the subsequent year.

Local variations in the response of a catchment to an El Niño or La Niña event will depend partly on how the event maps onto local climate, and partly on catchment characteristics. New Zealand, for example, experiences substantial variations in climate over a small geographical area, and the local consequences of El Niño and La Niña events vary considerably. Figure 4.26b shows average low flow anomalies for the years during and after El Niño and La Niña events for two catchments in New Zealand (Mosley, 2000). In both catchments streamflows are higher during La Niña events, but at different times.

Although there is an association between ENSO and European weather (a tendency for above average precipitation in winter and spring following an El Niño (Wilby, 1993; van Oldenbergh *et al.*, 2000)), the effect of ENSO on European streamflows remains unclear. However, river flows in both northern and southern Europe are influenced by the North Atlantic Oscillation. During winters with a positive NAO index (i.e. more vigorous depressions) winter streamflow is above average in western Britain, the Nordic countries and northern Germany, but below average in Spain and Portugal (Shorthouse & Arnell, 1997). Below average streamflows in high NAO years extend through the Mediterranean basin to Turkey: flows in the Tigris and Euphrates are low during high NAO years (Cullen & de Menocal, 2000).

Streamflow variability in the western part of North America is influenced not only by ENSO, but also by the Pacific Decadal Oscillation. Differences between decade-long average streamflow during warm and cool PDO phases

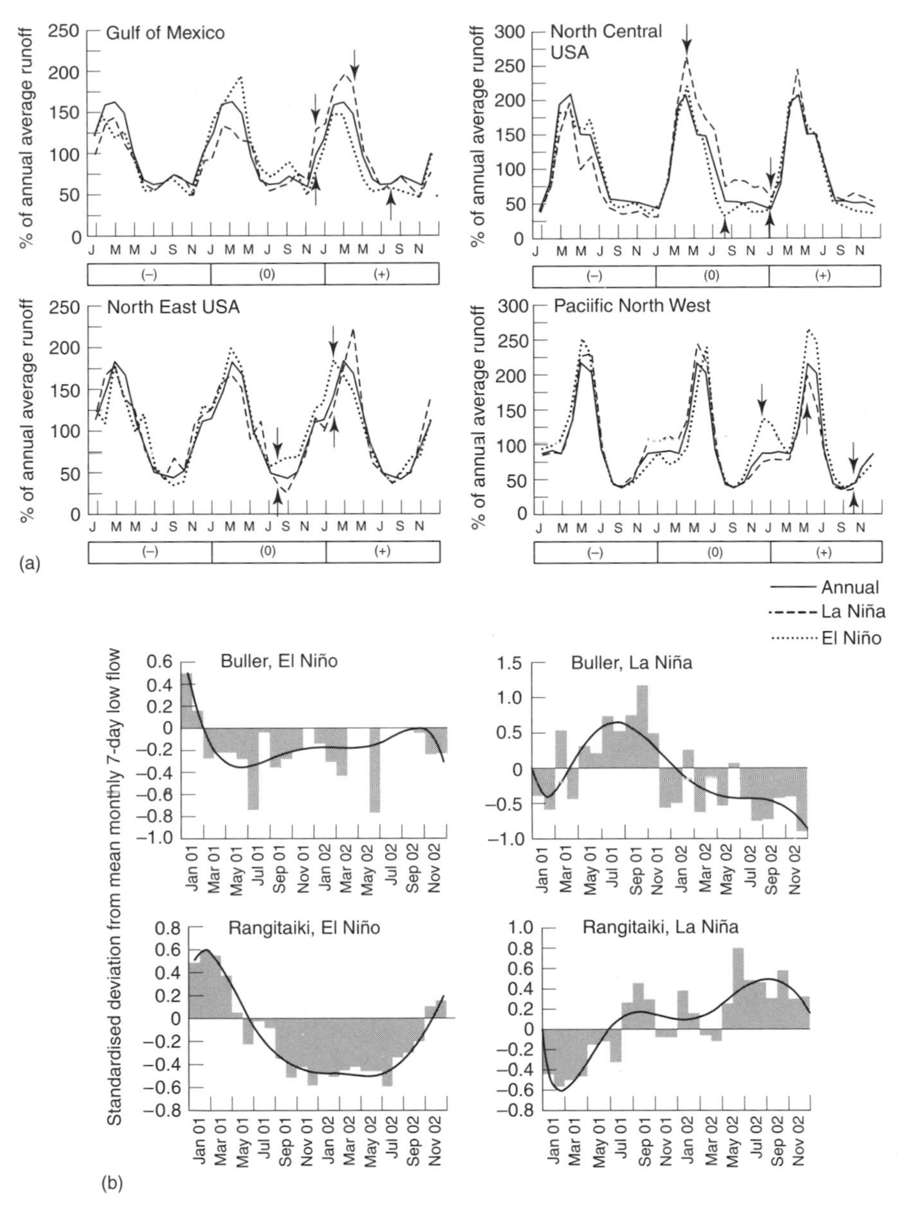

Figure 4.26 Sequencing of flow and streamflow anomalies during El Niño and La Niña events. (a) Regional patterns for four US regions during "normal", El Niño and La Niña conditions (Dracup & Kahya, 1994): the arrows denote the beginning and end of the El Niño and La Niña periods; (b) 24-month composite anomalies in low flows for two New Zealand catchments during El Niño and La Niña events (Mosley, 2000: the dates show time from the start of the event)

Table 4.6 Probability (in per cent) of "low" and "high" streamflow in the Columbia River, associated with ENSO and PDO (adapted from Miles *et al.*, 2000). Low and high flows are defined to be more than 1.5 standard deviations from the long-term mean

Low flows				
	El Niño	ENSO neutral	La Niña	*All years*
Warm PDO	36	0	8	*14*
Cool PDO	11	0	0	*4*
All years	*21*	*0*	*3*	*8*
High flows				
	El Niño	ENSO neutral	La Niña	*All years*
Warm PDO	0	0	0	*0*
Cool PDO	0	18	19	*12*
All years	*0*	*9*	*12*	*7*

in the Columbia River, for example, are typically around 15% (Hamlet & Lettenmaier, 1999a), with the lowest flows during the warm phase. Table 4.6 shows the probability of "low" and "high" streamflows in the Columbia River during different combinations of PDO and ENSO (Miles *et al.*, 2000). In this case, there is an 8% chance, for example, that flows in any one year will be "low". In El Niño years, the probability rises to 21%, and during the warm phase of the PDO is 14%. In an El Niño year during the warm phase of the PDO the probability of experiencing low flows rises to 36%. Similarly, the probability of experiencing high flows is greatest in La Niña years which occur during the cold phase of the PDO.

The amount of rainfall in the Sahel region of west Africa and north east Brazil is significantly affected by the position of the Inter-Tropical Convergence Zone, which is at its northernmost limit in the Northern Hemisphere summer. The position of the ITCZ is itself related to sea surface temperatures in the tropical Atlantic Ocean. When the tropical Atlantic Ocean is warmer than average, the ITCZ is shifted to the north of its usual position: rainfall is higher than average in the Sahel, but drought conditions prevail in north east Brazil. The reverse holds when the tropical Atlantic Ocean is cooler than average. Streamflow, of course, is closely related to rainfall, and so streamflow anomalies should also follow the sea surface temperature anomalies in the Atlantic. Uvo & Graham (1998), for example, showed that streamflow anomalies in north east Brazil were strongly correlated with Atlantic sea surface temperatures, with a lag of between three and nine months.

Hydrological records with memory: lakes

Systems with large amounts of storage may take a long time to respond to anomalous inputs. The level of Lake Victoria in east Africa, for example, rose by several metres in the early 1960s, and this was subsequently blamed on influences as diverse as deforestation in the catchment (see Chapter 6) and a persistent change in climate. Detailed analysis, however, showed that the abrupt and long-lived increase in lake level was due to one particularly wet year (Sene & Plinston, 1994): it took several years for the "surplus" water to drain away through the River Nile. This happened again in 1997, when a large precipitation anomaly led to an increase in the level of Lake Victoria of around 1.7 m (together with increases in other east African lakes), and lake levels are expected to remain high for the first few years of the twenty-first century (Birkett *et al.*, 1999). This is an example of a hydrological system with built-in long-term persistence, leading to a long-duration, possibly abrupt, change in hydrological behaviour for a short-lived change in inputs.

4.5 Hydrological behaviour: an overview

This chapter has shown how hydrological behaviour varies across both space and time. Spatial variations in hydrological characteristics reflect variations in climate (particularly the amount and timing of precipitation and the presence or absence of snowfall) at the broadest scale, with catchment properties influencing the detail of hydrological regime in a given catchment. Properties of the flood hydrograph, such as the time to peak and the relative magnitude of the flood peak, are predominantly affected by catchment topography, whilst catchment geology and soil type affect the proportion of event precipitation that goes quickly to runoff, the rate at which flows decline, and low flow characteristics of a catchment.

However, climate varies at different time scales. Day-to-day, season-to season and year-to-year variability are incorporated within the concept of "climate", and are conventionally assumed to be constant over decadal and longer time scales. In statistical terms, it has traditionally been assumed that each year is a sample from an unchanging population. Increasingly, however, it is being realised that the characteristics of climate are not necessarily stable from one decade to another, and also that the likelihood of a particular set of hydrological characteristics may vary from year to year depending on the "mode" that the climate system is in. Flood risk, for example, may be substantially higher in an El Niño or a cool phase PDO year, for example, than in other years. Such climate signals are much stronger in some parts of the world than others, but current research into climatic variability is leading to increased understanding of how climate varies across the entire globe. The WMO's CLIVAR project (CLIVAR, 1997) seeks to use understanding of atmospheric and ocean processes to observe, describe, analyse and predict climatic variability. Increased understanding of the nature of climatic variability is already leading to improvements in the ability to forecast streamflows in some parts of the world one or more seasons in advance. Prototype forecasting methods have been developed for rivers in Australia, based on sea surface temperatures (Chiew *et al.*, 1998), the Columbia River basin in north west North America, based on ENSO and the PDO (Hamlet & Lettenmaier, 1999a), and rivers in the Amazon basin, based on sea surface temperatures in the Pacific and Atlantic Oceans (Uvo & Graham, 1998).

There are two final messages from this chapter. First, the effects of a given environmental change – in the catchment or the climate – will depend on the hydrological regime of that catchment, and different environmental changes will affect different dimensions of the hydrological regime. Second, it may be difficult to separate the effects of environmental change in a catchment from the effects of natural climatic variability operating over different time scales. These two issues will be explored in detail in Chapter 6.

Chapter 5

Water quality and the flux of materials

5.1 Introduction

The previous chapters have concentrated on the *quantity* of water and its variation over space and time. This chapter explores variations over space and time in the *quality* of water. Water quality is in fact rather difficult to define precisely, as "quality" is a value-laden term. Water that may be of acceptable quality for one use or function may be unacceptable for another. The term "water quality" therefore is usually taken to mean the physical and chemical characteristics of the water. Physical characteristics (or *determinands*) include the concentration of sediment in the water, the temperature of the water, and its colour. Chemical characteristics reflect the material dissolved in the water (as solutes) or held in the water on particles of sediment, and can be indexed by many different chemical determinands. Water has many unusual properties, including its ability to exist as vapour, liquid or solid, and its very high specific heat capacity. In terms of water quality, however, its most significant property is that virtually all substances can dissolve in it. This chapter first considers physical characteristics of rivers and lakes, before moving on to review chemical characteristics.

It is important to distinguish at this point between "water quality" and "pollution". A pollutant is simply a constituent that is present in a sufficiently large quantity to have an adverse impact on the use or appearance of water, or aquatic life in the water. Microbial pollutants include pathogens which can cause ill-health, such as cholera, faecal coliforms (including *E. coli*) and *Cryptosporidium*. An excess of biodegradable material (such as sewage effluent) can lower dissolved oxygen concentrations. Heavy metals, such as zinc and mercury, can be toxic. Excessive nutrients can lead to eutrophication, again leading to reductions in dissolved oxygen. Nitrate has been associated with human health problems. Organic micro-pollutants, including pesticides and some materials used in industry, may also be toxic, and endocrine disruptors (chemicals which affect the health, growth and reproduction of a wide range of organisms) are an increasing concern in many developed countries. High sediment concentrations can adversely affect the aquatic environment, as can high temperatures; "abnormal" water colour can be a problem for drinking water supplies. Excessive salinity limits the use of water and also pollutes soil. Many – but not all – of these pollutants have their origin in human activities in the catchment. This chapter concentrates on the physical, chemical and biological processes which determine water quality in general terms: the consequences of human interventions are explored in Chapters 6 and 7.

Hydrologists have conventionally taken what can be termed a *water-oriented* approach to water quality: their focus has been on the water body (river, lake, soil or groundwater), and they have been concerned with understanding how different types of material get into the water, how these materials affect the water body and determine its "quality", and how their characteristics and concentration change over time. This reflects concern over the pollution of the water environment. Another way of looking at water quality, however, focuses on the *materials* and the role of the hydrological system in the cycling of those materials from one store to another (as carried out for example by Berner & Berner, 1996). In the most general terms, the fluxes and changes in state of an element or compound can be described by its *biogeochemical cycle*. The hydrological cycle can in fact be interpreted as the biogeochemical cycle of the compound H_2O. More specifically, however, emphasis is usually placed on cycles of the biologically-important elements carbon, nitrogen, phosphorus, oxygen and sulphur. The hydrological system plays an important role on these biogeochemical cycles, partly by transporting material and partly by acting as a medium for chemical and biological reactions. This role is described in Section 5.4, and the implications of human-induced alterations in these cycles is explored in Chapter 8.

5.2 Physical water quality

Sediment load

Material held in moving water by upward turbulence is termed *suspended sediment*, and generally consists of particles less than 0.2 mm in diameter. The actual size distribution of suspended sediment varies depending on the geology of the catchment and river channel. Suspended sediment is made up of both inorganic matter (derived from soil and rocks) and particulate organic matter (POM: derived from vegetation and aquatic animals). Bedload is material rolling and bouncing along the river bed. It generally contributes only a small proportion of the total sediment transported, with its relative importance increasing as channel slope increases. Sediment concentration is, rather obviously, the amount of material held in a given volume of water, and is usually expressed in mg l^{-1}. Sediment load or yield is the total amount of material passing a particular point over a defined time period. It is usually expressed as either tonnes a^{-1} or tonnes ha^{-1} a^{-1}. Box 5.1 summarises the methods used to estimate sediment concentrations.

The sediment budget for a catchment is composed of a number of sources of sediment and a number of sinks, both on the catchment land surface and in the channel. There are three sources of sediment. The first is erosion from hillslopes. This can reach a river channel either through overland flow (which is, as shown in Chapter 3, most likely in semi-arid and arid areas) or by being

Box 5.1 Measuring sediment transport

The concentration of suspended sediment at a given instant can be measured by taking a sample of water, filtering the sample, and weighing the sediment left on the filter paper: concentrations are expressed in mg l^{-1}. The first technical problem lies in the selection of the sampling site, as sediment concentrations vary with turbulence. In deep rivers, where mixing is likely to be less complete, this problem is avoided by taking a depth-integrated sample (slowly lowering the sample bottle down through the river) or taking samples at defined depths. The second problem is, however, more significant, and relates to sample frequency. Suspended sediment concentrations vary rapidly over time, and a few "snapshot" samples may give very misleading indications of the total amount of sediment transported over a period. One response to this is to install an automatic sampler which takes samples once triggered by a rise in river level. Another is to continuously measure the turbidity of the river, using a photoelectric device, and estimate suspended sediment concentrations using a calibrated relationship between turbidity and sediment concentration. These relationships, however, may not be very stable.

Bedload is harder to measure and continuously monitor than suspended sediment. Most studies use bedload traps sunk into the bed of the river, which collect bedload and are emptied and weighed at defined intervals. This is often rather difficult, and the data give an integration of bedload transport over time. Continuous measurements of bedload transport can be made by installing pressure sensors under the bedload trap and monitoring the weight of material in the trap. Again, the traps need to be periodically emptied. Another approach uses tracers, which can range from simple painted pebbles to synthetic pebbles with embedded tracking devices which can be continuously monitored.

dumped directly into the watercourse. Much material eroded from hillslopes never reaches the river channel, and is deposited in sinks on the land surface. The second source is erosion from the bed and banks of the river channel. The relative significance of these two sources varies considerably. In Coon Creek, Wisconsin, for example, only around 6% of sediment is derived from the river channel (Trimble, 1981), and approximately 19% of the sediment output of the Culm River in south west England derives from the banks (Ashridge, 1995). As much as 65% of sediment in the St Lawrence River, Canada, comes from river bed and banks (Rondeau *et al.*, 2000). The third source of suspended material is autochthonous material formed in the water body, in two main ways. Particulate organic matter may be formed from phytoplanktonic algae or indeed from parts of any plants or animals living in the water. Particulate inorganic matter can also be formed by the chemical

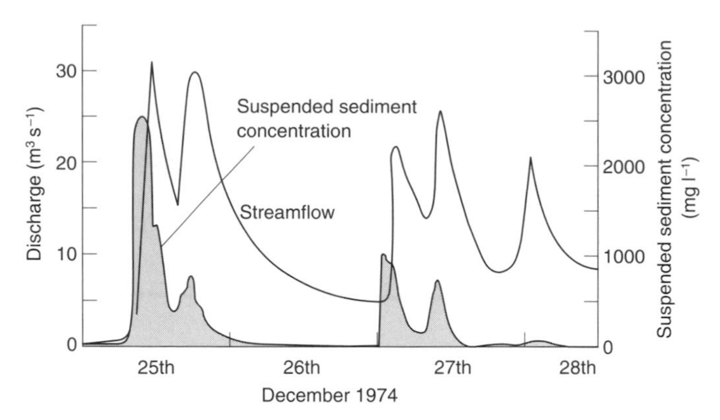

Figure 5.1 Variation in suspended sediment concentration and streamflow in the River Exe, Devon (adapted from Walling, 1977)

process of precipitation from a supersaturated solution. Calcium carbonate, for example, may precipitate out of solution, and can contribute a substantial proportion of the suspended sediment load during summer in areas draining limestone (up to 70% of the suspended material in the Loire River, France, during summer, for example, may be precipitated calcium carbonate: Meybeck *et al.*, 1989). Sediment that arrives in or derives from the channel, from whatever source, may be deposited in channel sinks. Only a fraction of the sediment generated within a catchment and river channel will therefore be carried as suspended sediment or bedload past a certain point, and this fraction is known as the *sediment delivery ratio*. This ratio may be as low as 10% (Walling & Webb, 1996). The composition of suspended sediment and bedload depends on catchment geology, and can be used to identify source areas.

The suspended load of a river varies strongly with streamflow, and may vary over several orders of magnitude. Figure 5.1 shows a typical sequence of streamflow and suspended sediment concentration over four days in the River Exe in Devon (Walling, 1977). The peak sediment concentration tends to occur before the peak streamflow, and sediment concentrations may be lower in subsequent events because there is less material available to transport. The relationship between streamflow and suspended sediment concentration is therefore non-linear (different on the rising and falling limb of the hydrograph) and variable over time. Estimates of total suspended sediment yield based on combining a sediment/streamflow relationship based on point samples with a time series of streamflow may therefore be very inaccurate.

Sediment transport tends to be concentrated in only a few days. In a small catchment in Devon, for example, 90% of the total long-term sediment yield was transported in just 5% of the time (Walling & Webb, 1996). Although the amount of sediment transported increased as discharge increased, moderate flood events made a greater contribution to the long-term sediment yield than big events because they occurred more frequently (Wolman & Miller, 1960). However, in other environments it is possible that very extreme "catastrophic" events could make a greater contribution to total sediment flux.

Variations in hydrological regime over time (Chapter 4) mean that sediment transport also varies from year to year and decade to decade. Flows in the Magdelena River in Colombia, for example, are higher during La Niña years than El Niño years, by a factor of around 1.5 (Restrepo & Kjerfve, 2000): mean daily sediment loads are twice as high during La Niña years than El Niño years. Similarly, Inman & Jenkins (1999) found that the average annual sediment flux from the 20 largest streams entering the Pacific Ocean from the central and southern California coast was five times greater during the "wet" phase of the Pacific North Atlantic pattern than during the dry phase, and in one catchment more sediment was transported in the abrupt transition from dry to wet phase during 1969 than in the preceding 25-year "dry" period.

Although sediment loads vary over time, there are clearly large variations in suspended sediment transport between rivers (Walling, 1996). High levels of transport can be expected where the catchment rock is highly erodible, where relief is steep (and hence flow velocities high), where there is little vegetation cover, where rainfall is intense and where there is tectonic instability. The global average sediment concentration is around 450 mg l^{-1} (Meybeck, 1982), but 50% of major world rivers have average concentrations less than 150 mg l^{-1}. Most sediment is carried in the few highly turbid rivers, located in south east Asia, around the Mediterranean, and parts of east Africa, western North America and western South America. It has been estimated that between 20 and 25% of the total world export of sediment from land to sea comes from the steep, wet islands of the East Indies (Sumatra, Java, Borneo, Sulawesi, Timor and New Guinea), which only make up around 2% of the global land area (Milliman *et al.*, 1999). In Britain, average concentrations are low in global terms, with averages typically less than 50 mg l^{-1}: maximum concentrations rarely exceed 5000 mg l^{-1} (Walling & Webb, 1996).

The amount of sediment that can be held in a river or transported as bedload depends on the river's velocity and turbulence, and the size distribution of the particles. Bedload is generally limited by velocity and turbulence (and is therefore often termed a "capacity load"), but the load of suspended material is usually determined by the amount of material supplied. However, it is important to recognise that increasing supply does not necessarily mean an increase in concentrations of material transported down-river, if sediment concentration is near capacity.

Sediment concentrations and yields can be significantly influenced by human activities in the catchment and river network: examples are given in Chapter 6.

Water colour

Although pure water is colourless, water in the natural environment is rarely totally clear. Different types of suspended sediment may give the water colours ranging from yellow (the Yellow River in China, for example, is named for the yellow silt it carries), through the turquoise/green colour of glacial outwash streams to dark brown, and algae may give a green hue. This coloration ("apparent colour") clears if the water is allowed to settle. Dissolved humic substances, resulting from the leaching of decaying organic material, may also colour water, and this coloration (known as "true colour") is not lost by settling. It occurs downstream of swamps and bogs, and tends to reach a maximum when humic material is flushed out after a long dry spell, often occurring a year after the dry spell (Naden & Macdonald, 1987). Rain water percolating through leaf litter can also pick up dissolved organic matter and become stained.

Water temperature

Temperature is perhaps the most important physical characteristic of water. It affects the rate of operation of biological and chemical processes, and has a large influence on dissolved oxygen concentrations (see below). It also influences the ecological characteristics of a river or lake.

The temperature of a water body is determined by the energy balance. Inputs of heat energy come from the sun (short-wave radiation), the atmosphere (long-wave radiation and sensible heat flux from the air), precipitation, condensation and advection from upstream inflows or groundwater. The water body emits heat energy through radiation, evaporation and advection downstream, and energy is exchanged with the atmosphere and sides of the water body by conduction. In practice, the radiation terms are dominant. The water body attempts to maintain an equilibrium temperature with the atmosphere. Air temperature of course varies over time, and so, therefore, does the temperature of the water body. The lag between air and water temperature change largely depends on the size of the water body, but may also be influenced by the degree of mixing of the water body due either to the downstream movement of water (in rivers) or wind and thermally-induced currents (in lakes).

River water temperature varies more rapidly with air temperature than lake water temperature, because of the smaller volume of water. The temperature of a particular stretch of river therefore shows diurnal, day-to-day and seasonal variability. Figure 5.2 shows the variation through the year in water temperature in and air temperature above a small stream in the New Forest, Hampshire, southern England. Water temperature in the river may also be affected by the dominant source of streamflow at a given point in time. Snowmelt, for example, may lower water temperature, and temperature frequently falls when flows rise following rainfall.

The primary reason for differences in water temperature between catchments is variation in air temperature. For a given climate regime, water temperature and its variability depend on the source of water (groundwater-dominated rivers tend to have cooler water with less diurnal variation than rivers with little groundwater), the volume of flow (larger rivers tend to show less variation in temperature over time) and, at the smallest scale, the nature of the river corridor (water temperature is lower in rivers shaded by overhanging vegetation): see Walling & Webb (1996).

Turbulence within a river generally means that water temperatures are broadly similar throughout a river reach. Lake water temperature, however,

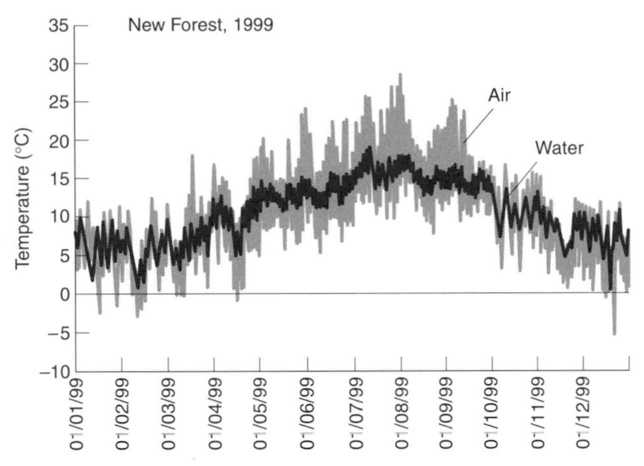

Figure 5.2 Variation in air and water temperature through a year in the New Forest, Hampshire

not only shows less variation over time than river water temperature and a more lagged response to air temperature, but may also vary with depth. The density of water increases as temperature falls, reaching a maximum at 4 °C. Deep lakes therefore exhibit *thermal stratification*. In summer, the water at the surface of the lake will be warmer and therefore less dense than water at greater depth, and two distinct layers tend to form. The surface layer (epilimnion) is generally between 5 and 8 m deep, and overlays the denser, cooler hypolimnion. The surface layer cools in winter and becomes denser, thus enabling greater mixing. This mixing may be induced by wind action, and is important because the two layers have different properties: the lower layer contains less dissolved oxygen and more nutrients (Section 5.3) than the surface layer. Deep tropical lakes are generally thermally-stratified throughout the year.

When water cools to 0 °C, ice may form. Initially, ice forms at the margins of a river or lake when the surface layer reaches freezing point, and where flow is slow and laminar (Gray & Prowse, 1993). In turbulent parts of a river, however, water is constantly mixed, and ice formation can only occur once the entire river cross-section has cooled to 0 °C. As temperature falls further, small ice particles (frazil ice) less than 0.5 mm in diameter form within the flowing water, increasing its viscosity. Frazil ice crystals continue to grow, and eventually produce large frazil floes, which float on the flowing river water. As winter progresses, a solid ice cover may be formed by outgrowth of river margin ice and the accumulation of frazil floes. Small rivers may freeze to the bed, whilst larger rivers will continue to have flowing water under the ice layer. The duration of ice cover is obviously an important influence on the use and ecological characteristics of a river or lake, but it also has implications for the hydrological regime of a catchment. For example, river flows may be at a minimum in early winter whilst the ice cover is forming from water in the river, rather than later in the winter when the amount of runoff being generated is at a minimum (Gray & Prowse, 1993). Also, the peak flow during spring in some rivers or years may be influenced more by the characteristics of the break

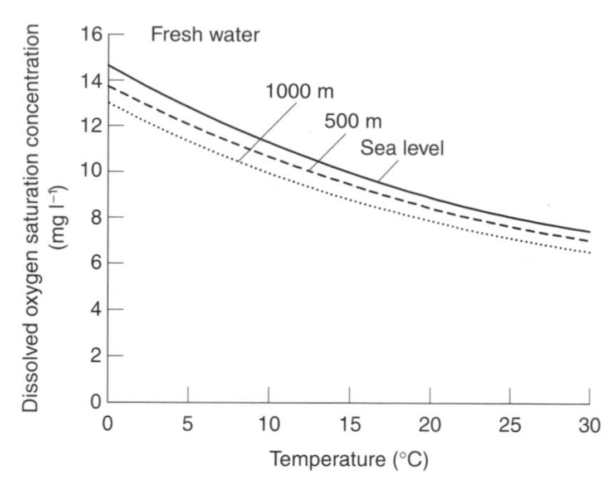

Figure 5.3 Variation in the saturation dissolved oxygen concentration with temperature and altitude

up of the ice cover and the breaking of ice jams than by the melting of snow stored on the catchment surface. Ice cover is rare on British and most maritime temperate rivers.

Dissolved oxygen

Water contains dissolved solids (Section 5.3) and also some dissolved gases. These include nitrogen and carbon dioxide, but the most important is oxygen because it is fundamentally important to aquatic organisms. Dissolved oxygen levels are therefore usually taken as indicators of the physical, rather than chemical, characteristics of water.

The amount of dissolved oxygen that can be held in water varies with temperature, atmospheric pressure and, to a lesser extent, the concentration of dissolved solids in the water. Figure 5.3 shows that warmer water can hold less dissolved oxygen than cooler water, with the saturation dissolved oxygen concentration decreasing by between 0.2 and 0.3 mg l^{-1} for every degree Celsius increase in temperature. Dissolved oxygen concentrations are often expressed as a percentage of the saturated dissolved oxygen concentration. Fish can generally only survive in water with more than 3 mg l^{-1} of dissolved oxygen.

There are two primary sources of oxygen in a river or lake: re-aeration across the water surface, and photosynthesis by plants. Re-aeration is influenced by the turbulence of the water and the oxygen deficit (the difference between actual and saturated dissolved oxygen concentrations), and is greater in rivers than lakes. Oxygen is released by fixed plants, phytoplankton and algae within the water body through photosynthesis, which occurs during daylight hours. Oxygen is consumed by the respiration of plants and animals in the water, which continues throughout the day. Oxygen is also consumed by the bacterial decomposition of organic material (the amount required for this is defined as the biochemical oxygen demand (BOD)) and in the oxidation of ammonia to nitrate and nitrite (Section 5.4). The concentration of dissolved oxygen in a water body is therefore very dependent on biological processes, which are themselves influenced by the amount of nutrients available (Section 5.3), and the amount of organic material available for decomposition.

Average dissolved oxygen concentrations in UK rivers tend to be around 10 to 11 mg l^{-1}, with BOD typically 1 to 2 mg l^{-1}, except in polluted rivers.

5.3 Chemical water quality

Water in the soil, river, lake or aquifer – and indeed in clouds – contains dissolved substances. The actual chemical properties of a water body depend on the sources of material, the chemical processes operating in the landscape and in the water body, and the biological processes in play. Solids that dissolve in water can ionise to form ions by the separation of the original molecules into electrically-charged components. The ions sodium (Na^+), potassium (K^+), calcium (Ca^{2+}), magnesium (Mg^{2+}), chloride (Cl^-), sulphate (SO_4^{2-}) and bicarbonate (HCO_3^-) are collectively known as the major ions and these, together with silica (SiO_2), present in non-ionic form, are the basic descriptors of chemical water quality. *Nutrients* are broadly defined as elements vital to the health of plants and animals, although a rather more narrow definition restricts the term to nitrogen (N), phosphorus (P) and organic carbon. An excess of nutrients can lead to *eutrophication*, or the excessive growth of algae. Two general indicators of dissolved material in water are the total dissolved solids (TDS, expressed on a mass per volume basis) and salinity (total dissolved inorganic matter, expressed on a mass per mass basis). As a general rule, saline water has a TDS of greater than 3000 mg l^{-1}.

Hydrogen ions (H^+) are present in small concentrations, but have an important effect on water quality. Concentrations of hydrogen ions – as measured in pH – control the speciation of other chemicals in the water, influence (chemical) precipitation and dissolution, and affect the aquatic life that can be supported. The pH of unpolluted water is typically around 7.7.

This section first looks at the sources of dissolved material and the processes they are subject to, before looking at variations between catchments and variations in a catchment over time in dissolved concentrations and load: it is important to note here that a large proportion of the total load of some chemical constituents may be carried as suspended sediment. Box 5.2 outlines the approaches used to measure the chemical characteristics of water. Hydrometric agencies in many countries are increasingly measuring a range of water quality determinands (in the UK's Harmonised Monitoring Network, for example, and the national stream water quality networks in the USA: Alexander *et al.*, 1998), and the water component of the Global Environmental Monitoring System (GEMS-Water, organised by UNEP and the World Health Organization) collates data collected by participating countries (GEMS-Water, 1999).

This chapter does not offer a detailed discussion of chemical processes. For more detail from a hydrological perspective see Webb & Walling (1996) and McCrutcheon *et al.* (1993): for a more chemically-oriented perspective see Andrews *et al.* (1996), Howard (1998) or vanLoon & Duffy (2000).

Box 5.2 Measuring chemical water quality

The concentrations of different chemical determinands in water can be undertaken in the laboratory or in the field. Laboratory methods are the most reliable, and include titration, colorimetry and spectrometry. They are undertaken on samples of water and, as with suspended sediment (Box 5.1), the major practical problem lies in the sampling frequency. Also, the laboratory methods may be cumbersome or require expensive equipment. During the past few years a number of electrodes have been developed to measure directly specific ions. These can be used in the laboratory, but increasingly instruments have been refined and made sufficiently robust that they can be left in the field to continuously record ion concentrations. pH probes have also been developed which can provide continuous measurements in the field.

The total dissolved solids (TDS) in a sample of water is much easier to measure than the concentration of specific ions or constituents: in the laboratory, a known volume of sample water is evaporated and the residue weighed. The electrical conductivity of water is related to the concentration of total dissolved solids, and conductivity is easy to measure both in the laboratory and, more importantly, in the field. With the aid of an empirically-calibrated relationship, it is therefore relatively easy to continuously monitor total dissolved solids concentrations. Salinity is strictly defined as the weight in grams of inorganic ions dissolved in 1 kg of water, but in practice this is difficult to measure, so salinity is usually assessed by measuring the conductivity of the water relative to the conductivity of a standard sea water.

Sources and processes

There are three broad sources of material which may be transported in solution: atmospheric inputs, products of chemical weathering of bedrock, and material derived from soil and vegetation. Suspended sediment sources provide the remainder of the total chemical load. Human activities also provide chemical load, as discussed in Chapters 6 and 7.

Material can move from the atmosphere to the land surface in three ways. *Wet deposition* occurs during precipitation, and consists of the particles that have acted as condensation nuclei at the heart of raindrops (Section 3.2) together with atmospheric particles below cloud level washed out by falling raindrops. Concentrations in newly fallen snow are higher that those in rain because snowflakes are effective at entraining material as they fall, and concentrations in the snowpack change through the season. As snow particles coalesce and recrystallise, dissolved ions migrate to the crystal surfaces. During the winter these crystals will melt from the surface inwards, and meltwater will have higher solute concentrations than the original snow. This meltwater will subsequently

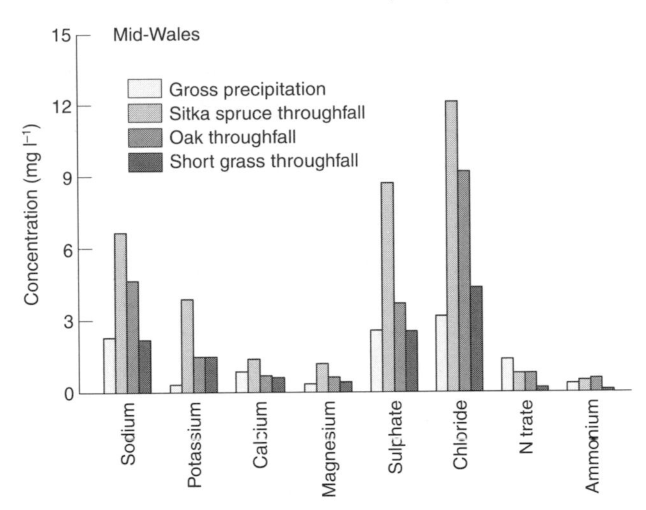

Figure 5.4 Chemical composition of precipitation and throughfall under different vegetation types, mid-Wales (data from Soulsby, 1997)

produce a flush of solutes into the soil or stream. *Occult deposition* occurs when mist and cloud droplets are intercepted by vegetation (Section 3.3). This usually has higher solute concentrations than precipitation, because the water droplets are smaller and closer to the ground (Wilkinson *et al.*, 1997). Some of this intercepted water is evaporated, increasing the concentration of solutes. *Dry deposition* is fallout of material from the air either as particles falling against the force of gravity or as the absorption of gases directly by the land surface – particularly vegetation. The relative importance of these different mechanisms of deposition depends on the frequency and type of rainfall, and also distance from the source of the material. Much of the material in the atmosphere has a marine origin, but some comes from blown soil and dust, biological emissions from living organisms and – as outlined in Chapter 7 – human activities. Atmospheric inputs provide the bulk of the sodium (Na) and chloride (Cl) loads in rivers, with their importance obviously increasing with proximity to the sea. The vegetation cover clearly has a very big influence on the amount of chemical material and its concentration in water that reaches the ground, as illustrated in Figure 5.4. Concentrations of all but nitrates are considerably higher in throughfall from the Sitka spruce than in either the gross precipitation or the throughfall from oak, and are little affected by the short grass: all the vegetation types, however, "strip" nitrates from the incoming precipitation. The "enriched" water will be concentrated in certain areas under the vegetation canopy by the processes of throughfall and stemflow, leading to substantial spatial gradients in soil water chemistry (Chang & Matzner, 2000).

Once the water reaches the ground, it infiltrates into the soil and may percolate to groundwater. The soil is a highly reactive zone (Soulsby, 1997), where chemical, physical and biological processes can operate under saturated (i.e. without oxygen) and unsaturated (i.e. with the presence of oxygen) conditions. The chemical processes include mineral weathering, the details and products of which depend on the parent material, and cation exchange. Clay minerals and humic substances caused by decomposition of plant and animal litter both have large areas of negatively-charged surface, forming the soil cation exchange complex, which allows soil particles to adsorb cations from percolating water. The total quantity of cations which can be exchanged (the cation exchange capacity) varies with soil properties, and the higher the organic and clay contents the higher the cation exchange capacity. In calcareous soils calcium tends to move into the water, whilst in acidic soils aluminium may be released. Decomposition not only increases cation exchange capacity, but microbial activity also breaks organic materials, such as nitrogen, into other compounds (microbial decomposition of organic nitrogen, for example, creates ammonia, which can then be oxidised to form nitrate) releasing nutrient ions into the soil water. Other processes operating in the soil to alter soil water chemistry include chelation, anion adsorption, oxidation and reduction. Soil water will therefore contain a mixture of inorganic and organic ions, and its composition will vary through the soil profile. Vegetation will not only release organic matter through decomposition which is subsequently leached into soil water, but also extract nutrients from the soil and store them: this will vary through the year. The storage and processing of material by vegetation may be more important in regulating stream chemistry than inputs from the atmosphere or soil weathering (Webb & Walling, 1996).

The pathways that soil water follows to get into the river channel (Section 3.7) have a very significant effect on the chemistry of the receiving river. If the predominant pathway is through macropores, then water passing through the soil will undergo limited chemical change, whereas if the water is percolating slowly through unsaturated soil it will acquire a greater chemical load. "Old" water, pushed out of a hillslope by "new" water entering from above, will also have a relatively high chemical load.

Water that does not flow laterally to the river channel percolates down to bedrock, where it enters a much less chemically-active zone than the soil. Groundwater chemistry is dominated by chemical processes operating in saturated conditions, but – unlike in soils – over very long time periods. Weathering and reduction processes are most important, with the resulting composition of groundwater depending on the nature of the parent material. Figure 5.5 shows the chemical composition of groundwater from three different aquifers in Britain. The chalk groundwater, for example, has high calcium concentrations due to carbonate dissolution, whilst the granite has higher concentrations of magnesium and sodium due to silicate weathering. It is, of course, the distinctive chemical properties of different aquifers that (apparently) allow consumers to distinguish between different brands of bottled mineral water.

Water reaches the river channel with a particular chemical signature determined by its passage through vegetation, the soil and bedrock. A small number of materials pass through the river system with no further change, including sodium, chloride and most heavy metals. The rate at which they move through the catchment therefore depends on transit times, mixing and diffusion

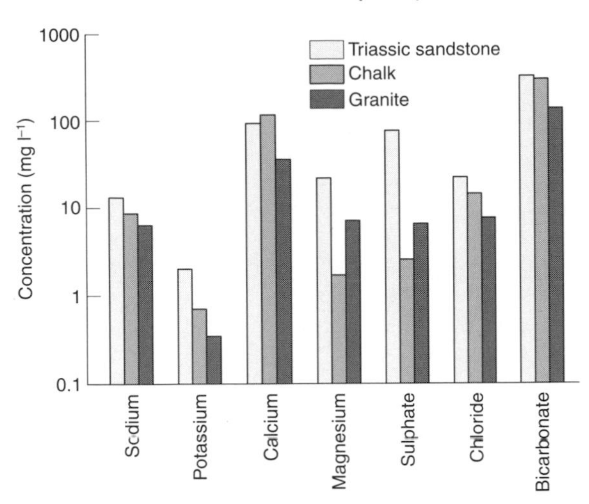

Figure 5.5 Chemical composition of groundwater in three British aquifers (data from Soulsby, 1997): note the logarithmic scale

processes within the water body, and the amount of evaporation from the river: solute concentrations in the River Nile, for example, increase downstream as river water evaporates. Most materials, however, undergo further changes in the river or lake. Nutrients continue to be cycled in rivers and lakes, for example (Newbold, 1996). Nitrification and denitrification through microbial processes both continue, and nutrients and silica will be harvested by aquatic organisms. Chemical interactions between suspended sediment particles and solutes may also lead to changes in the materials "attached" to sediment particles. The rate at which biological and chemical processes operate is dependent on, amongst other things, the temperature of the water and the amount of nutrients available for plant, animal and microbial growth, and the velocity of the water has a significant effect. The slower the water, the longer the time available for biological and chemical processes to operate before material is dispersed. In-stream changes in water quality are therefore most likely to occur in slow-moving rivers, particularly in so-called "dead-zones" where water is virtually still.

The sources and pathways of chemicals in groundwater, rivers and lakes are therefore very complex. Vegetation plays an important role, as do *biological processes* generally: nutrient status is therefore not just a measure of water quality, but also an important determinant of it. *Hydrological processes* determine the actual pathways that water follows to get to the river, and *hydraulic processes* affect mixing within rivers. *Chemical processes* control the reactions between the various elements and compounds in water.

Variations over space

The chemical composition of water varies geographically, reflecting variations in bedrock geology, climate (and hence weathering), and vegetation characteristics. Figure 5.5 gave an indication of the variability in chemical composition

Table 5.1 Solute concentrations in catchments underlain by different rock types (adapted from Ward & Robinson, 2000)

Rock type	Total dissolved solids (mg l^{-1})	Principal ions
Igneous and metamorphic	<100	Sodium, calcium, bicarbonate
Sedimentary	50–250	Varies depending on composition
Limestone and dolomite	100–500	Calcium, magnesium, bicarbonate
Evaporites	<10 000	Sodium, calcium, sulphate, chloride

Table 5.2 Human influences on ionic and nutrient loads

Constituent	Sources
Sodium	Industrial sources, mining of rock salt
Chloride	Urban and industrial sources, mining, water treatment
Potassium	Mining
Sulphate	Industrial and agricultural sources, acid deposition
Nitrogen	Agricultural fertiliser, animal wastes, municipal sewage
Phosphorus	Agricultural fertiliser, animal wastes, municipal sewage
Heavy metals	Industry and mining

of groundwater in different British aquifers, under similar climatic conditions. For a given rock type, the concentration of a particular constituent may vary with climate. Silica concentrations, for example, increase with temperature as chemical weathering processes become more effective. Table 5.1 summarises solute concentrations in streams in catchments underlain by different rock types (Ward & Robinson, 2000).

Calcium, magnesium, bicarbonate and silica concentrations are little affected by human activities. The other major ions (sodium, chloride, potassium and sulphate), together with nitrogen and phosphate-based compounds and many heavy metals, can be very affected by human activities (Table 5.2; Chapters 6 and 7). The seven major ions plus silica make up the vast majority of the total dissolved load in most rivers.

Table 5.3a shows the average dissolved concentration in six major world rivers of the seven major ions and silicon, together with nitrate and phosphate concentrations. Bicarbonate, deriving from the weathering of carbonate rocks, is the major constituent in all of the rivers except the Murray-Darling, where it provides only just under a quarter of the load: here, chloride is the dominant solute. Calcium carbonate is also the major single solute in most British rivers (Table 5.3b).

Spatial variations in stream water chemistry can also be seen at the catchment scale, largely reflecting differences in bedrock, soil type and vegetation. Figure 5.6 shows the variation in several indicators of chemical water quality

Table 5.3 Solute concentrations in six world rivers and six British rivers. World data from http://www.cciw.ca/gems; British data from Hydrological Data UK yearbooks

	Concentration mg l^{-1}										
	Total dissolved solids	Calcium Ca	Magnesium Mg	Bicarbonate HCO_3	Silica SiO_2	Sodium Na	Potassium K	Chloride Cl	Sulphate SO_4	Nitrate/ Nitrogen NO_3/N	Phosphate/ Phosphorus PO_4/P
(a) World rivers											
Yukon, Alaska	183	31.8	7.2	109	7.7	2.6	1.2	1.1	7.7	0.01	0.12
Huang He, China	460	47	20.6	205	7.7	54.5	4.1	54.7	7.7	0.016	0.17
Chari, Chad	68	3.9	1.6	29.8	24	2.8	1.6	2	24	–	–
Amazon	43.5	5.4	0.9	21	6.9	1.9	0.8	2.2	6.9	0.02	0.17
Murray-Darling, Australia	453	21	17	94	5	101	6	171	5	0.1	0.03
Po, Italy	354	62.1	11.9	178	4	16.8	3	18.1	4	0.084	1.4
	Total dissolved solids	Calcium Ca	Magnesium Mg	Calcium carbonate $CaCO_3$	Silica SiO_2	Sodium Na	Potassium K	Chloride Cl	Sulphate SO_4	Nitrate/ Nitrogen NO_3/N	Phosphate/ Phosphorus PO_4/P
(b) British rivers											
Trent at Nottingham	566	106.1	22.1	159.3	7.18	73.8	9.9	98.9	169.6	8.6	1.53
Derwent at Wilne, Yorkshire	422	73	17	155.7	5.27	50.9	5.2	67.6	103.2	4.4	0.89
Avon at Evesham	596	119.9	28.4	195.6	10.74	57.9	9.9	78.6	196	10.6	1.8
Stour at Langham, Dorset	586	134.5	8.8	246	7.76	43.7	7.6	69.4	104.1	7.8	0
Tamar at Gunnislake, Cornwall	117	17.3	4.8	36.3	4.79	12.6	3.2	22.9	15.6	2.7	0.09
Exe at Thorverton, Devon	109	16.6	4.1	40	3.99	10.8	2	17.8	13.8	2.5	0.11

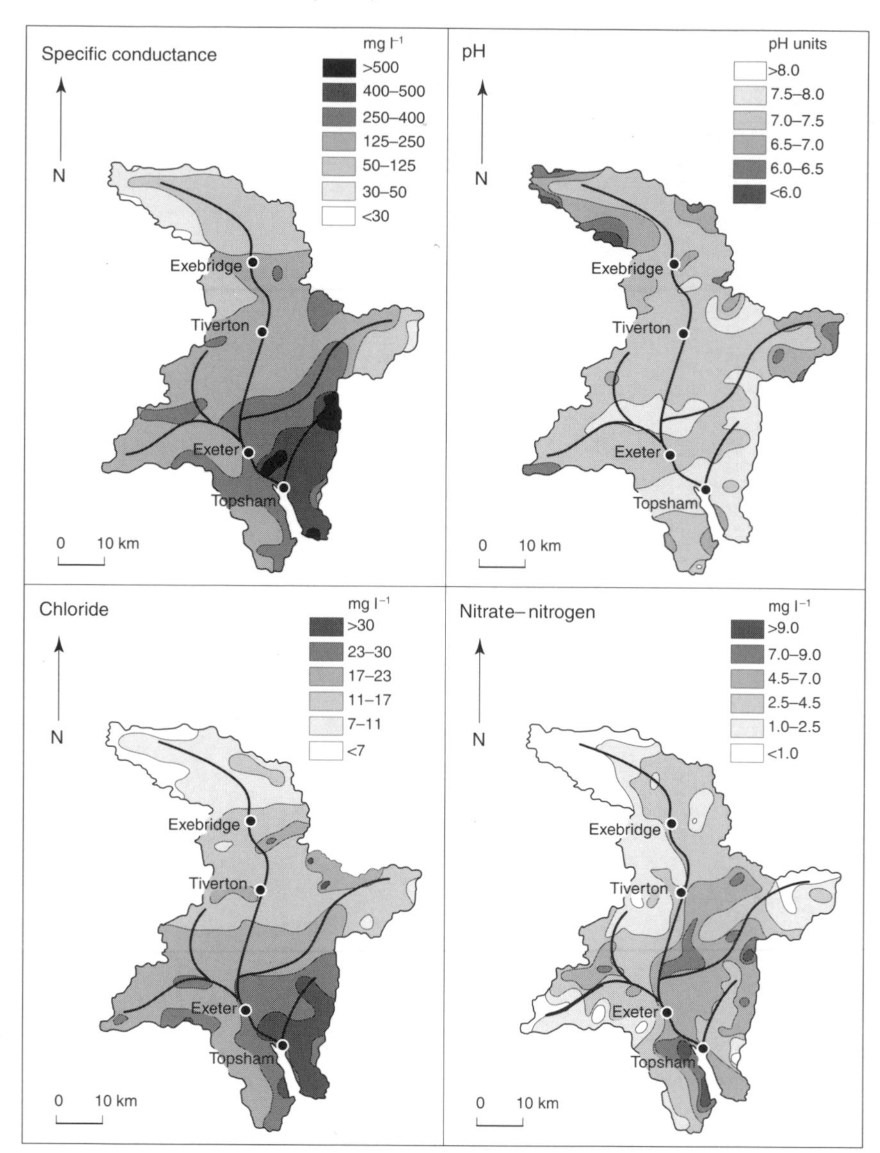

Figure 5.6 Variation in pH, specific conductance, chloride and nitrate concentrations across the Exe catchment, south west England (Walling & Webb, 1981)

in the Exe catchment, south west England (Walling & Webb, 1981): chloride concentrations reflect distance from the sea, pH varies with bedrock, nitrate concentrations vary with land use (Chapter 6), and specific conductance reflects the sum of all the different controls.

The chemical composition of river water in a catchment will reflect the mixture of bedrock, soil type and vegetation. Calcium and carbonate in the

Average annual runoff by the 2050s

HadCM2-x

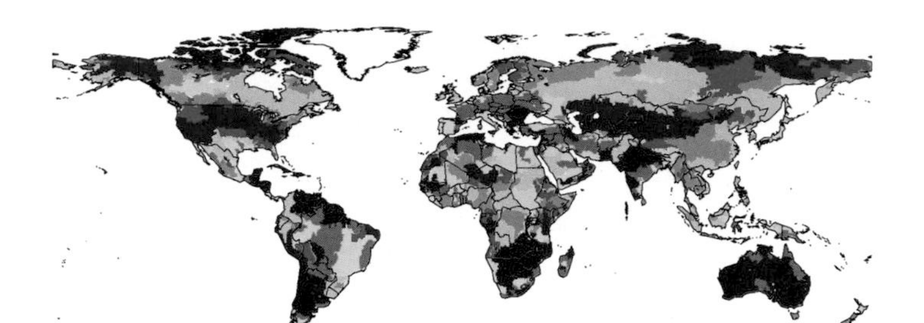

HadCM3

Plate IV Percentage change in average annual runoff, by the 2050s, under two climate change scenarios (Arnell, 1999b)

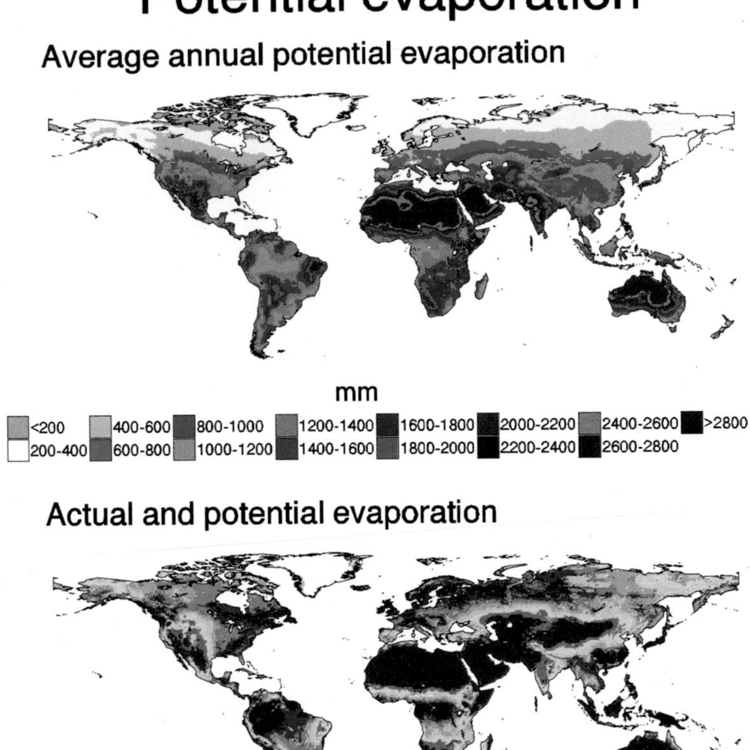

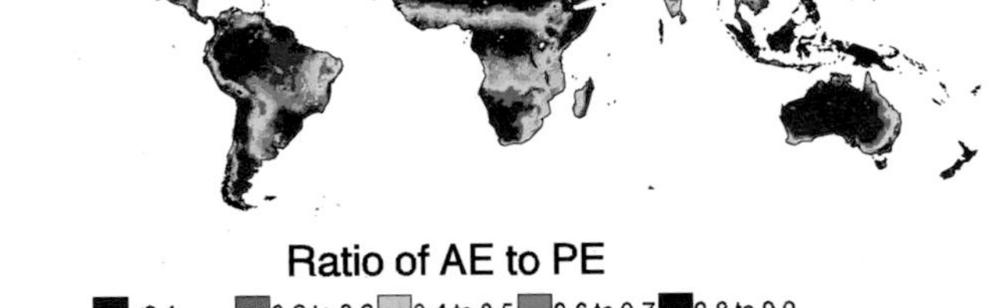

Plate III Global distribution of potential evaporation (Penman–Monteith), and the estimated ratio of actual to potential evaporation

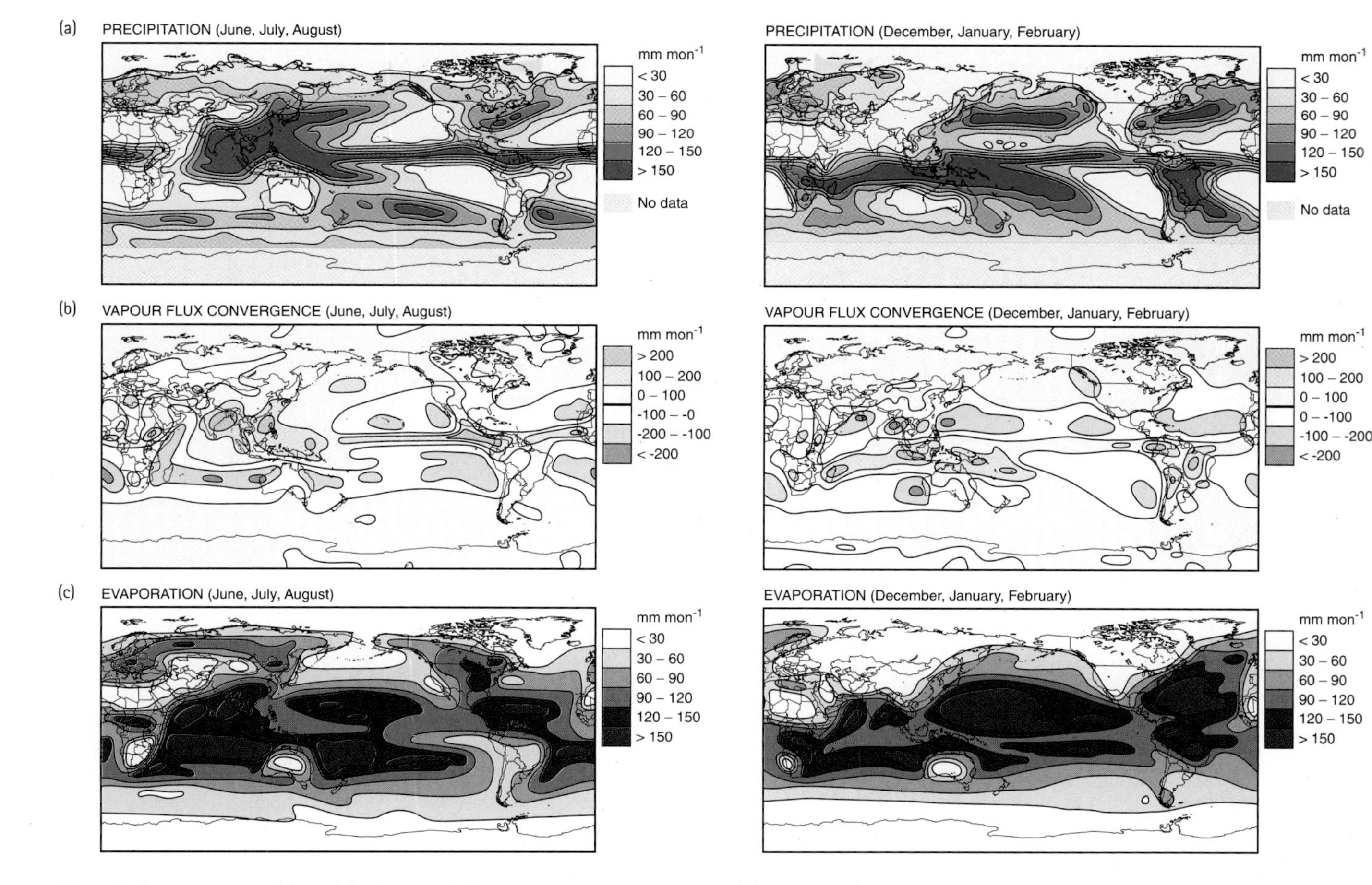

Plate II Components of the global water balance over the entire globe: (a) precipitation, 1988–98, from the Global Precipitation Climatology Project (http://orbit-net.nesdis.noas.gov/arad/gpcp); (b) vapour flux convergence (Oki, 1999); (c) evaporation (Trenberth, 1998)

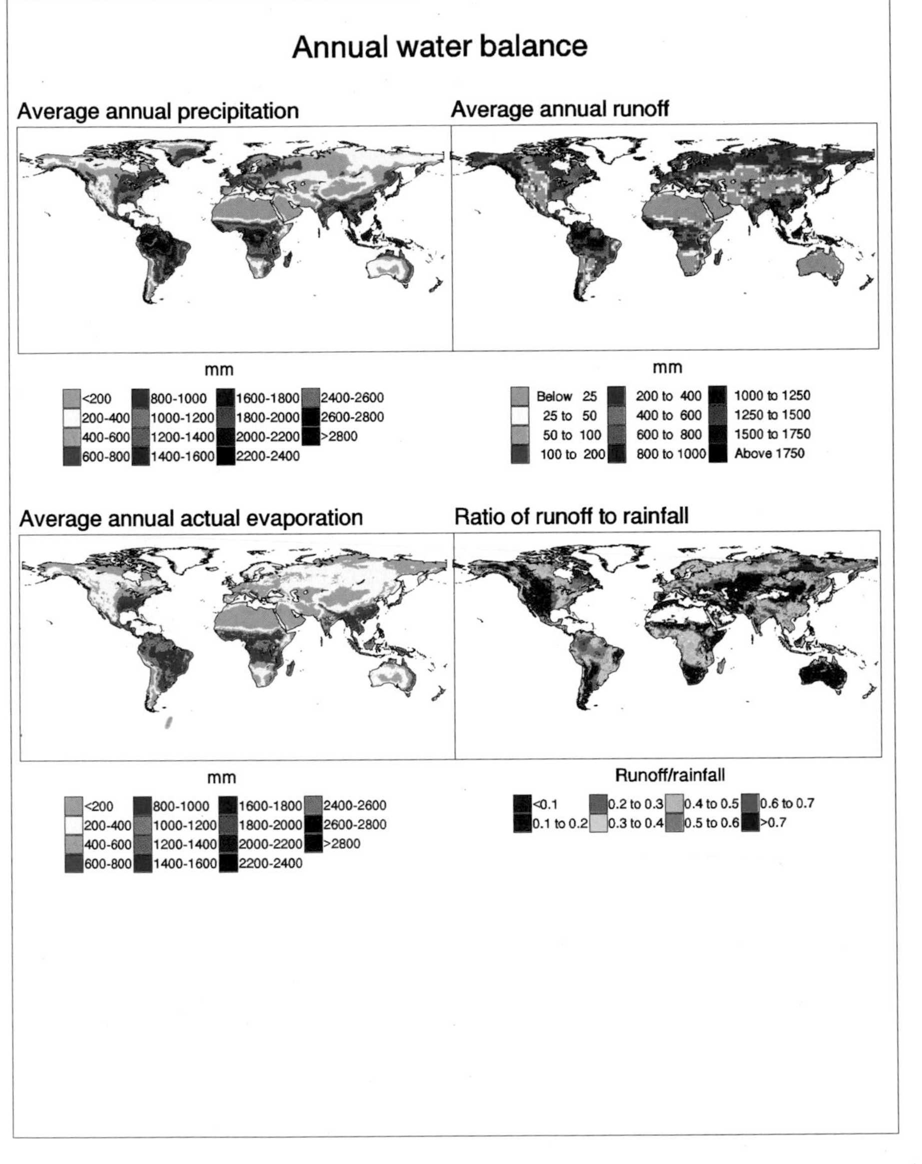

Plate I Components of the water balance over land (a) precipitation over land over the period 1961–90 (New *et al.*, 1999); (b) average annual runoff (Korzun, 1978; Cogley, 1994); (c) average annual evaporation over land, and (d) percentage of annual precipitation that becomes runoff

River Amazon, for example, primarily derives from headwaters in the Andes, whilst silica, potassium, carbon and nitrogen largely come from the lowland forest areas (Devol *et al.*, 1995).

Variations over time

Like suspended sediment concentrations, solute concentrations vary as streamflow varies. Total dissolved solids concentrations tend to reduce as streamflow increases in response to a precipitation, but different constituents respond in different ways. The concentrations of material deriving from chemical weathering in bedrock (such as calcium, magnesium and bicarbonate) and transported to the stream by groundwater flow, tend to decrease as streamflow increases, because the groundwater is diluted. Solutes associated with the upper soil layers, such as aluminium, dissolved organic carbon and hydrogen ions, increase with increasing flow as they are flushed out of the soil: this has been observed in many catchments, including Plynlimon in upland Wales (Neal *et al.*, 1997b). The flushing of hydrogen ions in some upland and heathland catchments produces an acid "pulse". Flushing of material generally tends to be greatest when a rainfall event follows a long dry period. Concentrations of nitrates and phosphates also tend to rise following rainfall as material is flushed through the catchment, but concentrations of nutrients also exhibit strong seasonal cycles associated with vegetation growth and decay (e.g. Arheimer *et al.*, 1996; McHale *et al.*, 2000). Snowmelt too can dilute groundwater, leading to lower concentrations of weathering-derived constituents (e.g. in the Fraser River, Canada (Cameron, 1996) and upland Scotland (Soulsby *et al.*, 1997)). River water is therefore a mixture of water with different chemical signatures. In fact, this recognition allows the identification of the different sources of streamflow on the basis of their chemical characteristics (Section 3.7).

Figure 5.7 shows the variation in calcium concentration with flow in the Schuylkill River, Pennsylvania (showing the dilution effect), together with the variation during an event in a small upland Welsh catchment of silica and dissolved organic carbon (Soulsby, 1995): dissolved organic carbon concentrations increase with streamflow. Figure 5.8 shows the range in major ion and silica concentrations over the long term in the River Avon in the English Midlands. Silica and sodium concentrations can be less than 30% of the mean, and sodium and chloride concentrations can be more than 75% higher than the mean.

Variations over time in the chemical characteristics in a lake reflect not only variations in the chemical inputs to the lake from rivers, but also physical, chemical and biological processes taking place in the lake itself. Mixing, for example, is influenced by wind and also by changes in the thermal stratification of the lake. Evaporation leads to increased concentrations of chemical species, and the precipitation of calcium carbonate can be enhanced when water temperatures and evaporation are at their highest (see for example Rosen *et al.*, 1996).

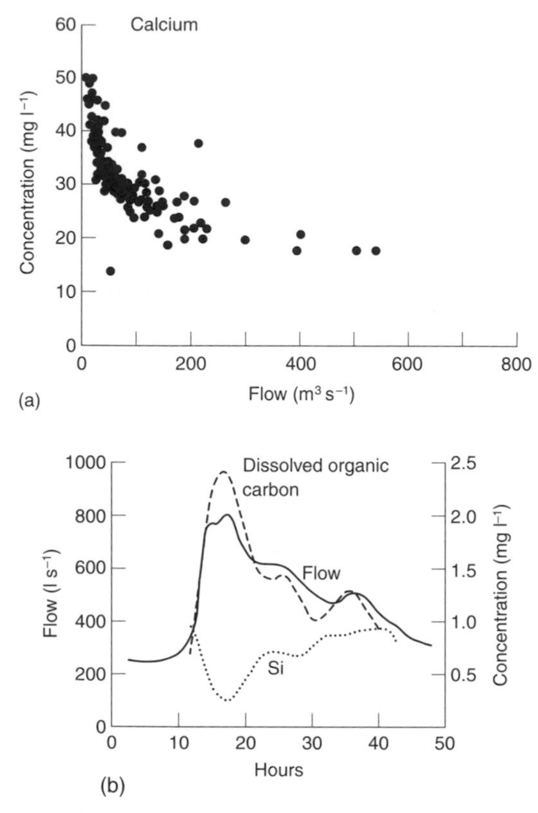

Figure 5.7 (a) Variation in calcium concentrations with streamflow in the Schuylkill River, Pennsylvania, and (b) variation in silica and dissolved organic carbon concentrations in a small afforested catchment in mid-Wales (adapted from Soulsby, 1995)

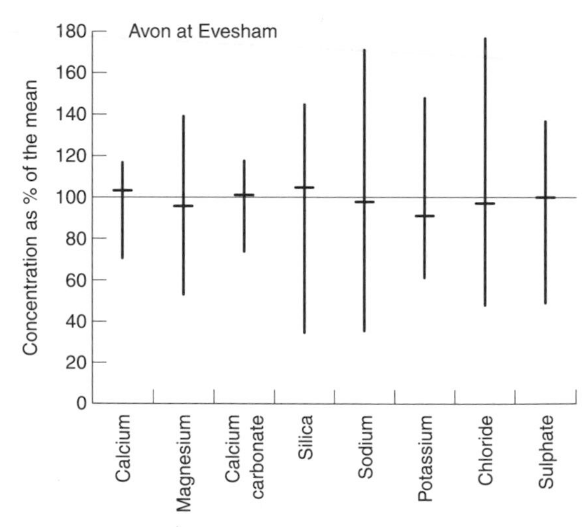

Figure 5.8 Range in major ion and silica concentrations in the River Avon, England: the lines show the 5%, median and 95% values, expressed as a percentage of the long-term mean

Table 5.4 Relative importance of sediment and solute loads (adapted from Gregory & Walling, 1973)

River	% of total load	
	As suspended sediment	As solutes
Colorado River, Grand Canyon, Arizona	94	6
San Juan River, Bluff, Utah	97	3
Wind River, Wyoming	73	27
Iowa River, Iowa	83	17
Corey Creek, Pennsylvania	56	44
Pond Branch, Maryland	16	84
Volga, Russia	36	64
Wieprz, Poland	5	95
Pilica, Poland	7	93
Tyne, northern England	65	35
East Devon, southern England	23–55	45–77

Total chemical load: solute and suspended sediment transport

The total chemical load of a river is made up of the dissolved material together with the material carried in suspension and as bedload. Suspended sediment and bedload are of course made up of organic and inorganic material washed into the river and generated within the river, and substances initially held in solution may bond with sediment particles through physical or chemical sorption processes. Suspended sediment can account for very high proportions of the total transport of some elements from the land to the sea. More than 90% of all aluminium, iron and silicon that reaches the sea, for example, is transported as suspended sediment, and more than three-quarters of the total phosphorus yield may be in the suspended sediment phase (Webb & Walling, 1996). The vast bulk of the trace elements and heavy metals transported in rivers are carried as solid suspended sediment (Meybeck *et al.*, 1989). In contrast, 99% of chloride is carried in solution. With a global average suspended sediment concentration of around 450 mg l^{-1} (Meybeck, 1982) and a global average total dissolved solid concentration of 120 mg l^{-1} (Walling & Webb, 1986), approximately one-quarter of all material moved from land to sea is carried in solution. This figure will, of course, vary substantially from catchment to catchment, as indicated in Table 5.4. The lowest proportion of load carried as solutes tends to be found in arid areas – where most streamflow is generated from overland flow – with the highest in lowland catchments in temperate areas.

5.4 Biogeochemical cycles

The previous sections have concentrated on the physical and chemical characteristics of water from the perspective of the water body. This section focuses on the movement of materials through the earth system – movement in which the hydrological cycle plays a very significant role. The cycling of material

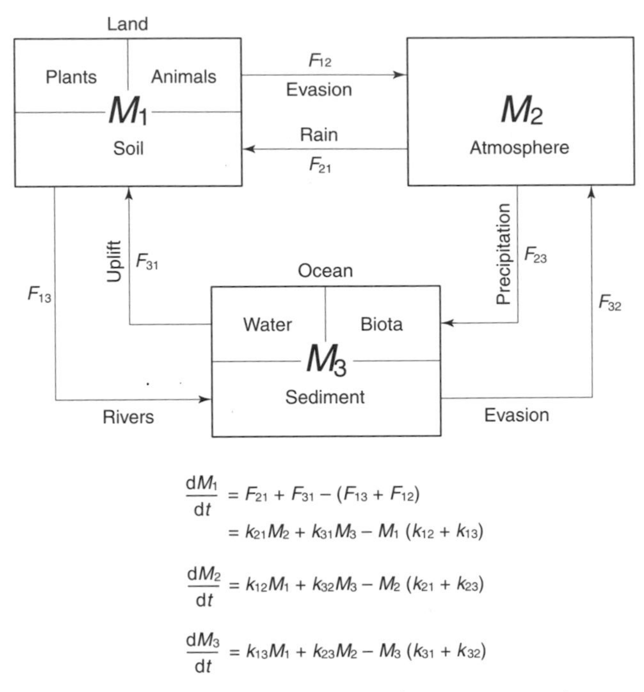

Figure 5.9 A schematic global biogeochemical cycle (Mackenzie, 1998)

from one state or store to another is described by its biogeochemical cycle. Figure 5.9 shows a schematic global biogeochemical cycle (Mackenzie, 1998). A biogeochemical cycle consists of a number of stores or reservoirs, together with transport paths and fluxes. The turnover time of a reservoir gives an indication of the time material spends in that reservoir, and is defined as the time it will take to empty the reservoir in the absence of any new inputs: it is calculated as the ratio of the mass of the store to the sum of either the sources or sinks of the substance.

The sediment cycle is conceptually a biogeochemical cycle, and geomorphologists and geologists have long been interested in determining the sources, sinks and fluxes of particulate sediment. Milliman and Meade (1983), for example, produced initial estimates of the total flux of sediment from land to sea, and these estimates have subsequently been refined: current estimates are that between 15 and 30 billion tonnes of material are moved from land to sea each year, two-thirds of which come from south east Asia (but note that human activities greatly affect these fluxes: Chapters 6 and 8). A substantial proportion of this total load of material is transported in dissolved form.

Conventionally, however, biogeochemical cycles are constructed for individual elements or compounds, although of course many cycles interact. Most attention has been directed to the cycles of carbon, nitrogen, phosphorus, sulphur and oxygen. The hydrological system affects these cycles in four main ways. First, fluxes of these elements from land to atmosphere, often through

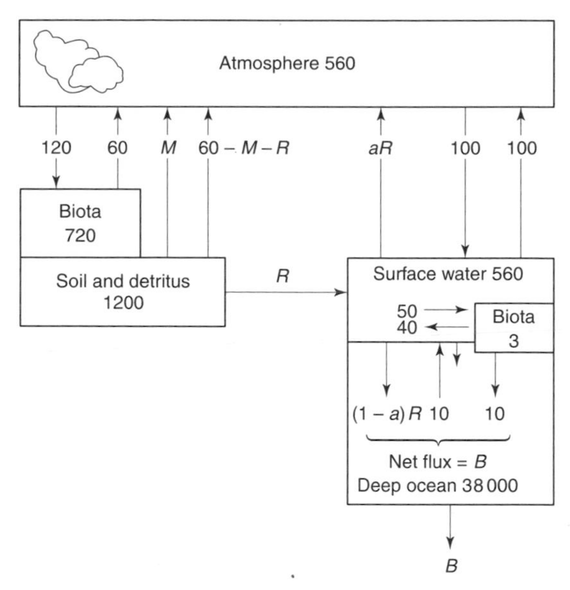

Figure 5.10 The global carbon cycle (Harvey, 2000). *M* is the methane flux, *R* is the riverine flux of particulate carbon, *B* is the deep burial flux, and *a* is notionally the proportion of the riverine inputs to the ocean that is emitted to the atmosphere. The numbers in the boxes are stores in Gt C (one gigatonne is equal to 1000 million tonnes), and the fluxes are in Gt C a^{-1}.

vegetation, are affected by soil moisture status and the amount generally of water at the land surface. Second, freshwater can be a source or sink. Third, groundwater can act as stores of material. Fourth, rivers are transport pathways, carrying material from land to sea. The remainder of this section briefly summarises the main biogeochemical cycles, focusing in particular on the role of the hydrological system.

Biogeochemical cycles can be constructed for any element or compound. Strictly, a complete cycle can only be constructed at the global scale. However, it is possible to produce catchment-scale biogeochemical *budgets*, which identify the inputs and outputs of the material, together with the fluxes and stores of the material within the catchment.

The carbon cycle

Figure 5.10 shows the global carbon cycle. The vast bulk of carbon lies in storage in sediment, with the next largest store – the deep ocean – holding less than 1% of the amount stored in sediment. Terrestrial biota, the soil and the surface of the ocean hold even less carbon, but this is the store that interacts rapidly with the atmosphere.

Plants absorb CO_2 in photosynthesis. Some of this carbon is returned to the atmosphere through the respiration of plants and animals which feed on them, but much is stored as carbon in plant (and animal) material. Plant growth is a function of, amongst other things (including nutrients, temperature and light availability) water availability, and soil moisture dynamics can influence the uptake of carbon at the land surface. The variation of soil moisture over time under vegetation depends on the timing of inputs of water, the ability of water to move through the soil, and evaporation and transpiration – which is itself a function of soil moisture. Transpiration and carbon fluxes are therefore closely linked, and indeed much of the current research into transpiration is being conducted in association with research into carbon fluxes.

As plants and animals decay, carbon is stored in the soil. This organic matter is decomposed by bacteria, releasing both CO_2 and methane (CH_4) into the atmosphere: rates of bacterial composition are affected by temperature and soil moisture. Methane is produced when organic matter decomposes in the absence of oxygen (anaerobic conditions), in wet soils, wetlands and thawing permafrost. In South Africa, savanna soil is a source of methane when wet and a sink when dry, constituting a small net sink (Otter & Scholes, 2000). Flooded areas, however, emitted methane at around 40 times the rate of even wet soil, and although only a relatively small proportion of South Africa is composed of seasonally-flooded wetland, these areas produce considerably more total methane than the savanna. The extent and duration of inundation in a year are, of course, determined by the amount of streamflow generated in the catchment and routed along the channel network to the floodplain: hydrological processes therefore have a clear control on total methane emissions from South Africa.

Organic material is flushed into rivers, and then transported to the oceans. Organic carbon is transported in dissolved (dissolved organic carbon: DOC) and particulate (particulate organic carbon: POC) forms. DOC and POC concentrations in a catchment are dependent on the catchment vegetation. In north east Scotland, for example, the highest concentrations of both DOC and POC come from catchments with large areas of peat (Hope *et al.*, 1997). Both DOC and POC are flushed out during storm events, so concentrations are higher in storm flows than in base flows. In the Plynlimon catchments in mid-Wales, for example, DOC concentrations are twice as high in quickflow than in baseflow. The relative magnitudes of DOC and POC vary considerably. In two major rivers in north east Scotland, POC contributes between 10 and 25% of the total organic carbon load, although it makes a greater contribution in some tributaries (Hope *et al.*, 1997). In the Rhone River, however, POC makes up around 60% of the total organic carbon flux (Sempere *et al.*, 2000), and the global average contribution is around 47% (Meybeck, 1983). The amount of organic carbon in water depends on the time of year and amount of decaying organic matter available. Fresh leaf litter is an important source of DOC in forested catchments (Hongve, 1999), and in deciduous forests concentrations will therefore be at their highest in autumn and early winter. Variations in organic carbon concentrations between rivers in the same geographical area are largely explained by vegetation differences, and Dillon & Molot (1997) showed how DOC concentrations in forested Ontario catchments varied with the proportional area covered by peat and beaver ponds.

Water also contains dissolved *inorganic* carbon, derived either from the atmosphere – from atmospheric CO_2 – or by the weathering of carbonate rocks. This dissolved inorganic carbon (DIC) is held primarily as dissolved CO_2 gas or as dissolved bicarbonate (HCO_3). The relative proportions held in these two forms varies with pH. At an average pH of 7, bicarbonate makes up around 84% of the total carbonate species, with the proportion decreasing to around 50% as pH falls to 6. The DIC concentration of river water is largely a function of catchment geology, and concentrations are higher in baseflow than quickflow.

The relative magnitudes of organic and inorganic carbon fluxes varies in a catchment over time, depending on the source of streamflow and time of year, and varies between catchments depending on vegetation and catchment geology. In eastern Canada, with few carbonate rocks, DOC fluxes are nearly seven times those of DIC fluxes (Clair *et al.*, 1994), whilst in the Rhone basin, with large areas of limestone, total inorganic carbon fluxes are more than six times higher than total organic carbon fluxes (Sempere *et al.*, 2000).

Organic carbon within a water body can be incorporated into the food web and ultimately respired as inorganic carbon (i.e. CO_2). In practice, most organic carbon that enters a river is transported to the sea (Newbold, 1996), because the "turnover" time for DOC and small POC – which provide the bulk of the organic carbon load – is long relative to river lengths. The turnover time for larger POM, such as algae or invertebrates, is short, but this contributes relatively little to total organic carbon load.

Global fluxes of organic and inorganic carbon from land to sea can be estimated by extrapolating from observations made at the mouths of major rivers or by developing models predicting carbon concentrations from catchment vegetation and geology. Estimated fluxes are uncertain, but it is estimated that under natural conditions (i.e. with no human disturbance), approximately 400 million tonnes a^{-1} (0.4 Gt a^{-1}) of carbon is transported from land to sea as organic carbon, and a similar amount as inorganic carbon (Meybeck, 1983; Mackenzie, 1998), making a total "natural" flux from land to sea of around 800 million tonnes a^{-1}: this is small compared to the other fluxes shown in Figure 5.10.

The nitrogen cycle

Nitrogen, like carbon, is an essential element for life. Figure 5.11 shows a simplified version of the nitrogen cycle in water, in which the role of biological processes is emphasised.

Although nitrogen is the most abundant element in the atmosphere, it cannot be used directly by plants. Bacteria and blue-green algae in the soil and attached to plant roots convert nitrogen to ammonia (NH_3) by the process of nitrogen fixation, and other bacteria convert this ammonia to nitrite (NO_2) and nitrate (NO_3) by the process of nitrification. Nitrate and ammonia are then taken up by plants. Denitrification is the conversion of nitrate back to

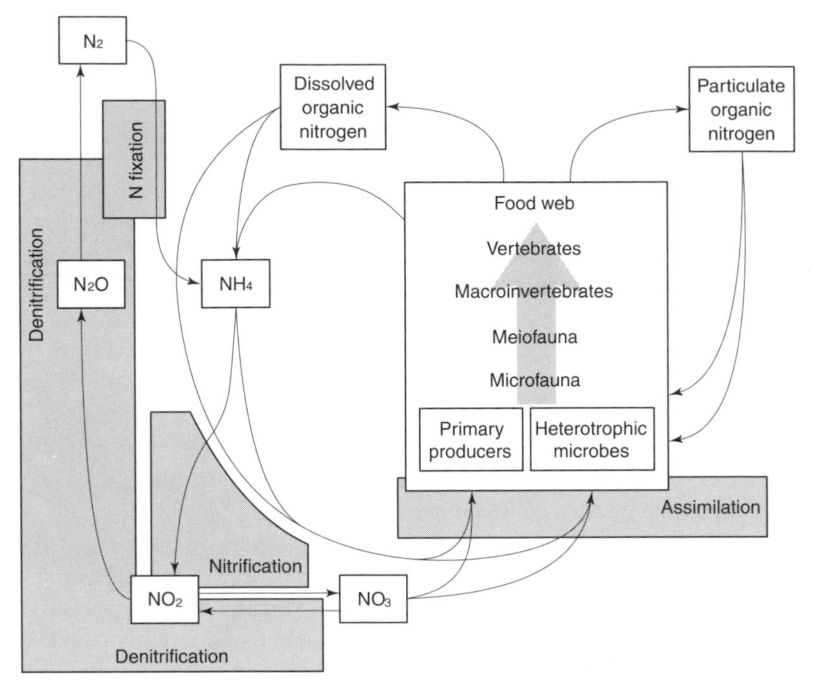

Figure 5.11 The nitrogen cycle in streams and rivers (Newbold, 1996)

nitrogen and nitrous oxide (N_2O) by microbial organisms. The nitrous oxide is released to the atmosphere, but the nitrogen can be converted again into ammonia – and the cycle continues. These various species of nitrogen can be leached into rivers and groundwater, and the processes of nitrification and denitrification operate not only in the soil but also in water bodies.

Soil water and rivers can contain nitrogen in five forms: ammonia, nitrite, nitrate, dissolved organic nitrogen and particulate organic nitrogen. The organic nitrogen comes from plant material (organic matter typically contains between 2 and 5% nitrogen): typically around two-thirds is in particulate form. Groundwater tends to contain only ammonia, nitrite and nitrate. Concentrations of ammonia, nitrite and nitrate are usually reported in terms of the equivalent nitrogen concentrations.

The amount of nitrogen held as nitrate in water tends to be between 10 and 100 times that held as nitrite, and between 5 and 10 times higher than that held as ammonium (an ion of ammonia formed when ammonia dissolves). Nitrate-nitrogen concentrations in unpolluted streams are typically of the order of 0.1 mg l^{-1}. Dissolved organic nitrogen concentrations, however, are generally higher than nitrate-nitrogen concentrations, and particulate organic nitrogen is usually larger still.

Ammonia, nitrite and nitrate concentrations in river water vary seasonally with the growth of vegetation, and are generally at their maximum in winter: they may vary too over longer time scales as vegetation in the catchment changes composition (as forest reaches maturity, for example). Concentrations are higher in quickflow than in baseflow (because the material is held in the soil), and peak concentrations may be particularly high after dry spells. Most nitrogen is therefore transported during high flow periods: McKee *et al.* (2000) found in the subtropical Richmond River in Australia, for example, that 96% of nitrogen was transported in less than 6% of the time. Ammonia, nitrite and nitrate peaks also tend to be high during spring snowmelt floods, as material stored over winter is flushed into the river. Year-to-year variations in total exports of inorganic nitrogen species may be considerable, due largely to variations in streamflow volume. The coefficient of variation of annual fluxes of nitrate into the Atlantic coastal zone of North America, for example, ranges from 0.03 to 2, although in most catchments the CV is between 0.2 and 0.3: the greatest variability from year to year occurs from the drier catchments draining the western Gulf of Mexico, where the variability in flows is greatest (Alexander *et al.*, 1996). Monteith *et al.* (2000) found synchronous patterns of variation in nitrate concentrations across much of upland Britain, and attributed these patterns to the North Atlantic Oscillation.

Variations in concentrations and total loads *between* catchments are primarily associated with variations in the amount of nitrogen available in the upper layers of the soil and the rate of nitrification (Williams *et al.*, 1997).

The nitrification of ammonia to nitrite and nitrate occurs in rivers, keeping ammonium levels in unpolluted streams low (Newbold, 1996). In small rivers, the bacteria involved in nitrification tend to live in the sediments at the river bed, but in larger rivers the bacteria are largely associated with suspended sediments. Nitrification consumes dissolved oxygen, thus lowering dissolved oxygen levels in the water. Denitrification also occurs in rivers, again largely within the sediment, and requires an organic carbon source. Nitrogen fixation by blue-green algae can also take place in rivers, but rates are small compared to other nitrogen fluxes (Newbold, 1996).

Catchment-scale nitrogen budget calculations (e.g. by Dillon *et al.* (1991) in Ontario) show that the exports of nitrogen, in all its forms, from a catchment may be very different from the inputs. In some cases nitrogen accumulates in the catchment, but in others may be released from long-term stores (such as peat bogs).

It is estimated (Berner & Berner, 1996) that the "natural" average annual flux of dissolved inorganic nitrogen to the ocean is around 4.5 million tonnes a^{-1} (mostly as nitrate), and around 10 million tonnes a^{-1} is transported as dissolved organic nitrogen. The particulate organic nitrogen flux is around 21 million tonnes a^{-1}, making a total "natural" flux to the sea from the land of 36.5 million tonnes a^{-1}. This is a small proportion of the nitrogen assimilated annually by the terrestrial biosphere (Berner & Berner, 1996), implying that the biological recycling of nitrogen is very efficient. Whilst the flux to the sea is small compared to nitrogen fixation processes operating in the ocean, rivers transport significant amounts of nitrogen nutrients to the estuaries and the coastal zone (Chapter 8). Nitrogen cycles in many catchments have been very significantly affected by human activities (Chapter 6).

The phosphorus cycle

Phosphorus is an integral part of the DNA and RNA of living organisms, and an essential nutrient. All phosphorus ultimately derives from rock weathering. Bacteria break down phosphorus in the soil into orthophosphate (PO_4), which is then taken up by plants and subsequently passed on to animals. Phosphorus is released back into the soil when plants and animals decay, to be recycled.

Phosphorus in water can be held as dissolved or particulate material, and in organic or inorganic forms. Dissolved inorganic phosphorus is essentially orthophosphate, and dissolved organic phosphorus is made up of complex organic material. Particulate organic phosphorus consists of particles of dead organic matter, whilst particulate inorganic phosphorus comes from material incorporated into mineral structures and adsorbed onto the surface of suspended sediment particles. Phosphorus is leached into a river through the soil, and undergoes significant changes within a water body. Dissolved inorganic phosphorus becomes adsorbed onto both organic and inorganic particle surfaces, increasing the fraction of phosphorus that is transported in particulate form: this is particularly significant where there are large numbers of fine inorganic sediment particles (Newbold, 1996). Phosphorus is also removed from the water by algae and higher-order plants in the water and on the banks of the water body, and added to the water body by excretion and death.

The vast proportion of phosphorus (generally over 90%) is transported in particulate form. Concentrations are higher in quickflow than baseflow, because much of the material derives from the soil, and most of the phosphorus exported from a catchment moves in only a few days (98% of the phosphorus load in the Richmond River, Australia, was carried in less than 6% of the time: McKee *et al.*, 2000). As with nitrogen, catchment phosphorus budgets can show that phosphorus is accumulating in stores in a catchment, or being gradually released (Dillon *et al.*, 1991).

Like the nitrogen cycle, the phosphorus cycle has been significantly affected by human activities. Fertilisers containing phosphorus are widely applied to farm land, and sewage treatment effluent contains phosphates (Chapter 6).

The sulphur cycle

Sulphur is another important nutrient, and as sulphate (SO_4) it is the second most abundant anion in rivers and the major cause of acidity in water (Slaymaker & Spencer, 1998). Biological processes in the ocean and on land result in the production of a number of sulphur-based gases, including hydrogen sulphide and dimethyl sulphide (DMS). Sulphate in the atmosphere, formed from these sulphur-based gases, constitutes one of the most common cloud condensation nuclei. Most of the sulphur-based gases are emitted from the oceans, but a smaller proportion are emitted from water-logged soils – with a clear hydrological control.

Sulphur gets into the catchment system partly from atmospheric deposition and partly from rock weathering. Relatively little is taken up by vegetation, and a large proportion of sulphate is leached into watercourses. Sulphate concentrations in river water in a specific catchment tend to vary with the source of incoming precipitation rather than the streamflow generation process, as indeed do concentrations in different catchments. Although estimates of magnitudes vary, the "natural" annual average sulphur flux from land to sea is around 100 million tonnes, approximately half of the input to the oceans from precipitation (Andrews *et al.*, 1996).

The sulphur cycle has been heavily perturbed by human activities, and in particular the burning of fossil fuels. This has resulted in the problem of "acid rain" (Chapter 7) – and may have more than doubled the flux of sulphur from land to sea via rivers (Andrews *et al.*, 1996).

The oxygen cycle

The oxygen cycle is closely coupled to the carbon cycle. Oxygen is produced by plants during photosynthesis, when CO_2 is consumed, and is extracted from the atmosphere by respiration and decay. Plant growth is of course influenced by water availability, and this is perhaps the major contribution of hydrological processes to the oxygen cycle. Rivers, however, also transport organic matter from land to sea, and some of this organic matter is made up of unrespired oxygen.

5.5 Concluding comments

This chapter has explored in some detail the processes influencing the physical and chemical characteristics of water bodies, from the perspective both of the catchment and the element. Total fluxes of materials through the catchment system depend on the amount of material available, the amount of water available to move material, and the pathways followed by water to the river channel.

Throughout the chapter, the emphasis has been on processes and fluxes in "natural" catchments. Human activities have had three very significant effects on the quality of water bodies. First, they have altered the *amounts* of material moving through the system. Agricultural fertilisers, for example, increase nitrogen and phosphorus loads, and clearance of vegetation may lead to increases in sediment supply. Second, they have altered the *types* of material moving through the system: new synthetic chemicals have frequently been released into the river environment. Third, they have affected the *transfer* of material along the river network by, for example, building dams to obstruct flows or altering flow regimes by changing catchment properties. These human impacts are explored in the next two chapters, and Chapter 8 looks at how changes in the physical and chemical characteristics of water may impact upon global processes.

Chapter 6

Change in the catchment

6.1 Introduction

There are two components to global environmental change (Turner *et al.*, 1990). *Systemic changes* operate at the global scale, affecting large areas at the same time: global warming is the classic example. *Cumulative change* describes the net effect of local-scale changes occurring in different areas, perhaps for different local reasons: deforestation is a good example. This chapter focuses on the cumulative change component of global environmental change, looking in detail at how changes in the catchment affect hydrological behaviour at the catchment scale. Chapter 7 explores the effects of systemic changes – largely changes in the *inputs* to the catchment – whilst Chapter 8 examines the effect of changes in the catchment on the global environmental system *beyond* the catchment.

Figure 6.1 provides a conceptual overview of catchment-scale changes in the hydrological system. Three broad groups of *changes* have three types of *effect* on the quantity and quality of water in the catchment, which have a set of *consequences* or *impacts*. It is clear that a particular "impact" can have several different contributing factors, and it may therefore be difficult in practice to work backwards and attribute an observed hydrological change to any one catchment change. Also, a given change can have several different effects and consequences. The centre of Figure 6.1 shows three broad types of effect of change in the catchment on hydrological behaviour in the catchment: a change in the volume of flow or groundwater storage, a change in the timing of flow, and a change in the quality of water in soils, rivers and groundwater. The changes shown around the outside of Figure 6.1 can affect one, two or all of these dimensions of the hydrological system.

Changes in the catchment are divided into three groups. "Land cover changes" essentially incorporate changes to the vegetation covering the catchment. Each of the five types of land cover change shown in Figure 6.1 – deforestation, afforestation, agriculture, urbanisation and mining – have many facets, but as a general rule all are done for reasons other than deliberately to alter hydrological behaviour (although, as will be emphasised at the end of the chapter, these changes are now frequently being done in ways which seek to minimise their hydrological consequences). The effects of these changes are therefore not necessarily anticipated or indeed predictable.

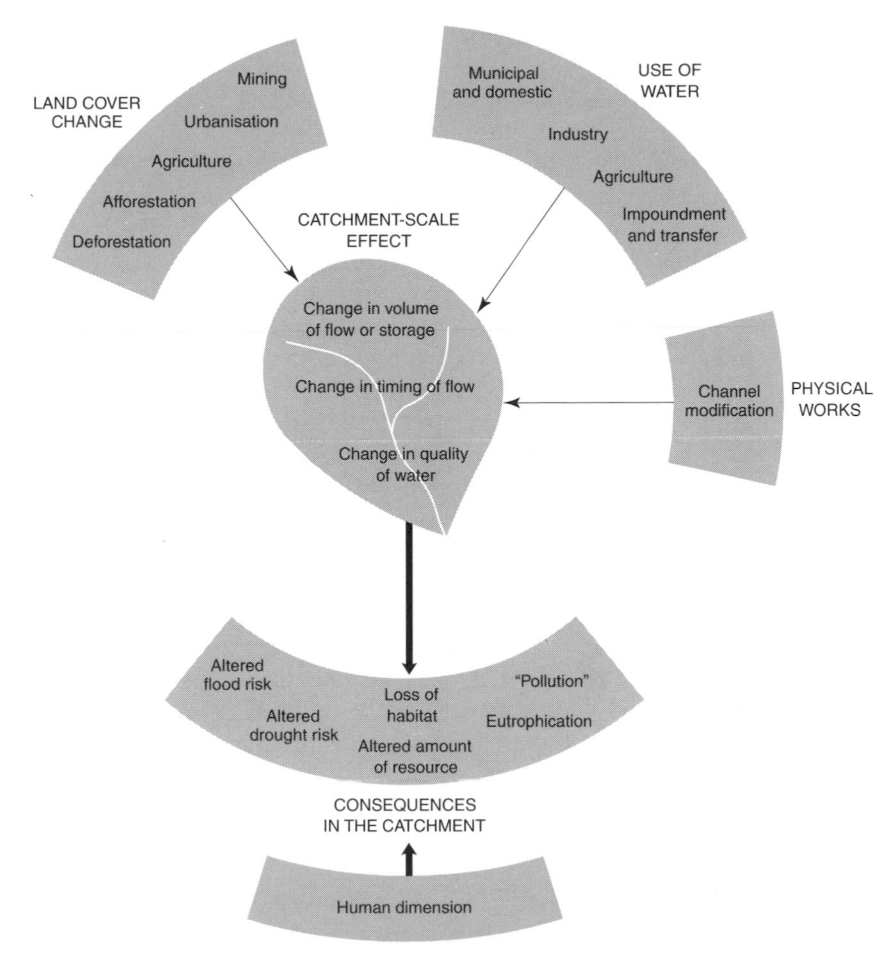

Figure 6.1 Effects and consequences of changes in the catchment

The second group of changes relates to the use and exploitation of the water in a catchment. Three of these interventions refer to particular uses of water, both as a resource and as a means of waste disposal. The fourth, "impoundment and transfer", covers the construction of reservoirs to regulate rivers for flood control, navigation, water supply or power generation, and the import of water from other catchments. These are all deliberate interventions in the catchment hydrological system with, in principle, identifiable and predictable effects. However, some effects may be unanticipated, or even if anticipated of no concern to those responsible for the intervention – and therefore ignored by them.

The third group of changes shown in Figure 6.1 cover physical changes to the river network, either through works in the channel (such as dredging, embankment, realignment and the construction of locks and weirs) or on the floodplain to prevent inundation. These changes primarily affect the timing of flow, and can be seen as having a largely *hydraulic*, rather than hydrological, effect.

It is important to distinguish here between "land cover" and "land use": the two terms are related, but not synonymous. Land cover describes what covers the land, whilst land use describes how the land is used. A forest, for example, can be used for many different purposes. The use of land can affect much more than just land cover. Agriculture as a land use, for example, involves not only changing the vegetation in a catchment, but also the abstraction of water from rivers, digging of ditches, disposal of waste and perhaps the construction of major impoundments. Similarly, urbanisation affects land cover, water use, waste disposal and is often associated with physical changes to the river network.

The bottom part of Figure 6.1 shows a number of the potential consequences in a catchment of the hydrological effects of catchment change. Altered flood and drought risk are self-explanatory, and are a function of changes in the quantity of water (and, of course, changes in the exposure to hazard). The other consequences, however, can be a result of changes in both quantity and quality. The amount of water resource available, for example, can be altered not only by the volume of flow but also by the quality of that water: the resource can be reduced by a decline in quality with no change in flow. Similarly, habitat loss may be due to a reduction (or indeed increase) in flows or a degradation in water quality. "Pollution" is broadly defined in Figure 6.1 to mean a degradation in the quality of water such that some use (human or environmental) is adversely affected. It is affected both by the amount of contaminants in the water and the volume of flow. There are many different types of pollution, including pollution by organic material (which consumes oxygen), excessive nutrients, microbial pathogens, heavy metals, organic micro-pollutants (such as pesticides and fertilisers) and suspended sediment: excessive colour and heat can also be pollutants. A distinction is made between *point* and *non-point* sources of pollution. A point source comes from a specific pipe at a specific location, whilst non-point pollution comes from a large area, such as an expanse of farmland. Chapter 5 introduced the wide range of chemical parameters that characterise water characteristics. In operational water management it is necessary to reduce these to indicators of water quality. Box 6.1 summarises the

Box 6.1 Chemical and biological indicators of water quality in the UK

The General Quality Assessment (GQA) of rivers in the UK has four components: chemical, biological, nutrient and aesthetic (Environment Agency, 1998a). The chemical GQA assesses dissolved oxygen concentrations, biochemical oxygen demand (BOD) and ammonia concentrations against standards expressed as percentiles. Ammonia and BOD are general indicators of pollution from sewage or agricultural land. The table below shows the classification:

(continued)

(continued)

Chemical GQA grade	Description	Dissolved oxygen (% saturation) 10-percentile	Biochemical oxygen demand (mg l^{-1}) 90-percentile	Ammonia (mg N l^{-1}) 90-percentile
A	Very good	80	2.5	0.25
B	Good	70	4	0.6
C	Fairly good	60	6	1.3
D	Fair	50	8	2.5
E	Poor	20	15	9
F	Bad	<20	>15	>9

In order to bc graded "very good", no more than 10% of samples must have a dissolved oxygen concentration less than 80% of saturation, and at least 90% of samples must have BOD less than 2.5 mg l^{-1} and ammonia concentrations less than 0.25 mg N l^{-1}. The biological GQA uses macroinvertebrates in the river as indicators of water quality, and compares observed populations with those which would be expected if the river was in a pristine condition:

Biological GQA grade	Description	Outline description
a	Very good	Biology similar to or better than expected. High diversity. No one group dominating
b	Good	Biology falls a little short of that expected
c	Fairly good	Biology worse than expected for an unpolluted river
d	Fair	Pollution-sensitive groups scarce. Range of pollution-tolerant groups present
e	Poor	Biology restricted to pollution-tolerant species
f	Bad	Biology limited to small number of very tolerant groups of species, or in the worst case no life present

Techniques have been developed to express the biological GQA classes in terms of quantitative ecological indicators. The nutrient GQA is currently not used operationally, and a future scheme is likely to be based on phosphate concentrations. The aesthetic GQA identifies four classes, based on a number of indicators including litter, odour, colour and the presence of scum.

Environmental quality standards setting maximum levels of a wide range of substances have been set by many agencies involved in water regulation. In the European Union, for example, cadmium concentrations must be less than 5 μg l^{-1}, and rivers designated as salmonid fisheries must have a BOD less than 3 mg l^{-1} and ammonia concentrations less than 0.005 mg l^{-1}.

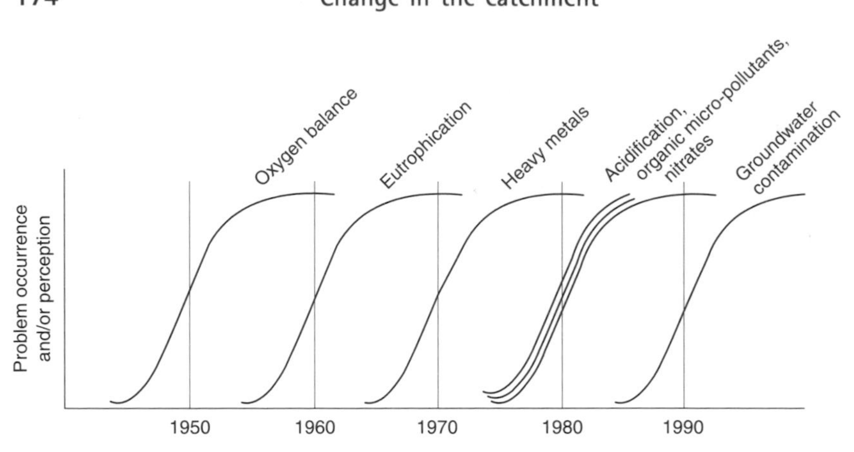

Figure 6.2 Evolution of water quality concerns (extended from Meybeck *et al.*, 1989)

classifications of water quality used in the UK. Eutrophication is excessive plant and algae growth, and is generally triggered by changes in nutrient inputs, although it can be exacerbated by low flows. Eutrophication is damaging to the water environment in three ways: when the plants and algae decay, large amounts of oxygen are consumed, algal blooms may be toxic, and blooms are unsightly.

Whilst there has been concern over the pollution of watercourses for several hundred years, real scientific concern over the effect of catchment changes on the hydrological system really began around the end of the nineteenth century. Since that time the focus of attention has shifted substantially. Initial concerns were over the effects on floods and erosion following deforestation, with scientific studies commencing in the United States and Switzerland. During the 1930s research in the United States started to look at how agricultural practices affected runoff and erosion, and more particularly at how to farm dry land in a way that minimised erosion. Interest in the effects of *afforestation* developed in the UK from the mid-1950s, leading to a major suite of experiments in upland areas. During the 1960s research began to show the hydrological consequences of urbanisation. The last few decades of the twentieth century, however, were dominated by concerns over the quality of water in rivers and aquifers. Figure 6.2 (developed from Meybeck *et al.*, 1989) summarises the evolution of water quality concerns since the middle of the twentieth century.

It is possible to explore the hydrological effects of catchment change from either the top or the bottom of Figure 6.1, focusing either on the change or the consequence. This chapter is organised by the type of change, but seeks to emphasise both throughout the chapter and in the final overview section the multiple causes of some important hydrological changes in a catchment. First, however, it is important to explore in detail the problems and challenges associated with detecting and estimating the effects of catchment change.

6.2 Detecting and estimating the effects of catchment change

Hydrological behaviour varies over time, at hourly, daily, seasonal, annual and inter-annual scales. Some of this variability in a catchment will be due to variations in climatic inputs and seasonal cycles of vegetation growth and decay, but some will be due to human-induced changes to the surface of the catchment and the way water moves through the catchment. How is it possible to detect human-induced change, or indeed to attribute a change in hydrological behaviour to human activity? There are three groups of approaches: analysis of observed time series, catchment experiments and computer simulation.

Analysis of observed time series

There are two ways of looking at observed time series when considering the impacts of potential human interventions. The first looks to see if a known human intervention is having a demonstrable effect on the hydrological system: "is urbanisation having an effect on flood peaks in this catchment?", for example. This involves constructing a time series of the intervention (dates of urban development, for example) and seeking to find a corresponding pattern in a time series of observations (magnitudes of flood peaks, for example). The main technical problem with this approach lies in the separation of any human-induced trend from the effects of inherent climatic variability. Hydrological time series, especially indicators of extreme flow behaviour such as flood peaks or low flows, are very "noisy", and the signal of some intervention may be small in comparison to the background noise. A second conceptual problem is that even if a pattern is found in the hydrological data which is correlated with the pattern of the intervention, it does not prove a causal link.

The other way of looking at time series seeks to quantify an apparent trend and identify its cause: "have there really been an unusually high number of floods in this catchment in the last few years, and if so are they due to the recent urbanisation?", for example. This approach obviously tends to be applied when it looks as if hydrological characteristics may be changing, and in fact is much more widely applied than the first approach. Again, however, the main problem lies in distinguishing statistically the effects of natural climatic variability from a human intervention in the catchment. A large number of tests to evaluate changes in time series have been applied in hydrological studies (Salas, 1993), including the Mann–Kendall trend test (e.g. Hirsch *et al.*, 1982) which tests for the presence of a continuous trend, and tests looking for shifts in the mean. However, all such techniques assess trend in the context of a time series in which it is assumed that there is no variability at time scales longer than one year: Chapter 4 showed how this assumption may not be true. Another problem with trend detection in hydrological data is that the data collection techniques and procedures may have varied over time, leading to discontinuities in the time series. Finally, the effects of one sort of change, such as catchment urbanisation, may be hidden by another, such as river channelisation.

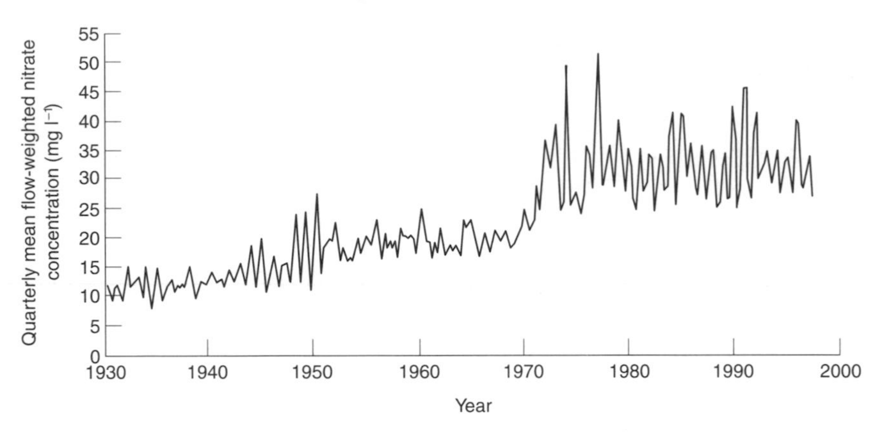

Figure 6.3 Increasing concentration of nitrate in the River Thames (Environment Agency, 1998a)

There are, however, many examples of clear, unambiguous trends in hydrological data resulting from human interventions, with perhaps the clearest occurring in the field of water quality. Figure 6.3, for example, shows the increasing concentration of nitrate in the River Thames.

Catchment experiments

Catchment experiments are controlled attempts to explore the implications of a particular intervention (generally a land use change) on hydrological characteristics, and have a long history: the Wagon Wheel Gap experimental catchments in Colorado were established in 1909. There are two types of catchment experiment, although in practice the two are often done together.

The first type uses sets of paired catchments, in which everything is the same apart from the parameter of interest. Typically, paired catchments will be adjacent with the same climate regime and underlying bedrock, but different land covers. The Plynlimon experimental catchments in mid-Wales are a classic example (Box 6.2). Fluxes and storages in the two (or more) catchments are continually monitored, and differences attributed to the difference in land cover.

Box 6.2 The Plynlimon experimental catchment

During the 1950s, concern was raised in Britain that upland afforestation was reducing the amount of water entering water supply reservoirs. In response, a paired catchment experiment was established in the late 1960s on Plynlimon Mountain in mid-Wales, based on the adjacent headwaters of the River Severn and the River Wye (Kirby *et al.*, 1990; Neal, 1997). Both catchments have an area of around 10 km^2, but whilst the Wye catchment is covered with moorland the Severn is largely covered with plantation Sitka spruce. Precipitation is measured at a

(continued)

(continued)

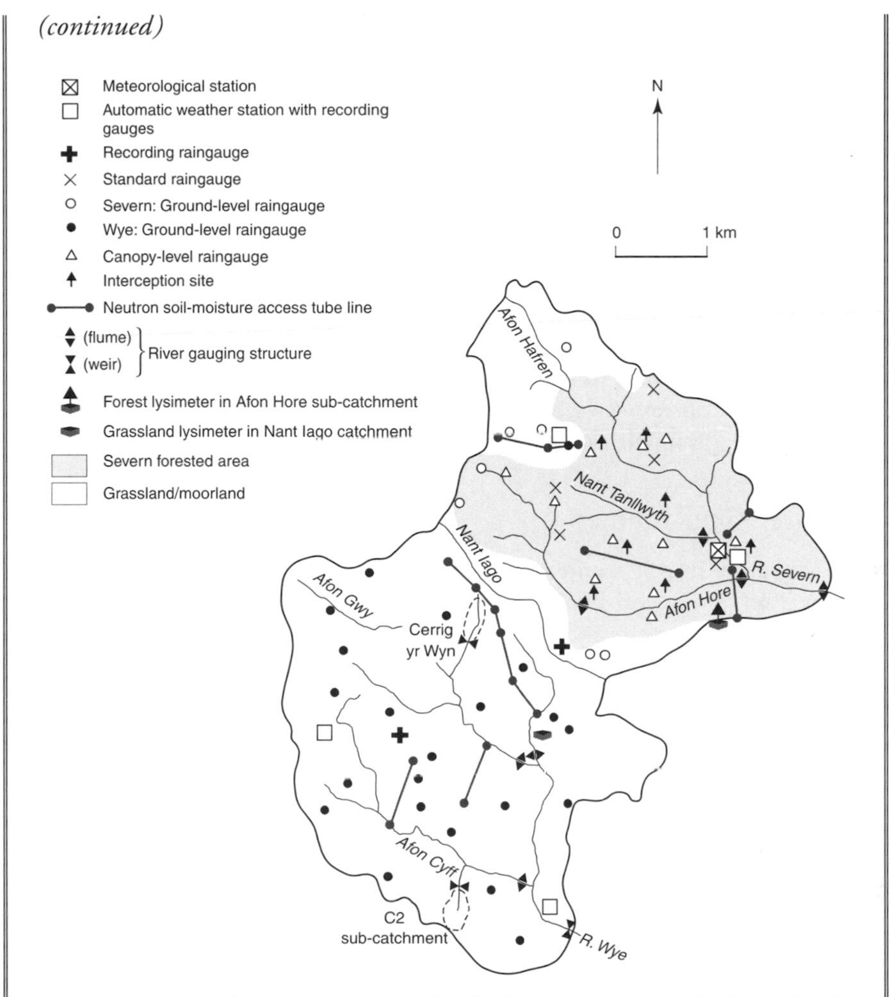

Figure 6.4 An experimental catchment: the Plynlimon catchments, mid-Wales (modified from Hudson *et al.*, 1997b)

large number of ground and canopy level gauges, and streamflow recorded not only at the catchment outlets but also on small tributary streams. Soil moisture is routinely monitored, and short-term experiments examined interception, throughfall and stemflow (Figure 6.4). During the 1970s water quality began to be measured in the catchments – reflecting changing scientific priorities – and in the mid-1980s part of the Severn catchment was clearfelled. The experiment faced many technical and practical difficulties associated with the measurement of fluxes in a harsh environment with a precision enabling any differences to be identified with confidence, and many innovative instruments and techniques had to be developed. By 1998, the Plynlimon experiment had generated around 300 publications (Kirby *et al.*, 1997).

The second type of catchment experiment is "longitudinal". A catchment is monitored for a number of years, until some controlled intervention is made, and the hydrological characteristics during and after the intervention are compared to those before. Often this is done with just one catchment (or indeed study plot – the same principles apply), but it may be difficult to attribute the hydrological change to the intervention, especially if the records are short. An exceptional flood after a plot of land has been drained may have occurred even if the land had not been drained: it could be impossible to tell. Many longitudinal studies therefore initially make measurements in more than one catchment, leaving one or more of the catchments unchanged as a control. Some studies make different changes in different catchments, allowing the effects of different types of land use change to be evaluated in a controlled way. These experiments at catchment scale are obviously expensive and must continue for a long time; "plot-scale" experiments are easier to implement.

Computer simulation

This approach uses a computer simulation model to simulate the effects of human intervention on hydrological characteristics within a catchment. One variant of this approach calibrates a hydrological model on data measured before an intervention takes place, and then simulates how the catchment would have responded to subsequent inputs if there had been no land use change (e.g. Lorup *et al.* (1998) in Zimbabwe). The difference between observed and simulated behaviour is then assumed to reflect the impact of human intervention. In practice, of course, the difference will reflect not only the human intervention but also the model errors, which may be particularly significant if the latter period contains inputs outside the range of those experienced during the initial calibration period. Another version of the approach uses a hydrological model to simulate the change in hydrological behaviour in a catchment as defined human interventions take place. It is often used where a trend has been identified in the hydrological data, but the cause of this trend is unclear. If the simulated changes match the observed changes, then it can be inferred that observed changes in hydrology can be attributed to the interventions incorporated in the simulation. This approach can be used to test different potential explanations for an observed trend.

A third application of computer simulation is to assess the effects of human intervention where there have been no direct observations. This uses a model which does not need to be calibrated on site data, together with estimates of input data and trends in human intervention, to simulate hydrological characteristics in a catchment or across a region with and without the intervention. This approach tends to be used for regional-scale, rather than catchment-scale assessments of the impact of human activities (e.g. the entire Columbia River basin in North America: Matheussen *et al.*, 2000), or to simulate changes over very long time scales (e.g. Coulthard *et al.*, 2000).

The final use of computer simulation is to conduct "virtual experiments" in order to assess what would happen under different hypothetical interventions, or what will continue to happen in the future following an intervention that has already taken place (for example Kamari *et al.*'s (1998) simulation of the effects of deforestation on nutrient fluxes in a small Finnish catchment). Predictions of change will, of course, be subject to uncertainty because of model formulation and parameterisation, and given these uncertainties it is possible that only the effects of very large interventions could be detectable. Nandakumar & Mein (1997), for example, applied a model to experimental catchments in Victoria, Australia, and showed that uncertainties in data inputs and model parameters meant that up to 65% of the forest area would need to be cleared (in the model) in order to produce flow increases detectable with 90% confidence.

6.3 Effects of changes in catchment land cover

Deforestation

Deforestation is the removal of forest cover, and has been practised for a very long time. Much of the primary forest in Europe was removed before the twentieth century, and most of Europe's forests – with the exception of the boreal forest in northern Europe – is secondary regrowth, often of alien species planted for forestry purposes. It is estimated that 85% of the primary forest in eastern North America was removed over the space of two or three hundred years by European settlers (Mackenzie, 1998), and again much of the current forest in this area is secondary regrowth. Although deforestation in some areas continues – approximately 240 km^2 of primary forest is harvested each year in the Pacific North West of the USA, for example – in developed countries there has been a net increase in forest cover since the latter part of the twentieth century (Mackenzie, 1998). The situation is rather different in the tropics, however. Half of the original tropical forest area has been cleared in the last 200 years, with most removed since 1950 (Mackenzie, 1998). It is difficult to get precise estimates, but around 1.5 million km^2 was lost during the 1980s (Slaymaker & Spencer, 1998): the rate of loss slowed slightly during the 1990s.

The vast bulk of deforestation is deliberate, either to clear land for farmland or to harvest the trees and replant. Forest fires, however, also remove trees, and these fires may be started naturally (by lightning) or (inadvertently or maliciously) by people. Acid rain can also lead to deforestation: this is explored in Chapter 7.

There have been many catchment experiments examining the effects of deforestation on the quantity of streamflows. Sahin & Hall (1996) collated results from 145 separate experiments. Studies from western Europe have tended to focus on the removal of plantation forest – because the removal of the primary forest took place centuries ago – as have some of the studies from the United States. However, most American studies have focused on the consequences of removing the primary forest in the Pacific North West, partly in order to develop forestry practices which attempt to have minimum impact. The increasing number of studies from open and closed tropical forests have looked at the removal of primary forest. Table 6.1 lists some of the studies from different regions: a few examples are discussed in more detail below.

Table 6.1 Some studies into the effects of deforestation on water quantity

Region	Reference	Comments
Europe		
UK: plantation forest, Plynlimon	Roberts & Crane (1997)	Longitudinal study
Belgium	Bultot *et al.* (1990)	Model study
France: Mediterranean pine	Lavabre *et al.* (1993)	Longitudinal study: fire
Israel: Mediterranean woodland	Inbar *et al.* (1998)	Longitudinal study: fire
North America		
Pacific North West: Oregon	Jones & Grant (1996)	Paired catchments
Pacific North West: Columbia River	Matheussen *et al.* (2000)	Model study
North west California	Wright *et al.* (1990)	Paired catchments
Colorado	Troendle & Reuss (1997)	Paired catchments
California pine	Keller *et al.* (1997)	Longitudinal study: fire
Tropical forests		
Malaysia	Malmer (1996)	Paired catchments
Indonesia	Asdak *et al.* (1998b)	Paired catchments
Nigeria	Lal (1996)	Longitudinal study
Subtropical forests and woodland		
Lake Malawi catchment	Calder *et al.* (1995)	Model study
South Africa	Scott (1997)	Paired catchments
South east Australia	Croke *et al.* (1999)	Paired catchments
Western Australia	Bari *et al.* (1996)	Paired catchments

The major effect of deforestation is to increase the total amount of runoff (Bosch & Hewlett, 1982). The removal of trees reduces the amount of water intercepted, and hence the amount evaporated back into the atmosphere, and more water reaches the ground more quickly. The effect of the removal of trees depends on the area of forest cleared, the type of forest, and the climate. Stednick (1996) collated results from catchment studies in the United States, and showed that the effects of deforestation were detectable in the Rocky Mountains once more than 15% of the catchment was cleared, but more than 50% of the catchment needed to be cleared to produce a detectable change in the Great Plains. This arises partly because of the different mix of trees in the two areas, but partly because of the difference in climate. Much of the precipitation in the Rocky Mountains falls in winter as snow, and snow accumulation is significantly increased if forest is cleared (Troendle & Reuss, 1997): removal of forest here has a big effect on runoff. In the Great Plains, most of the precipitation falls during summer, and snow storage is less important. Differences in climate make it difficult to compare the effects of removing different types of forest. Both Bosch & Hewlett and Sahin & Hall (1996) collated the results of large numbers of catchment experiments, and attempted to derive generalised relationships between the amount of clearance and the increase in runoff: Table 6.2 shows estimates of change in runoff for a 10%

Table 6.2 Effect of a 10% reduction in forest cover on average annual runoff (Bosch & Hewlett, 1982; Sahin & Hall, 1996)

Forest type	**Change in average annual runoff (mm)**	
	Bosch & Hewlett (1982)	**Sahin & Hall (1996)**
Coniferous forest	+40	+20 to 25
Deciduous hardwood	+25	+17 to 19
Eucalyptus	+40	+6
Scrub	+10	+5

reduction in cover. There are some significant differences between the two sets of relationships, reflecting different data used and different methods. Removing coniferous forest has the greatest absolute effect on runoff, but the percentage increase may be smaller than that after the removal of deciduous hardwood because coniferous forests tend to be located in areas with greater rainfall, and hence runoff. Clear-felling approximately 29% of the coniferous forest in a sub-catchment of the Plynlimon catchment led to an increase in average annual runoff in that sub-catchment of between 130 and 160 mm, or between 6 and 8% (Roberts & Crane, 1997), close to the rate of change estimated for coniferous forest by Bosch & Hewlett. There are particularly large differences in the estimated effect of removing eucalyptus forest, probably reflecting the wide variety of tree cover under different species of eucalyptus. The climate of an area will also influence the effects of deforestation through variations in the frequency with which rain falls. As shown in Chapter 3, a greater proportion of annual precipitation tends to be intercepted the more frequently rain falls, and hence the longer the canopy is saturated. It can therefore be expected that removal of a particular type of tree from a wet catchment (such as in upland Wales) will have a greater effect than in a dry catchment (such as southern England).

The actual effects of deforestation on total runoff depend, however, not only on the type of forest cover that is removed but also on the method of removal (Table 6.3). As a general rule, the greater the amount of mechanical intervention and the more "brutal" the method, the greater the increase in runoff.

Deforestation increases peak flows as well as total runoff, with small floods more affected than large floods. This is because the smaller the total volume of interception storage in a catchment relative to the total amount of precipitation, the less the effect of that interception storage on the amount of water reaching the ground. "Small" floods may become larger with deforestation, and "new" floods may be generated by rainfall that previously would not have produced a hydrological response. Lavabre *et al.* (1993), for example, found that the 10-year return period flood occurred three times in the year after a wildfire destroyed 85% of the forest in a small Mediterranean catchment. Increased flood frequency at Lake Babati in Tanzania has been attributed to deforestation in the catchment (Sandstrom, 1995a), and one of the possible influences on

Table 6.3 Effects of different methods of deforestation on runoff and water quality, relative to undisturbed catchments: tropical forest, Malaysia (adapted from Malmer, 1996)

Method of deforestation	Increase in runoff (%)	Total suspended sediment ($t\ ha^{-1}\ a^{-1}$)	Total N flux ($kg\ ha^{-1}\ a^{-1}$)	Total P flux ($kg\ ha^{-1}\ a^{-1}$)
Conventional methods: clear-felling, extraction of logs by tractor, and clearance by burning	+20	+1.9	+39.9	+1.3
'Minimum impact' methods: clear-felling, manual extraction, no burning	+8	+3.6	+27.0	+0.8

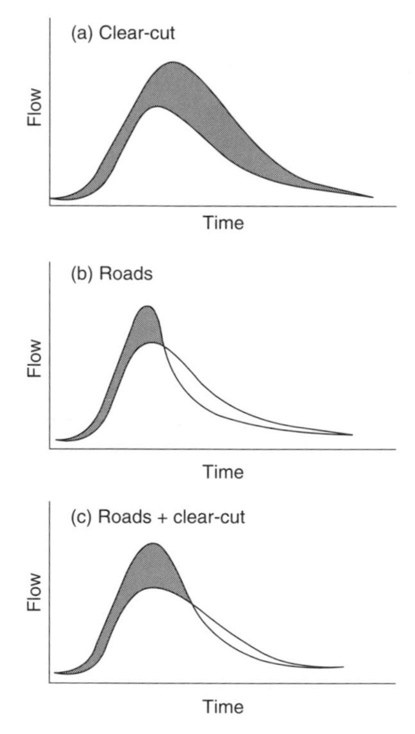

Figure 6.5 Effect of removal of trees and introduction of logging roads on the storm hydrograph: western Cascades, Oregon (Jones & Grant, 1996)

an apparent increase in flood frequency in the rivers draining the Himalayas is widespread deforestation of hillslopes in upland runoff-generating regions.

However, whilst the removal of trees affects the volume of water discharged in response to a precipitation event, the *timing* of the flood water is more affected by the infrastructure introduced in order to cut and remove the trees. Logging roads, for example, have a particularly large effect on the rate of response of a deforested catchment to precipitation. Figure 6.5 shows schematic

Table 6.4 Atmospheric deposition (g m^{-2} a^{-1}) over forest and moorland: Kielder, northern England (adapted from Fowler *et al.*, 1989)

Constituent	Wet deposition	Dry deposition		Cloud water deposition		Total deposition	
		Moorland	Forest	Moorland	Forest	Moorland	Forest
Sulphate SO_4^{2-}	1.31	0.31[1]	0.31	0.13	0.65	1.75	2.27
Nitrate NO_3^-(N)	0.35	0.22[2]	0.42	0.02	0.09	0.59	0.86
Ammonium NH_4^+(N)	0.45	0.18[3]	0.93	0.02	0.1	0.65	1.48
Chloride Cl^-	4.98	0.04	0.2	0.17	0.86	5.19	6.04
Sodium Na^+	2.81	–	–	0.09	0.46	2.90	3.27
Magnesium Mg^{2+}	0.33	–	–	0.01	0.04	0.34	0.37
Calcium Ca^{2+}	0.62	–	–	0.01	0.07	0.63	0.69

[1] as SO_2
[2] as NO_2 and HNO_3
[3] as NH_3

hydrographs estimated from experimental catchments in Oregon (Jones & Grant, 1996), illustrating the relative effects of removal of trees and introduction of logging roads on the storm hydrograph.

Effects of deforestation on low flows are less clear. In most cases, deforestation tends to increase low flows in line with increases in total runoff. However, in some circumstances low flows may be *reduced*: exposure of the soil and removal of vegetation may reduce infiltration capacities, meaning that water runs off quickly – creating greater floods and "wet season" runoff – and does not infiltrate into the soil or groundwater to sustain flows during drier periods. This is most likely to occur where soils are heavily disturbed during the deforestation process.

The duration of the deforestation effect obviously depends on what replaces the forest cover. If the forest is replaced by agricultural land, then the increased runoff will continue. If the forest is left to regrow – as would happen after a forest fire, for example, or when new trees are planted – then the rate of recovery depends on the rate of regrowth.

Whilst deforestation generally increases runoff, there are however some exceptions. The principal exception is where cloud forests are removed. These forests are characterised by large amounts of horizontal interception (Chapter 3). If the forests are removed, the amount of horizontal interception declines, and the input of water to the catchment falls. Hunzinger (1997) showed that winter season flows decreased in an Argentinian montane cloud forest after deforestation because of the virtual elimination of horizontal interception.

Removal of forests affects not only the quantity of water, but also its quality, in two main ways: the water quality of streams draining forested land is different from that of nearby streams draining non-forest land, and the actual process of deforestation leads to the flushing of nutrients and sediments. The quality of forest streams is different from those of non-forest streams because forests tend to "scavenge" more material from dry and cloud deposition (Chapter 5 and Table 6.4), have a different effect on nutrient cycles from short vegetation, and flow pathways may be different.

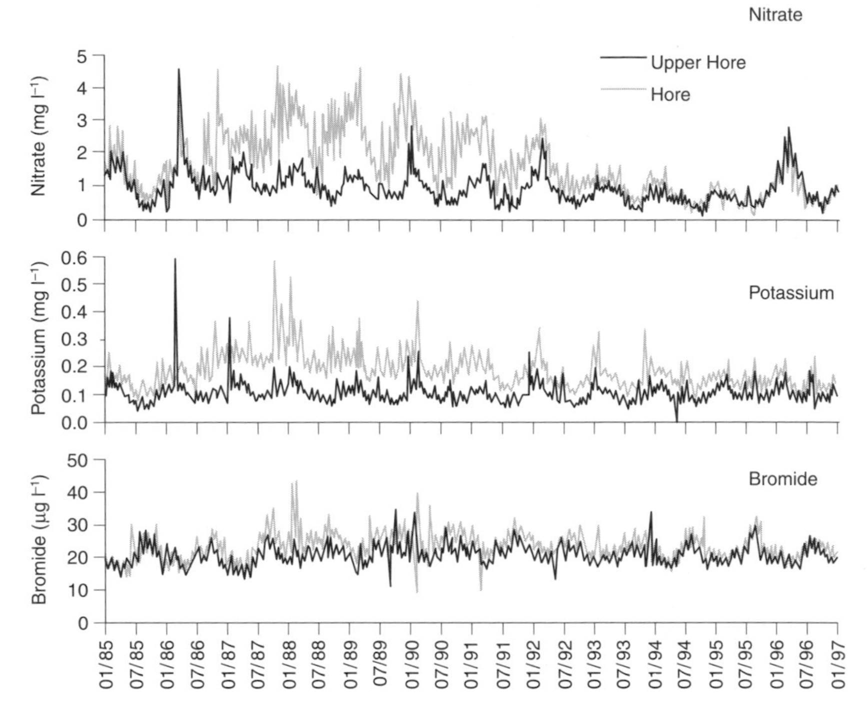

Figure 6.6 Time series of the nutrients nitrate, potassium and bromide in two sub-catchments of the Afon Hore, Plynlimon: the Hore sub-catchment was clear-felled between 1985 and 1988 (adapted from Neal *et al.*, 1997b)

The process of deforestation releases nutrients into the stream, and this has been observed in many environments (including tropical forests (Williams & Melack, 1997) and subalpine conifer forest (Reuss *et al.*, 1997): see also Table 6.3). Figure 6.6 shows time series plots for the nutrients nitrate, potassium and bromide in two sub-catchments of the Afon Hore, Plynlimon, one of which was clear-felled (Neal *et al.*, 1997b), showing the increase in nutrient concentrations following clearance. This arises for several reasons, including the decomposition of cut material and stumps, a break in the nutrient cycle and increased mineralisation of organic matter (Neal *et al.*, 1997). Although not obvious from the figure, peak nitrate and potassium concentrations tended to occur in early, rather than late, winter after clear-felling. Aluminium was also leached out of the catchment in the first few years after clear-felling, because the additional nitrate anions released aluminium from the soil's cation exchange store. After a few years, nutrient concentrations return to previous levels. Removing the forest cover leads to increases in both soil and water temperature (of around 3 °C in the Plynlimon example), which in turn leads to increased decomposition of organic matter: dissolved organic carbon concentrations therefore also tend to increase following felling. Kamari *et al.* (1998) estimated that it would take up to 40 years for nutrient levels to recover to pre-felling levels after deforestation and subsequent regrowth in a small catch-

Table 6.5 Effect of deforestation on stream salinity: Australia (adapted from Schofield & Ruprecht, 1989)

Average annual rainfall (mm)	Stream salinity (mg l^{-1})		
	0% of catchment cleared	0 to 50% cleared	50 to 100% cleared
>1100	144	176	233
900–1100	260	500	697
700–900	386	1095	756
500–700	70	1272	3488
<500	no data	11 988	19 255

ment in eastern Finland, although recovery would be more rapid in areas with faster regrowth.

Suspended sediment concentrations also generally increase after deforestation, for several reasons. The soil surface is exposed to rainfall and disturbed, increasing the likelihood of soil erosion, sediment may be generated from roads, and the stream channel and banks may be disturbed by machinery, releasing pulses of sediment (Leeks & Marks, 1997). Table 6.3 shows the effects of different deforestation processes on total sediment yield from an area of primary tropical forest in Malaysia (Malmer, 1996): yields were increased by between 7 and 13 times. Deforestation in part of the Plynlimon catchment increased the sediment yield by 39%, or 9 tonnes km^{-2} a^{-1} (Leeks & Marks, 1997: note the far higher yields in tropical Malaysia than Plynlimon). During deforestation the amount of woody debris in river channels increases. This debris causes blockages which impound sediment, and sediment yields may therefore initially be reduced immediately after deforestation – until the debris dam bursts. Not all the sediment generated from deforested land reaches the basin outlet. Some will be deposited in land and channel sinks, particularly if sediment concentrations are "naturally" high and little more can be carried. The "buffering capacity" of a catchment is related to the sediment delivery ratio (Chapter 5), with the lower the ratio, the greater the likelihood that sediment generated in part of the catchment will be stored on the surface or in the channel (Walling, 1999).

Finally, deforestation has been associated in Australia with dryland salinisation, which is the increasing concentration of salts in the soil (Smith, 1998). This arises because removal of forest increases recharge, leading to a rise in groundwater: if this groundwater is saline, as it is across large parts of Australia, then soil will become saline and so will water draining into rivers. In one catchment experiment, removal of eucalyptus forest led to an increase in recharge from 25 mm a^{-1} to between 60 and 100 mm a^{-1}, with an initial rise in groundwater of 2.6 m a^{-1}. Table 6.5 shows the generalised effects of clear-cutting on stream salinity in Australia (Schofield & Ruprecht, 1989): in humid areas the effects are small, but impacts increase as rainfall reduces. Dryland salinity is severely increasing the salinity of the Murray-Darling River in Australia (Smith, 1998).

Table 6.6 Average annual precipitation, runoff and evaporation: Plynlimon 1972 to 1985 (adapted from Hudson *et al.*, 1997b)

	Precipitation (mm)	Runoff (mm)	Evaporation (mm)
Wye (grassland)	2500	2008	491
Severn (forested portion)	2487	1795	692
Difference (mm)	−13	213	−201
Difference (%)	−0.5	−11	41

Deforestation, then, can have wide-ranging effects on water quantity and quality. It is, however, difficult to generalise quantitative impacts from one case study to another, and the impacts of deforestation clearly depend very much on how the deforestation takes place.

Afforestation

In the broadest sense, afforestation means the replacement of a non-forest land cover with forest, and thus includes natural regeneration of forest in abandoned land. A stricter definition of afforestation, however, restricts the term to the deliberate planting of trees for commercial forestry purposes.

In the late nineteenth century, the expanding cities of the United Kingdom outgrew their local water resources, and began to look for supplies from further afield. Reservoirs were built in upland Britain, and water piped to the growing cities or transported by rivers: these upland catchments were largely covered with moorland. The First World War revealed a "timber crisis" in Britain, and the Forestry Commission was established to expand considerably Britain's production of timber. From the 1920s, large parts of upland, moorland, Britain were covered with plantation forest, typically non-native Sitka and Norway spruce: many of the reservoired catchments were afforested. During the 1950s, however, evidence began to emerge that the forests were reducing the yield of water supply reservoirs (Law, 1956). The hypothesis that forests increased the "losses" of water from catchments ultimately led to the establishment of the Institute of Hydrology by the Natural Environment Research Council, and the initiation of the Plynlimon experiment (Box 6.2). The results of the Plynlimon experiment have shown that mature coniferous forest uses more water than grass or moorland, at least in upland Britain (Table 6.6): evaporation is 41% higher, and runoff is 11% lower.

Afforestation is, however, a process. Land must be prepared for planting, and trees take time to mature. Figure 6.7 shows in conceptual form the change in streamflow (and nutrients) over time in an afforested catchment (the figure excludes the effect of any initial forest clearance that may be necessary).

Before trees are planted it is often necessary to prepare the ground, particularly in wet upland Britain. Drains are dug into the moorland, and these drains lead to short-term increases in peak flows and reductions in lag-times (Robinson, 1986; 1998). Figure 6.8 shows the change in time to peak at Coalburn in

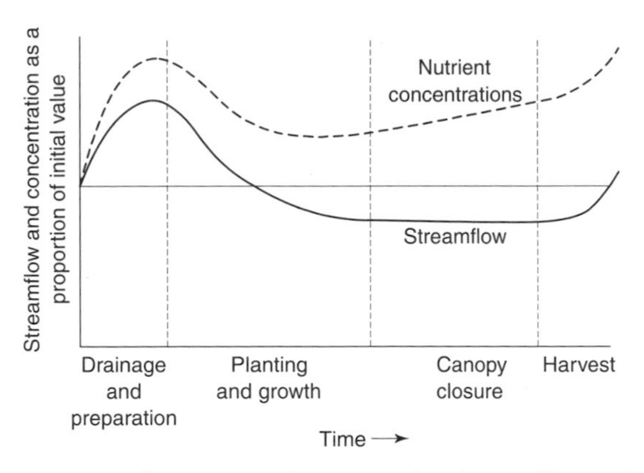

Figure 6.7 Change in streamflow and nutrients over time in an afforested catchment

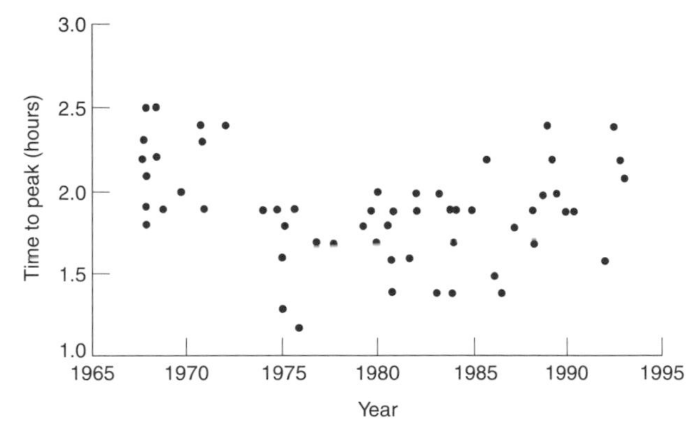

Figure 6.8 Hydrograph time to peak in the Coalburn catchment: drainage took place in 1972 (Robinson, 1998)

northern Britain: response times fell immediately after drainage in 1972, and rose steadily thereafter. This expansion of the drainage network also results in increased sediment concentrations and yields. Also, for the first one or two years total flow may increase as waterlogged soils are progressively dewatered. In some cases, these drains may lower the table sufficiently to produce longer-term reductions in runoff, as evaporation and transpiration may be suppressed by longer periods of soil moisture deficit. As time progresses, the drainage network becomes less effective as it is blocked by vegetation, evaporation is increased as trees grow, and total streamflow declines. The rate of change in evaporation, and hence streamflow, depends on the rate of growth of the trees. In a South African catchment, runoff began to decrease three years after eucalyptus trees were planted, and after nine years runoff had fallen to zero; pine plantation had a slower response, and it took 12 years before the runoff disappeared (Scott & Lesch, 1997). Evaporation reaches a peak once the canopy is closed – after about 20 years – and may actually thereafter decline as trees become more mature (Hudson *et al.*, 1997b).

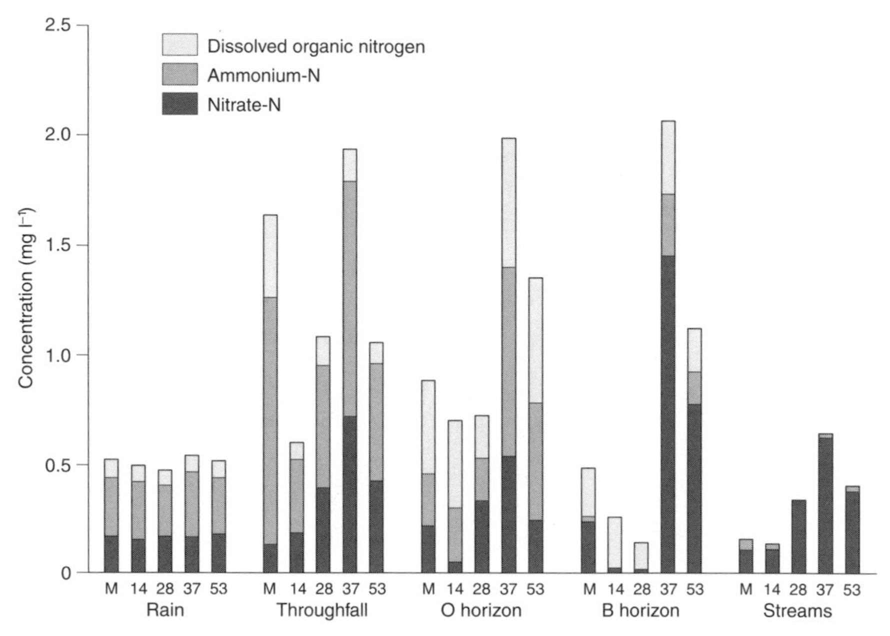

Figure 6.9 Annual mean concentrations of dissolved organic nitrogen, ammonium and nitrate, in moorland (M) and four forested catchments (14, 28, 37 and 53 years old), in rainfall, throughfall, two layers in the soil and streamflow (Stevens *et al.*, 1997)

Most work on afforestation in the UK has focused on upland conifers, the major form of afforestation during the twentieth century. Towards the end of the twentieth century two new trends emerged. The first, which is likely to be widespread in many developed countries at least, is an increasing interest in planting trees in order to maintain and increase biodiversity, and to sequester carbon in order to help slow the increasing concentration of carbon dioxide in the atmosphere. Across much of the temperate world this afforestation is likely to be of "native" hardwoods rather than conifers, and relatively little is known about the effects of broadleaf hardwoods on catchment water use (see Chapter 3). Evidence suggests that, as in coniferous forests, evaporation is higher from broadleaf woodlands than surrounding grassland, particularly in dry summers (Harding *et al.*, 1992). Widespread afforestation for biodiversity or carbon sequestration reasons may therefore impact upon the amount of water resources available. The second trend is the increased planting of short-lived, fast-growing trees for use as fuel. This is known as short-rotation coppice, and the trees commonly used – willow and poplar – can have extremely high transpiration rates because they prefer wet conditions (Hall & Allen, 1997).

Stream solute and nutrient concentrations also change over the course of forest growth. Concentrations increase as the initial drainage ditches are dug and then fall back, but continue at levels greater than nearby non-forested land (Stevens *et al.*, 1997). Figure 6.9 shows the variation in dissolved organic nitrogen, ammonium and nitrate with stand age in upland Wales. The inputs of nitrogen from rain to the moorland and forested sites are very similar, but

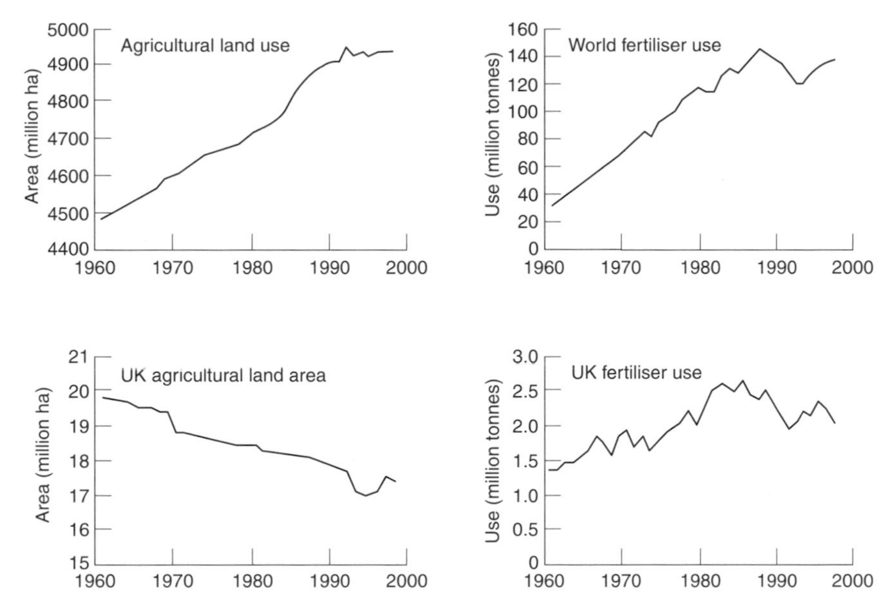

Figure 6.10 Area under agriculture and fertiliser application: world and UK

nutrient concentrations are considerably higher in the streams, and generally increase with age.

Like deforestation, afforestation therefore can have a very significant impact on the catchment system (drying up catchments in South Africa, for example). Also as with deforestation, the way in which afforestation is implemented has a substantial effect on its impact.

Agriculture

Land has, of course, been farmed for a very long time, and across much of intensively-farmed Europe, North America and Asia river flow and groundwater recharge regimes have long been different from what they would have been under "natural" undisturbed conditions. Agricultural practices are, however, changing very significantly, and these changes are having impacts on the hydrological system. In the developed world these changes include increasing use of agricultural chemicals and, at least in some areas, return of agricultural land to non-agricultural uses (known in Europe as "set-aside"). Agriculture is also changing in parts of the world currently under less intensive farming. Here, the crops grown, the methods of cultivation and the use of chemicals are changing. Figure 6.10 shows trends in the area under agriculture and fertiliser application at the global and UK (indicative of many developed countries) scales.

Agricultural practices can affect both the quantity of water and its quality. There are three main dimensions of agriculture which influence the quantities of water cycling through a catchment: the type of crop grown, the physical

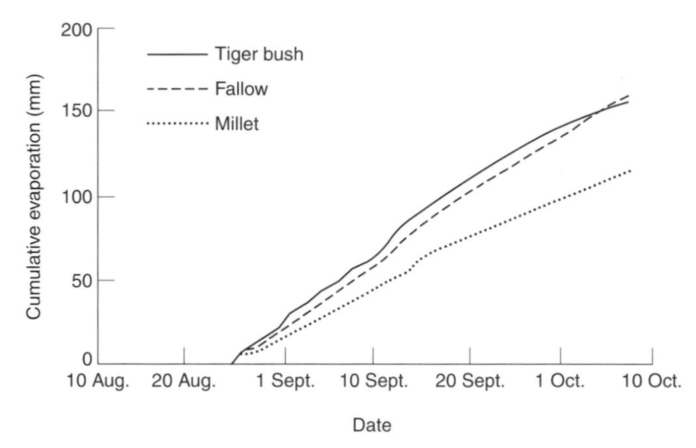

Figure 6.11 Cumulative evaporation from fallow land, tiger bush and cultivated millet: Niger (Gash *et al.*, 1997)

activities and works associated with cultivation, and the abstraction of water for irrigation (this is explored in Section 6.4). The type of crop grown has two primary effects. First, different crops use different amounts of water over different growing seasons. Transpiration rates are therefore different, and seasonal transpiration and evaporation totals may also be different. Figure 6.11 shows the cumulative evaporation during the growing season from fallow land, tiger bush and cultivated millet at a site in Niger, west Africa (Gash *et al.*, 1997). Over the growing season, evaporation from the millet was 22% less than that from the other vegetation types, and the extra water infiltrates into the soil or runs off to become streamflow. Second, land may be left bare for part of the year, exposing the land surface to rainfall. Transpiration will of course be zero at these times, and infiltration capacities may be reduced, encouraging overland flow.

Cultivation also involves a number of physical activities and interventions. Ploughing alters soil structure, and compaction by machines reduces infiltration capacity and increases the frequency of overland flow. In fact, in temperate environments infiltration-excess overland flow virtually only occurs from disturbed agricultural land, particularly during winter. van Wesemael *et al.* (2000) showed how frequent tillage of almond plantations in south east Spain has led to spatial variability in soil thickness and structure, and therefore patterns of variability in infiltration and evaporation. Farm roads and paths, too, can be significant areas of runoff generation. Ziegler & Gaimbelluca (1997) found, for example, that the 0.5% of catchment area covered by rural roads in northern Thailand contributed a very large proportion of the rapid event runoff during small events. In many parts of Europe there has been an increasing "rationalisation" of small fields into smaller numbers of large fields. This has involved the removal of field boundaries and the construction of new farm tracks, and enabled the use of machinery. Figure 6.12 shows the effects of such changes near Freiburg, Germany (Bucher & Demuth, 1985): there are more short-duration flood peaks in the modified catchment, and low flows are substantially

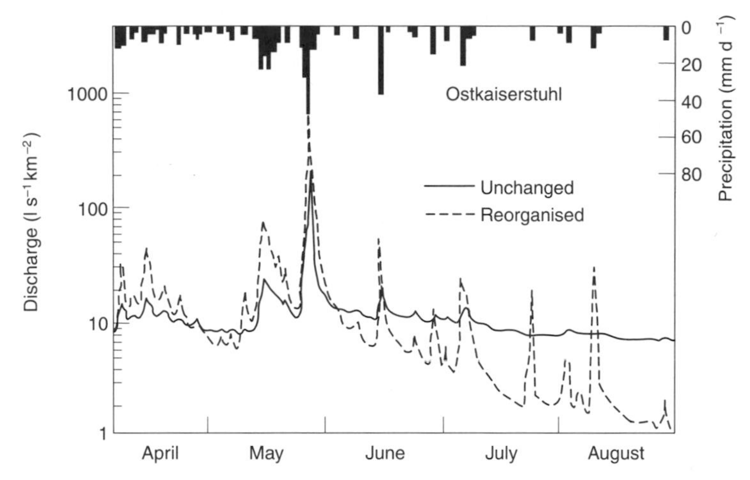

Figure 6.12 Effect of field reorganisation on runoff, Freiburg, Germany (adapted from Bucher & Demuth, 1985)

reduced. Cultivation practices do not always, however, lead to increases in rapid runoff. Manual weeding around millet in Niger, for example, breaks up the crusty soil surface which develops after soil is wetted, so rain falling immediately after weeding infiltrates rather than runs off the surface (Peugeot *et al.*, 1997). The adoption of conservation tillage techniques in the mid-west of the United States, designed primarily to reduce soil erosion, has also led to reductions in flood peaks and a lengthening of the time to peak (Potter, 1991).

However, one of the biggest effects of agricultural practices on runoff rates and recharge across much of northern Europe is through field drainage. This is necessary in order to lower field soil moisture contents to increase the growing season and allow access of machinery during the critical autumn ploughing season (Robinson *et al.*, 2000). The drains are designed to move water quickly from soil to watercourse. In the early 1980s upland field drainage was blamed by some of the inhabitants of York for causing flooding, and this claim triggered a comprehensive review of the downstream effects of field drainage. This review found that the effects of drainage depended very much on the type of soil that was drained (Robinson, 1990). Drainage of deep, permeable soils tends to increase storm flows, as water is more rapidly transmitted to the drainage network. Drainage of shallow soils, however, tends to decrease storm flows, as the drainage increases soil moisture deficits by lowering the water table.

Agricultural activities have many effects on water quality. Direct and indirect tillage practices, such as ploughing and compaction by machinery, affect soil erosion by both disturbing soil and increasing the frequency of overland flow. However, whilst erosion rates may be high, not all of this eroded material will reach a river channel, as much is stored on hillslopes and the floodplain. Much of the sediment that does reach the river channel comes from roads and paths, as found in Thailand by Ziegler & Giambelluca (1997), or through drainage ditches.

Four aspects of chemical water quality are affected by agricultural activities: nutrients, organic matter, pathogens and organic micro-pollutants (pesticides). The increasing intensity of agriculture particularly since the Second World War has led to increasing concentrations of nutrients, particularly nitrogen and phosphorus, in many rivers and aquifers in many countries: there have been many case studies and reviews (e.g. Carpenter *et al.*, 1998). A very high proportion of nitrogen in rivers and aquifers is in the form of nitrate (NO_3), and high concentrations of nitrate have been linked with human health problems (blue-baby syndrome and stomach cancer). High concentrations of phosphates can lead to eutrophication. Agriculture is not the only source of nitrogen (there are rain water inputs and contributions from sewage effluent: Section 6.4), but in lowland catchments it is typically the most important (Burt & Johnes, 1997). Nitrogen, and hence nitrates, come not only from the application of chemical fertilisers but also from animal waste, and nitrogen is released when grassland is ploughed. Only a relatively small proportion of the potentially available nitrogen is actually removed from the soil by water, as most is taken up by plants or immobilised on soil particles. Nitrate concentrations increase with flow, as nitrates are washed out of the soil, and concentrations also tend to be higher in winter when plant uptake is at its lowest (Figure 6.13a): field drains increase the rate of removal of nitrates from fields to rivers. Concentrations are also high in the first high flow event after a prolonged dry spell. Where snowmelt is the dominant hydrological process, peak concentrations occur during spring. Nitrate concentrations in rivers with no intensive agriculture are generally less than 1 mg l^{-1}, but across much of lowland Europe and lowland agricultural North America average nitrate concentrations in rivers can be greater than 5 mg l^{-1}. Figure 6.13b indicates the range in nitrate concentrations found in five lowland British rivers: note that the European Union maximum permissible concentration for drinking water is 11.3 mg l^{-1}. Some of the nitrate available in the soil is leached to groundwater, and it can take nitrate-rich water between 10 and 50 years to reach the water table (Adams *et al.*, 2000). Nitrate concentrations in some British aquifers began to rise in the 1980s in response to the ploughing of grasslands during the 1940s. Figure 6.13c shows the profile with depth of nitrate concentrations beneath unfertilised, moderately fertilised and highly fertilised grassland (indicating that in this instance an application of 400 kg N ha^{-1} a^{-1} is clearly too much for the grassland system to assimilate), and Figure 6.13d shows the downward movement of nitrate-rich water (Adams *et al.*, 2000).

Phosphate fluxes have been less intensively studied than nitrate fluxes, and the relative contributions of agriculture and sewage effluent (see Section 6.4) vary substantially from catchment to catchment. Most agricultural phosphate is exported from a catchment in particulate form, with concentrations at their highest during high flows.

When it degrades, biodegradable organic matter consumes oxygen, and a large influx of organic matter can result in total oxygen depletion – and death to oxygen-breathing organisms (such as fish) within the watercourse. Animal waste products are high in biodegradable organic matter, and intensive livestock

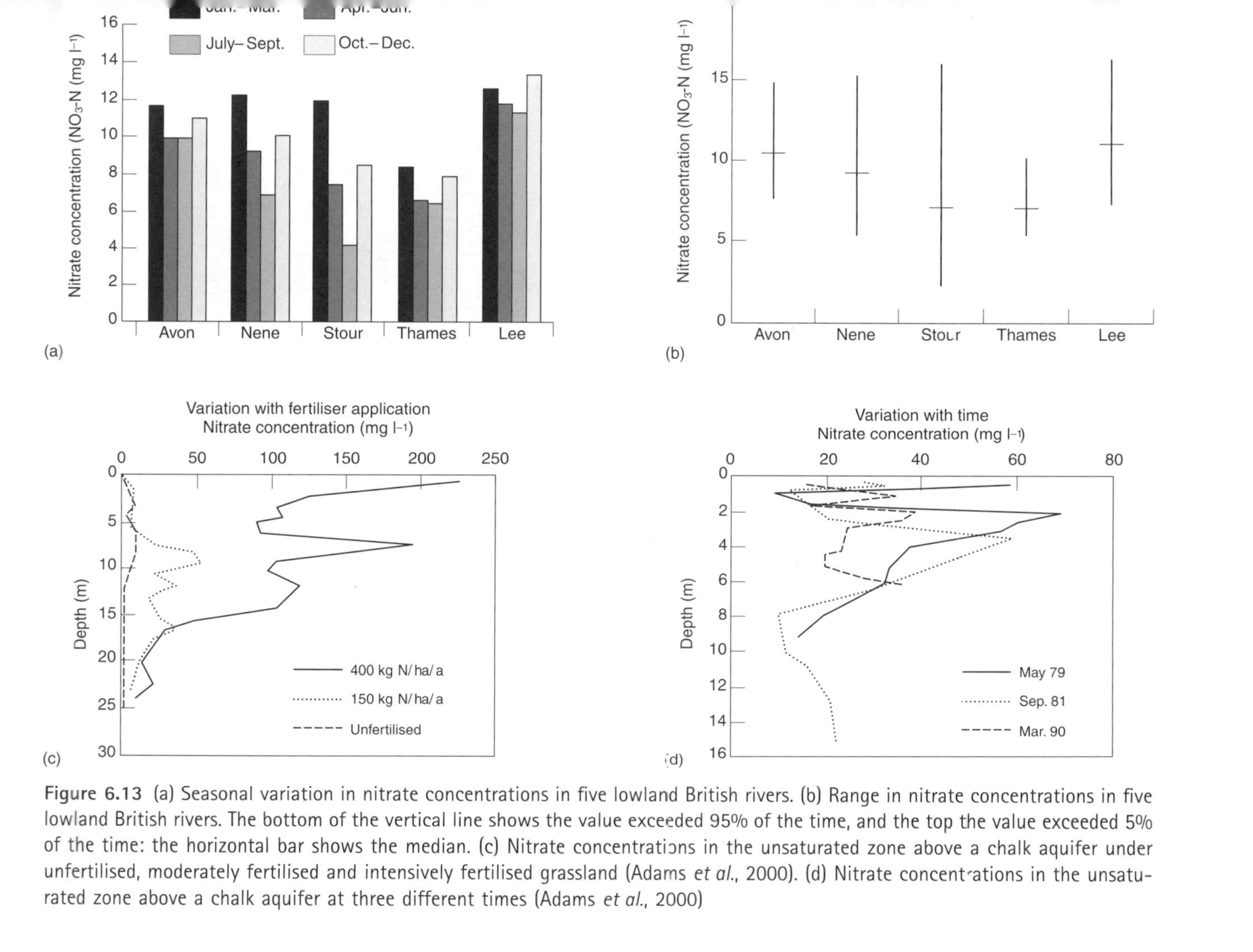

Figure 6.13 (a) Seasonal variation in nitrate concentrations in five lowland British rivers. (b) Range in nitrate concentrations in five lowland British rivers. The bottom of the vertical line shows the value exceeded 95% of the time, and the top the value exceeded 5% of the time: the horizontal bar shows the median. (c) Nitrate concentrations in the unsaturated zone above a chalk aquifer under unfertilised, moderately fertilised and intensively fertilised grassland (Adams *et al.*, 2000). (d) Nitrate concentrations in the unsaturated zone above a chalk aquifer at three different times (Adams *et al.*, 2000)

raising, when animals are indoors for all or part of the time, concentrates these wastes into a small area. Farmers store these wastes in slurry lagoons or tanks. The slurry is then often sprayed onto fields as fertiliser, and if it subsequently rains this material may be washed into a watercourse; slurry lagoons can also fail, releasing large volumes of slurry into a river. Slurry typically has a BOD (biochemical oxygen demand) of around 30 000 mg l^{-1} (BOD for natural water is usually between 1 and 2 mg l^{-1}: Chapter 5). Intensive livestock rearing also involves the storage of large amounts of animal feed, including silage (plant material compacted and stored in airtight conditions): silage often has a BOD of around 80 000 mg l^{-1}, and leakage of stored silage can have catastrophic effects on a watercourse. Dairy farming, with its high concentration of cattle, is responsible for the vast majority of agricultural water pollution incidents in England and Wales (Environment Agency, 1998a).

Microbial pathogens include bacteria, viruses, protozoa and more complex organisms that cause ill-health in humans (Meybeck *et al.*, 1989). Animal wastes (along with human wastes: Section 6.4) are a source of pathogens, again particularly where livestock are intensively reared. Faecal coliforms are indicators of faecal contamination, and Boyer & Pasquarell (1999) found concentrations in a karst aquifer in the Appalachian Mountains, USA, approximately 100 times greater below a dairy farm than beneath open pasture. A pathogen which has been causing concern in the UK and the USA is the protozoa *Cryptosporidium*, which is found in cattle waste and has been washed from pasture into streams feeding public water supplies.

Since the 1970s the application of pesticides increased substantially in many farms, and several studies in the 1980s and 1990s showed high levels of pesticides – particularly herbicides – in small agricultural catchments (e.g. Williams *et al.* (2000) in the UK, Scribner *et al.* (2000) in the mid-west of the USA, Garmouna *et al.* (1998) in France and Kreuger (1998) in Sweden). Concentrations are generally at a maximum in autumn, because herbicides are often used to clear the seedbed before planting. Pesticide concentrations are greatest, for a given rate of application, where rain water carrying pesticide residues can be transported rapidly to streams, and this happens where there are significant preferential flow pathways (Heppell *et al.*, 2000) or where drains have been laid in the field. Pesticides and herbicides have been found in shallow groundwater (e.g. in the United States: Kolpin *et al.*, 1998), but relatively small amounts of pesticide leach down to deep groundwater because the rate of transport is sufficiently slow for concentrations to be attenuated by degradation and absorption (Adams *et al.*, 2000). However, if the groundwater is recharged by rapid preferential flows, then the potential for contamination is increased. Predicting the concentration of pesticides in groundwater therefore requires an understanding of the recharge process. Pesticides are used not only in crop cultivation but also in livestock farming, particularly for dipping sheep. Until the mid-1990s organophosphorus-based sheep dip was widely used, which not only caused a number of pollution incidents but is also toxic to humans. Pyrethroid-based sheep dips have therefore become increasingly common, but whilst these are safer for humans they can be very toxic to stream fauna (Virtue & Clayton, 1997).

Urbanisation and industrial development

Urbanisation changes the physical characteristics of a catchment, and if the area urbanised is large enough, may also influence local climate (Barry & Chorley, 1998: Chapter 8). Changes in the characteristics of the catchment affect the quantity and quality of water downstream. Industrial development too – which does not necessarily take place in urban areas – impacts particularly upon water quality. The proportion of the world's population living in urban areas has increased from around one-third in 1960 to nearly 50% in 2000, and the actual area covered by urbanisation has increased correspondingly. In 1991, approximately 10% of the land area of England – or 13 780 km^2 – was covered by urban development.

Urbanisation essentially involves replacing a permeable vegetated land surface with a largely impermeable surface interspersed with small patches of vegetation. The drainage network is also substantially increased – and made permanent – and effectively includes not only storm drains, but also roads and the gutters of individual buildings. This change in land surface can have a very significant impact on the response to rainfall, and activities in an urbanised catchment affect water quality.

The most important hydrological effect of urbanisation is that rain runs off rapidly into streams rather than infiltrating, evaporating and moving gradually towards groundwater or rivers. Figure 6.14 shows a typical hydrograph upstream and downstream of an urban area: peak responses are faster and higher

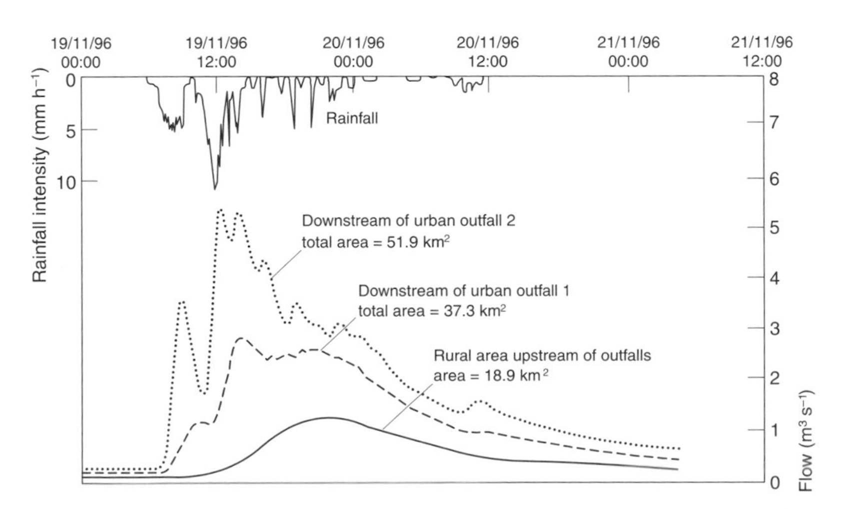

Figure 6.14 Response to a rainfall event upstream and downstream of an urban area (Robinson *et al.*, 2000)

Table 6.7 Accumulation rates for lead and zinc in two small lakes near Coventry, England (adapted from Charlesworth and Foster, 1999)

	Zinc (kg ha^{-1} a^{-1})		Lead (kg ha^{-1} a^{-1})	
	Urban lake	**Rural lake**	**Urban lake**	**Rural lake**
1850–1954	3.94	1.13	1.14	0.85
1954–91	29.37	6.23	5.27	2.02

(Robinson *et al.*, 2000). However, not all floods are affected in the same way, and the greatest effects are seen on the relatively small rainfall events, with return periods less than around 10 years (Hollis, 1975). A greater proportion of the event rainfall becomes streamflow, a higher proportion of the hydrograph is composed of new water (Gremillion *et al.*, 2000), and the number of rainfall events producing streamflow increases. Land cover has a lesser effect as rainfall magnitude increases, because the amount of catchment storage available (in soil, for example), becomes very small relative to the total rainfall: a similar effect is seen with afforestation. The difference between the urbanised and "natural" catchment in terms of available soil moisture storage capacity is greatest during summer (or the dry season), when soil moisture deficits are at their maximum, and urbanisation therefore tends to have a greater effect on the magnitude and frequency of summer floods. The effect of urbanisation depends on the nature of the catchment being urbanised, with the greatest effect the more permeable the catchment. In effect, urbanisation in a temperate environment makes the catchment respond much more like a semi-arid catchment: the response is rapid, and does not depend on antecedent catchment conditions. The same rainfall event will therefore produce a similar response at any time of the year.

Rivers downstream of urban areas tend therefore to have higher peak flows and lower low flows. However, the urban environment also contains networks of pipes transporting both potable water and sewage effluent. These pipes frequently leak (especially the potable water supply pipes, as they contain water at high pressure), and this leakage can contribute both to urban baseflows and to urban groundwater recharge. Yang *et al.* (1999), for example, found that around 65% of the annual recharge of the sandstone aquifer below Nottingham, England, came from leaking mains water pipes, and an additional 5% came from leaking sewers.

Many substances accumulate on the ground in an urban catchment. Hydrocarbons and heavy metals produced by motor vehicle emissions accumulate on roads, often in particulate form, as does salt spread in winter to reduce road icing and organic matter (Ellis, 1979). This material is washed off the roads by rainfall, through the drainage network, and into rivers which therefore experience a rapid pulse of potentially highly-polluting sediment. Charlesworth & Foster (1999) measured heavy metal concentrations recorded in sediments in two small lake catchments in Coventry (Table 6.7). Heavy metal concentrations (primarily zinc, lead, cadmium, nickel and copper) were between 3

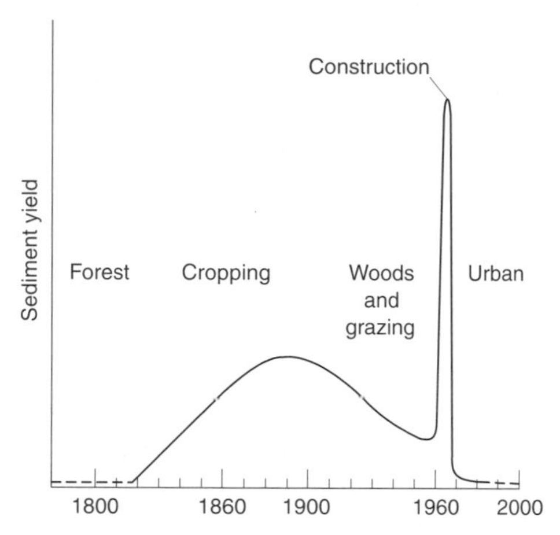

Figure 6.15 Sediment yield associated with land use change in the Middle Atlantic region of the USA (adapted from Wolman, 1967)

and 5 times higher in the more urbanised of the two lakes, but even in the "rural" lake heavy metal concentrations had increased substantially (by factors between 3 and 6) over the past few decades due primarily to material washing off roads. Material also builds up in the storm drainage network, and may be flushed out by particularly heavy rainfall. This may lead to sudden and catastrophic deoxygenation of the receiving waters. In most older urban areas, storm drains feed into the foul sewer network, and storm water goes to treatment works. However, if the volume of storm water is greater than the design standard of the combined storm and foul sewer system, then the sewer system will overflow ("surcharge"), and the storm water will be contaminated with raw sewage. Herbicides and pesticides are also used in urban environments, in private and public gardens and as part of roadside maintenance, and find their way to watercourses, as does the illegal discharge of waste down storm drains and material from spillages (Smith (1998) gives an example of leakage from a petrol station in Canberra, Australia, that contaminated urban groundwater) and road accidents. In urban areas in developing countries with inadequate sewerage systems, human waste and other organic material may get into rivers and groundwater. Finally, many urban and surrounding areas contain landfill sites, in which domestic and industrial waste is buried. Water may leach organic and toxic materials out of landfill sites, which make up around 30% of all confirmed and suspected sources of groundwater contamination in England and Wales, more than twice as many as the next most polluting activity (Environment Agency, 1998a). Urbanisation, then, can severely degrade water quality in many ways.

As is the case with afforestation and deforestation, the *process* of urbanisation has a significant impact on hydrological response, and particularly on water quality and sediment yields. In a classic study, Wolman (1967) produced a model of sediment yield associated with urbanisation (Figure 6.15): the construction process disturbs land and generates large sediment pulses.

Some of this degradation is due to industrial activities within the urban environment, but not all industrial development takes place in urban areas. Apart from the deliberate discharge of possibly contaminated water used in the industrial process (Section 6.4), water pollution from industry takes two forms: material washed off industrial sites and transported to the river either through storm drains or as overland flow, and materials leached from land contaminated by industrial processes. "Indirect" water pollution arising from industrial air pollution is discussed in Chapter 7. Very significant toxic heavy metal pollution occurred, for example, after the failure of an industrial storage pond in 1997 in the Guadiamar River, Spain (Cabrera *et al.*, 1999), and a spill at the Sandoz chemical factory in Basel, Switzerland in 1990 led to significant pollution in the Rhine as far downstream as the the Netherlands. Drainage of contaminated land and the subsequent pollution of aquifers is of increasing concern in many old industrial areas: contaminants often take the form of toxic heavy metals.

Mining and mineral extraction

These activities can have a very significant impact on water quantity and quality in many ways. It is useful to distinguish between deep mines and surface mineral extraction. Deep mines affect the quantity and quality of water in local rivers largely through the effects of mine drainage. During mining, it is often necessary to lower the water table by pumping water out. This can cause regional lowering of the water table, affecting groundwater discharge, and the pumped water is discharged directly to rivers. The quality of the pumped water is generally reasonable, as it is essentially ordinary groundwater. However, once mining has stopped and the pumps been switched off, the water table begins to rise again. This affects groundwater contributions to streamflow, but the greatest impact is on water quality. As the water table rises, water fills void spaces and the available soluble materials dissolve rapidly (Adams *et al.*, 2000), leading to increased solute concentrations and potentially toxic levels of some materials in water draining out of the mine. The most polluted mine water discharge in Britain came from the Wheal Jane tin mine in Cornwall in the early 1990s. The precise effect of ceasing pumping, however, depends significantly on the geological characteristics of the mine. The timing of the effect also varies, depending on the time taken for the water table to recover and reach soluble material. In the UK, this is typically a decade or more (Adams *et al.*, 2000), but can be shorter if large amounts of water were being pumped. In general, solute concentrations are highest immediately after mine water appears in streams, and decline after approximately five times as long as it took for the water to appear, to around 10% of the initial concentration, until the supply of soluble material is exhausted (Younger, 1997). Deep mining can also lead to land subsidence which alters surface topography and the drainage network, and sediments can of course be washed off the surface workings of deep mines – including spoil heaps – leading to degradations in stream water quality.

Shallow mining and gravel extraction have a much more obvious effect on the surface landscape, and therefore a much more direct impact on streamflows, local groundwater flows and water quality. As with deforestation – but more so – vegetation is cleared and the land surface exposed to rainfall. Downstream pollution may be very significant. In Latin America, for example, it is estimated that up to one million artisanal miners have together released around 5000 tonnes of mercury into the aquatic environment whilst mining for gold (UNEP, 1999).

Many of the sediments generated by mining activities are deposited in the river channel. These deposits may provide a source of polluting material for many decades after mining has ended.

6.4 Effects of changes in catchment water use

Water may be taken from a river or aquifer ("abstracted") for offstream use by domestic and municipal users, industrial users and agricultural users. A distinction is generally made between consumptive and non-consumptive uses. Non-consumptive users return the water to a river after use, whilst consumptive uses "consume" the water as evaporation. Returned water may contain waste materials, or may have different physical or chemical characteristics from the water originally abstracted.

At the global scale, abstractions of water are increasing rapidly (Figure 6.16), with agriculture – particularly irrigation – being by far the dominant user of water. The rate of change and the relative importance of different types of abstraction vary significantly between countries. Agricultural abstractions represent only 1% of the amount of water withdrawn in England and Wales, for example, with 53% being abstracted for domestic and municipal purposes (Environment Agency, 1998a). Total abstractions in England and Wales are

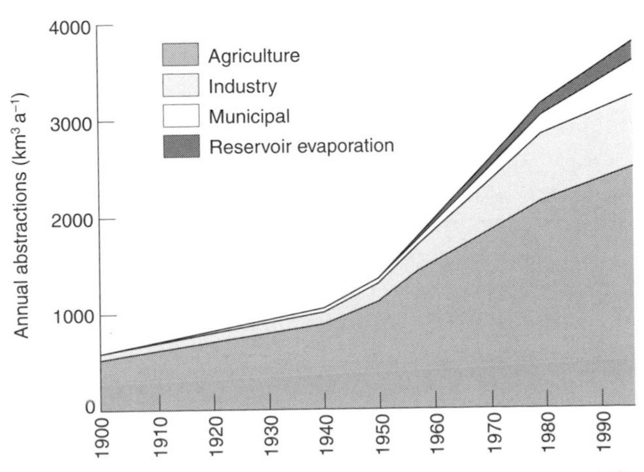

Figure 6.16 Global abstractions of water (adapted from Shiklomanov, 1998)

increasing, but the amount withdrawn by industry is falling, meaning that in some areas total abstractions are falling. The same pattern applies in many industrial countries.

Domestic and municipal abstractions and returns

"Domestic and municipal" abstractions provide water to households, shops and businesses for drinking, washing and flushing toilets. Virtually all of this water is returned to a river after use. In developed countries an increasing proportion of domestic and municipal supplies is used for garden watering: this is a consumptive use.

In developed countries, most domestic water is provided through public water supply systems, owned and operated by local authorities or the private sector, although some domestic consumers in rural areas abstract their own water from wells or springs. In developing countries, public water supply systems are often only found in towns and cities, and people in villages or small towns abstract their own water. As a general rule, individual abstractions have little effect on the quantity of river flow or groundwater recharge, as the amounts withdrawn are small. Abstraction for public water supply, however, has a greater potential for hydrological impact. Supplies can be taken either from reservoirs, direct from rivers or from groundwater (in England and Wales the three sources are used in approximately equal proportions). The effects of building reservoirs and abstracting water from them are relatively easy to identify, as the downstream effects will be determined by the operating policy of the reservoir operator and the total amount withdrawn. Similarly, the effects of direct abstractions from rivers are relatively easy to identify – and indeed avoid. It is, however, rather more difficult to predict and evaluate the effects on streamflows of abstraction from groundwater.

Although it has been known for many years that groundwater abstraction can have significant effects on river flows (Downing, 1993), awareness of the extent of the problem in Britain was dramatically increased during the drought of 1988–92 (Acreman *et al.*, 2000). Environmental pressure groups identified 78 rivers as being affected by "overabstraction", mostly for public water supply (Biodiversity Challenge Group, 1996). The National Rivers Authority identified a smaller number of "low flow problem" rivers across England and Wales (NRA, 1993) and instigated studies to ascertain the cause of recent low flows: were they due to drought or activities in the catchment (Box 6.3)? In most cases, groundwater abstraction was found to be increasing exposure to drought during dry periods, and programmes were put in place to reduce abstraction. Groundwater abstraction can lower streamflow by reducing recharge, but also affects flows by inducing lower groundwater levels around a well and thereby altering hydraulic gradients (Figure 6.17). The effects of a given abstraction therefore depend not only on how much water is being abstracted, but on the hydraulic properties of the aquifer and the position of the well relative to the river. If abstraction is greater than recharge, then water levels in the aquifer will progressively decline.

Box 6.3 A low flow problem: the River Pang, Berkshire

The River Pang drains a catchment area of around 170 km^2 just to the west of Reading, Berkshire. The catchment is almost entirely underlain by chalk, and water has been abstracted from the chalk for public water supplies since the mid-1960s: by the early 1990s abstraction averaged around 20 megalitres per day, or approximately 35% of the "natural" average streamflow. During the drought of 1988–92 river levels in the Pang fell to record levels and were low for prolonged periods. The upstream part of the river dried up, and the lower flows in the lower parts of the river meant that sediment, coming predominantly from small tributaries underlain by sands and clays, was deposited on the river bed, covering trout spawning grounds. Trout populations therefore reduced – the River Pang has an important trout fishery – and there were therefore not only hydrological impacts, but also ecological and indeed aesthetic impacts. The water supply company reduced abstractions substantially in 1992 to around a third of earlier levels, and flows in subsequent years recovered (abstractions ceased entirely in 1997 as nitrate concentrations in the aquifer became too high).

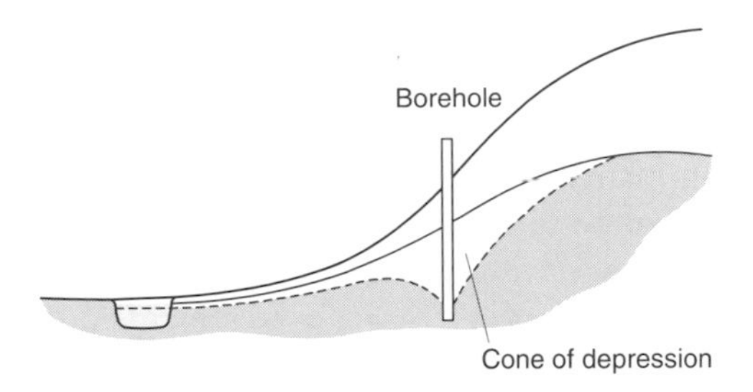

Figure 6.17 Effect of groundwater abstraction on streamflow

Water that is abstracted for public supply is rarely provided directly to consumers: it usually has to be treated in some way to remove various sorts of contamination. First, sediment must be settled out of the abstracted water, and the water may be disinfected using chlorine. Other chemicals may be added to aid clarity (aluminium sulphate) or dental health (fluorine). Excessive colour may need to be removed, as might potentially harmful chemicals such as nitrates and pesticides. Treated water is often termed potable water to distinguish it from raw water, and as a general rule groundwater usually needs less treatment than surface water.

Water abstracted for domestic and municipal use has to be returned to the catchment system, ideally in a way which does not degrade the receiving watercourse. Waste is usually discharged through a public foul sewer network or household systems such as latrines and septic tanks. Water passed through a

Table 6.8 Percentage of domestic and municipal waste water treated to primary, secondary and tertiary standards (adapted from World Resources Institute, 1996)

Country	Level of treatment			
	None	Primary	Secondary	Tertiary
Austria	28	2	42	28
Sweden	5	1	7	88
UK	13	14	62	12
Spain	41	15	40	0.4
USA	28	11	34	27
Turkey	94	1	5	0.5

foul sewer network is ultimately discharged to a river (or, in coastal communities, often to the sea). This sewage effluent will contain human wastes, organic material, washing powders, and a host of chemicals used around the house and in shops and businesses, together with traces of any chemicals added before the water is supplied. The effluent ("raw sewage") may be discharged directly into the receiving watercourse, degrading river water quality. Raw sewage can have a BOD of around 300 mg l^{-1}. In developed countries, most sewage effluent undergoes some form of treatment before being discharged. Primary treatment consists of just the removal of solid matter from the effluent (BOD lowered to around 150 mg l^{-1}). Secondary treatment uses various biochemical processes to remove much of the material held in solution and lower BOD (to around 20 mg l^{-1}), whilst tertiary treatment adds additional chemical processes to remove more contaminants (and lower BOD to around 10 mg l^{-1}). Table 6.8 shows the proportion of population served by primary, secondary and tertiary treatment at various dates during the 1990s in a sample of countries.

Immediately downstream of a sewage treatment outfall water quality deteriorates, depending on the amount of treatment the sewage has received and the dilution of the effluent in the receiving water. Quality improves with distance downstream as biochemical processes in the water degrade some materials (such as ammonia) and re-aeration causes oxygen levels to recover.

During the 1990s concern increased significantly over two components of municipal sewage. Whilst most nitrates in stream water come from fertilisers, substantial proportions of the *phosphorus* load may come municipal sewage. Phosphorus is the key nutrient in eutrophication, and is found not only in fertiliser but also in many domestic detergents. In some catchments, sewage effluent may be the major source of dissolved phosphorus (Williams *et al.*, 2000). In dry years, half the phosphorus load in the Murray-Darling catchment in Australia comes from sewage treatment effluent, but in wet years increased runoff from farmland means that sewage effluent only provides 10% of the phosphorus load (Smith, 1998). *Endocrine disrupters* are chemicals which adversely affect the health, growth and reproduction of a wide range of organisms by influencing the endocrine system. Since the early 1980s there have been concerns that oestrogenic substances in sewage effluent are leading to increased frequencies of "feminised" male fish downstream of sewage treatment works (Williams *et al.*, 2000).

Finally, sewage treatment works increase streamflow by discharging into rivers. Effluent discharges may represent a large proportion of total streamflow, particularly during dry periods – offsetting some of the more direct effects of urbanisation on low flows. For example, during dry conditions more than 60% of the flow of the River Soar in Leicestershire, England, is treated sewage effluent.

Industrial abstractions and returns

Industrial abstractions include water taken out in order to complete an industrial process (to make steel or process food, for example) or as a coolant (in power generation, for example). The proportion of "process" water that is consumed varies with industrial process, and the proportion of "cooling" water that is consumed depends on how the cooling occurs. Open cooling systems lose the water by evaporation to the atmosphere, whilst closed cooling systems recycle cooling water and return it, perhaps at higher temperature, to the river.

The vast majority of industrial abstractions are made by the users of the water directly, either from groundwater or from rivers. An increasing problem in many old industrial cities which overlay aquifers – including London and Birmingham – is rising groundwater tables caused by a *reduction* in industrial abstraction (Adams *et al.*, 2000). Industry in such cities previously abstracted groundwater from their own wells, and over decades groundwater levels fell. The groundwater level below Trafalgar Square in London fell by 65 m between 1846 and 1963 (Environment Agency, 1998a). However, during the late twentieth century much of this city industry closed down or moved away, and the amount of water abstracted fell. Groundwater levels therefore began to rise, and had risen by 20 m by 1986. This has implications for buildings and infrastructure in urban areas, but also has potential effects on streamflows and water quality: the rising water table may reach and mobilise pollutants.

Industrial processes release effluent into watercourses. Historically much industrial effluent has been released untreated into rivers, but this is increasingly rare. Effluent in developed countries at least is now either treated before discharge by the industrial user or sent as "liquid trade waste" through the foul sewer system to sewage treatment works. The contents of industrial effluent vary substantially between industries, but include heavy metals, and natural and synthetic organic pollutants. More than half the heavy metal load of the Ruhr River in Germany, for example, comes from industrial effluent, mostly treated in municipal sewage treatment works (Meybeck *et al.*, 1989). In Britain, high pesticide concentrations are found downstream of textile industries in the Pennines (Environment Agency, 1998a), and trade effluent in parts of the Midlands contains chemicals used in dyeing and the metal plating industry. Water used in cooling is frequently discharged back into rivers at higher temperature. This "thermal pollution" can raise water temperature by several degrees.

Agricultural abstractions and returns

Agricultural abstractions are virtually all consumptive. The vast majority of water abstracted for irrigation is evaporated (either from the crop or the soil), and does not return directly to the river. Irrigation schemes range from single-farmer wells to large reservoir systems on major rivers. Irrigation has had significant impacts on river flows and groundwater levels in many countries. Overabstraction of groundwater, for example, has led to significant wetland loss in Spain and many Mediterranean countries, and fossil groundwater is being "mined" in many areas, including Libya and the Great Plains of the USA. Major irrigation schemes have also led to the significant degradation of two major river systems, both frequently cited as major environmental disasters. The Colorado River in the western United States receives virtually all of its inflows from snowfall and snowmelt in the Rocky Mountains. The river then flows broadly southwards through a desert and into Mexico, where it enters the sea through an important and ecologically-rich delta. Throughout its course, however, the Colorado River is extensively used for large-scale irrigation, with major transfers out of the catchment. Figure 6.18a shows annual flows at three locations along the Colorado. The longest record shown is for a site on the river close to where it forms the border between California and Arizona. The first large decline in flows between 1935 and 1940 occurred because water was accumulating in Lake Mead behind the Hoover Dam; the decline since the 1950s has been due to increasing abstraction from the reservoirs along the Colorado, and flows have been very consistent since the completion of Lake Powell in 1963. The other two records are further downstream, above and below Imperial Reservoir, close to the border with Mexico: in most years, around 90% of the inflows to Imperial Reservoir are abstracted and used for irrigation in Arizona and California. The Aral Sea in central Asia is an enclosed sea with no exit, supplied with water by two rivers – the Amu Darya and Syr Darya – both fed, like the Colorado, by snowfall and snowmelt in their headwaters. During the 1950s and 1960s, irrigation, primarily for cotton and rice, increased significantly and the area under irrigation doubled between 1960 and 1990. Both cotton and rice require large volumes of water, and by 1975 only between 10 and 20% of the average inflows were reaching the Aral Sea (Micklin, 1988). The Syr Darya dried up completely between 1974 and 1986, and during the 1980s so did the Amu Darya. The Aral Sea has therefore shrunk significantly (Figure 6.18b, c), and is projected to continue to do so.

Once abstracted from the river or groundwater, irrigation water is distributed across the land surface by a variety of methods ranging from flooding through spraying to trickling. In arid and semi-arid areas an excess of irrigation water can lead to increased salinisation of the soil (known as wetland salinisation to distinguish it from the dryland salinisation associated with deforestation as discussed in Section 6.3). This primarily occurs because the irrigation water contains dissolved material, which is left behind when water is transpired by crops. The application of irrigation water with a salinity of 500 mg l^{-1} can add between 3 and 12 tonnes per hectare of salts per year (Smith, 1998). The problem is compounded by inefficient application of irrigation, which results

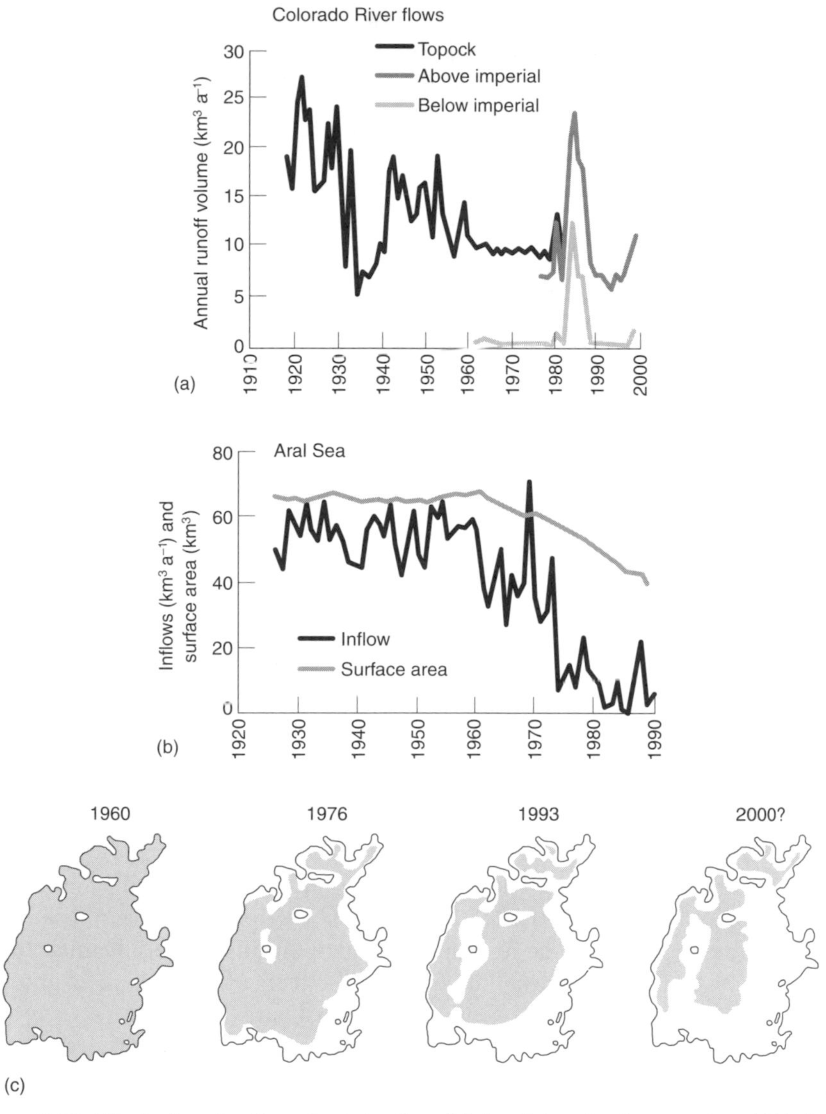

Figure 6.18 Effect of abstractions for irrigation: (a) the Colorado River, western USA, (b, c) the Aral Sea in south central Asia. Colorado River flow data from the US National Water Information System; Aral Sea inflows and surface area from Gleick (1993), and Aral Sea extent from Glantz *et al.* (1993)

in excess water infiltrating to groundwater and causing the water table to rise – bringing with it salts held at depth. Depending on the method of application, as much as 50% of applied irrigation water may add to groundwater recharge. Salinisation of irrigated soils has been recognised as a problem for hundreds of years, and a substantial proportion of semi-arid irrigated land is damaged by excessive salinity. This not only has significant impacts on crop productivity, of course, but also affects the quality of any water that drains from irrigated land into rivers. This "return flow" may also contain high concentrations of agricultural chemicals and suspended sediments. Smith (1998), for example, describes the high nutrient, pesticide and herbicide concentrations in water draining from irrigated areas in Australia, particularly in the Murray-Darling Basin.

Impoundment and transfers

The previous three sections have looked at particular uses of water – domestic and municipal, industrial and agricultural – and their impacts on the water balance, the distribution of flow over time and water quality. Some of these users take water from reservoirs or transfers between basins, but not all reservoirs and transfers are designed to supply water to offstream users. Reservoirs can be used to generate hydropower, control a river's flows to maintain navigation, or protect downstream communities against flooding. These functions impact upon streamflows and water quality. In the UK, the widespread and systematic development of impounding reservoirs began in the eighteenth century, reaching a peak in the 1960s (Gurnell & Petts, 2000). These reservoirs, of varying sizes, were largely built to provide water supplies to towns and industry, but a large number of relatively small reservoirs supply water to Britain's canal network. There are over 450 "large" dams in the UK, 80% of which are in the uplands (Gurnell & Petts, 2000), and several inter-basin transfers: Liverpool and Birmingham both receive water from Wales, and Manchester obtains its water from the Lake District. At the global scale there are currently around 45 000 large dams (World Commission on Dams, 2000), defined as dams more than 15 m high or storing over 3 million m^3 of water: almost half of these are in China. These dams store around 6000 km^3 of water (Shiklomanov, 1993: the estimated storage in natural lakes is of the order of 100 000 km^3: Chapter 2), and it is estimated that 46% of the world's 106 primary major basins are modified by at least one large dam (World Commission on Dams, 2000). An overview of the environmental and social effects of large dams is given by Gleick (1998) and the World Commission on Dams (2000).

The primary hydrological effect of a reservoir is to alter the timing of streamflows, with the effect depending rather obviously on the way the reservoir is operated. Reservoirs tend to lessen flow variability downstream, reducing the high flows and increasing low flows. Diversions from one river basin to another have a similar effect. Figure 6.19 shows the rather extreme effect of river regulation by impoundment and diversion on the River Murray in Australia (Smith, 1998). Thoms & Sheldon (2000) describe the effects of nine headwater dams and 15 main channel weirs (plus 267 agricultural water abstractors) on flows in the Barwon-Darling River, Australia: mean runoff has fallen by 42%, small floods (return period less than 2 years) have reduced in magnitude by between 35 and 70%, the seasonal variation in flow has altered, and flows have become less variable. In areas with high evaporative demand, substantial amounts of water may be evaporated from the open surface of reservoirs that would not necessarily have been evaporated from bare soil, or indeed may not

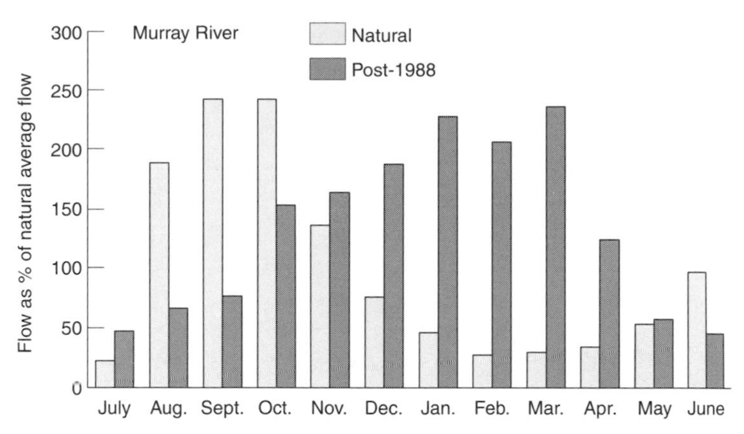

Figure 6.19 Average monthly runoff on the River Murray, Australia, before and after impoundment and diversion (adapted from Smith, 1998)

have been available for evaporation until it was transported from up-river. In the mid-1990s, evaporation from reservoirs at the global scale was 188 $km^3\ a^{-1}$, or approximately half the total amount of water abstracted for domestic and municipal use (Figure 6.16).

Evaporation and abstractions from reservoirs in semi-arid areas can together significantly reduce the volume of flow downstream. The Hadejia-Nguru wetlands in northern Nigeria, for example, are seasonally inundated by the Hadejia and Nguru rivers. During the 1970s, however, a major reservoir was built on the Hadejia River for irrigation purposes. Prior to the construction of this dam around 2000 km^2 of wetland was flooded annually: by the 1970s this had reduced to 1500 km^2 with even greater reductions during dry years (when even less was released from the reservoir). Full implementation of the proposed reservoirs and diversions would cause even greater reductions in wetland flooding – and the subsequent groundwater recharge (Thompson & Hollis, 1995).

Reservoirs have several effects on water quality. First, they trap suspended sediments, which not only fills the reservoir but also starves the river downstream, frequently leading to river channel change. It is estimated (Meybeck *et al.*, 1989; Milliman & Meade, 1983) that approximately 1500 million tonnes of sediment is trapped each year behind dams, representing approximately 10% of the global sediment discharge to the ocean, or between 0.5 and 1% of reservoir storage capacity lost each year (World Commission on Dams, 2000). This trapped sediment can hold adsorbed materials such as phosphates. Second, releases come from the bottom of the reservoir, and this water tends to be relatively cold and possibly with low oxygen concentrations.

6.5 Effects of physical changes in the river network

The preceding sections have shown how human activities affect the amount, timing and quality of water reaching the river channel. Changes to the river

Table 6.9 Percentage change in average annual flood peak magnitude after channel modification, and the capacity of the modified channel (adapted from Robinson, 1990)

	Increase in average annual maximum flood (%)	**Return period of new channel capacity (years)**
Ock, Oxfordshire	14	10
Barlings Eau, Lincolnshire	34	10
Witham, Lincolnshire	74	50
Ewenny, Mid-Glamorgan	88	100

Note: a "natural" channel has a capacity close to the magnitude exceeded once every two or three years

channel itself can also affect hydrological regimes. Channel modifications include channel realignment, relining the channel with some other material, and excavations within the channel (Sear *et al.*, 2000). These changes to the channel are done in order to prevent inundation of the floodplain, increase navigation potential, extract gravel and, increasingly, to restore river channels to their earlier form. The main physical effects of channel modifications include not only the obvious changes to the channel itself, but also the separation of the channel from the floodplain.

River channel modifications that change the shape of the channel are generally designed to change the shape of the flow hydrograph – to move water more quickly and to prevent it overflowing on to the floodplain. The channel works themselves alter the timing of streamflow, whilst the separation of river from floodplain alters the total volume of flood flows. As a general rule, the hydrograph peak increases, and the lag-time between rainfall and response falls. Unlike with deforestation, afforestation or urbanisation, the greatest effect of separating the river from its floodplain tends to be noticed for larger events (at least up to the design capacity of the modified channel), because the effect of the lost storage is greatest. The greater the change in capacity of the modified channel, the greater the effect on flood magnitudes. Table 6.9 shows the change in the average annual flood peaks in four different channel modification schemes in Britain (Robinson, 1990). The lag-time on the Witham reduced from between 20–26 hours before channel modification to around 16 hours after.

The process of channel modification creates increased sediment loads, due partly to the physical disturbance and partly to erosion by the modified channel. Brookes (1987), for example, found increases in sediment loads immediately downstream of channelised reaches of around 40%, and Sear *et al.* (1998) measured short-term increases of up to 150% downstream of a channel restoration project. However, a proportion of this extra sediment is deposited further downstream, and the effects of channel modification on catchment-scale sediment yields are currently not clear.

Flood protection is sometimes provided by building flood embankments out of the river channel and leaving the channel unmodified. This does not affect sediment loads and small high flows (those which do not cause out-of-bank flooding), but does affect the flows which would normally have spilled onto the floodplain.

Both channel modification and the construction of flood embankments alter the relationship between river and floodplain. Under "natural" conditions, the floodplain may be inundated to some degree most years, and some floodplain wetlands are much more closely connected with the river. Changing the frequency of inundation will affect floodplain vegetation, as will reducing river water levels and lowering the floodplain water table. Floodplain wetlands have many hydrological and ecological functions, including providing a storage for water and sediment, playing a major role in nutrient and carbon cycling within the catchment, and acting as a "buffer" zone between the hillslope and channel. Sear *et al.* (2000) claim that the loss of floodplain wetlands and associated storage functions, due largely to channel modifications, has profoundly affected the hydrology (in terms of quantity and quality) and geomorphology of UK river channels.

6.6 Overview

This chapter has reviewed the main effects of a wide range of human interventions in the catchment on water quantity and quality. The catchment system is exposed to many different pressures, and indeed many forms of environmental degradation are associated with hydrological processes in the catchment. There are two main points to make in conclusion.

First, many activities are going on in a catchment, and the hydrological characteristics of a catchment may also reflect the legacy of past activities. The catchment therefore integrates the effects of a range of human interventions over both space and time, and these effects are superimposed on catchment with geographically-variable properties exposed to variations in inputs over time. Different human interventions also operate at different time and space scales. Some interventions simply alter the timing of streamflow, and their effects are not seen when data are averaged over longer periods. Some interventions have greatest effect at the field or small catchment scale, but are swamped by others at larger scales. The effects of field drainage, for example, are often masked further down the river network by the effects of modifications to the river channel. More generally, the effects of land cover and land use changes on the volume and timing of streamflow tend to become less apparent the larger the catchment, as a proportionally smaller part of the catchment is affected. Changes in the hydrological regime of large catchments (greater than around 5000 km^2) tend to be more influenced by direct human interventions to the river channel, reservoir impoundment and abstractions and returns than by effects on land cover and use.

The effects of multiple interventions can best be seen in river and groundwater quality: each intervention adds its own contribution to the mixture of materials in the water. The chemical water quality of the River Thames, for example (Neal *et al.*, 2000), reflects the varying contributions of municipal and light industrial effluent and non-point pollution from agricultural land.

Box 6.4 Pressures and responses in the Soar catchment

The River Soar drains a catchment of 1380 km^2 in the Midlands of England, and has a population of 705 000 people (Environment Agency, 1998b). The catchment is mainly agricultural, but includes the city of Leicester (population 293 400) and some industrial development and mineral extraction. There are five public supply reservoirs in the catchment, and almost 400 licensed abstractions, most of which are very small. There are 28 landfill sites, and sewage effluent is treated at two treatment plants. Approximately a quarter of the total river length has been designated as a coarse fishery under European Union rules, and around 1% has "poor" river water quality (Box 6.1). The Environment Agency has produced a Local Environment Agency Plan (LEAP) for the catchment, identifying the key water (and waste) issues and potential responses. The key issues are (in no order of priority):

- coloured discharges from the textile industry: responses include to upgrade both on-site treatment and sewage treatment works, and also to reduce the use of dye;
- increasing eutrophication: monitor, then improve sewage treatment if justified;
- low flows below some public supply reservoirs: agree "compensation flows" with reservoir operators;
- loss of floodplain wetland: implement local water level management agreements;
- poor chemical and bacterial quality of some river reaches (below stated objectives): improve sewage treatment;
- dewatering for mineral extraction is lowering water tables: review planning permissions for new extractions;
- pesticides in public water supply reservoirs: liaise with farmers;
- copper contamination from circuit board manufacturers: increase on-site treatment and upgrade sewage treatment works;
- derelict and contaminated land: identify partners for land restoration.

The urban contribution tends to be diluted during high flows, but the agricultural contribution is increased as materials are flushed from fields. The water quality of the rivers entering the Humber estuary in eastern England reflects an even more varied mix of agriculture, industry (past and present) and urbanisation (House *et al.*, 1997). Phosphorus loads in the urbanised River Trent are dominated by dissolved phosphates originating from sewage effluent, whilst those in the agricultural River Swales are dominated by particulate phosphate derived from agricultural land. Heavy metals in the Humber estuary predominantly come from the Aire, Calder and Don, rivers draining old industrial areas. Box 6.4 summarises the multitude of pressures facing the water environment in the Soar catchment, Leicestershire.

Table 6.10 Some example actions to prevent or reduce human impacts on the water environment

Potential impact	Mitigating action
Increased flooding downstream of urban areas	Detention ponds to store flood runoff
Increased runoff and sediment transport following deforestation	Adopt minimum-impact methods to remove trees
High nutrient concentrations in aquifers and rivers in agricultural areas	Define Nitrate Sensitive Areas and restrict fertiliser applications. Maintain riparian buffer zones
High phosphate concentrations in sewage effluent	Install phosphate-stripping to remove phosphates
High concentration of heavy metals in industrial effluent	Install improved waste water treatment and reduce usage of heavy metals
Leakage of landfill	Seal landfill sites
Leaching from contaminated land	Implement bio-remediation

Second, the chapter has focused on the adverse effects of human intervention. In practice, of course, actions are taken to reverse, reduce or prevent adverse human impacts on the environment. Table 6.10 gives some examples of responses to human intervention: the list is not exhaustive. As a result, whilst there are many examples of continuing environmental degradation (e.g. UNEP, 1999), there are also many examples of either improvements or actions taken to prevent further deterioration. Between 1990 and 1995, for example, chemical water quality (Box 6.1) deteriorated in 12.6% of rivers in England and Wales, but improved in 40.2%, resulting in a net improvement of 27.8% (Environment Agency, 1998a). Nitrate concentrations in many British rivers and aquifers continue to rise, but concentrations of most heavy metals are declining. The number of major pollution incidents fell from around 600 in 1990 to 156 in 1996 (Environment Agency, 1998a). In other words, the overall picture is not necessarily as severe as implied by the litany of potential impacts reviewed in this chapter.

The next chapter explores the effects of changes in the climatic inputs to the catchment, and Chapter 8 looks at how human interventions may affect the global environmental system more generally.

Chapter 7

Changes to the inputs to the catchment: acid deposition and global warming

7.1 Introduction

Chapter 6 described the effects of cumulative global environmental change – similar types of changes happening in the catchment in many places for a range of reasons. This chapter looks at systemic global environmental change, which is defined as changes occurring at the global scale and affecting large geographical areas at the same time. Such changes affect the inputs of mass and energy to the catchment (Figure 7.1). The two dominant examples are acid deposition and global warming. They both directly affect mass and energy fluxes, and also indirectly affect catchment land cover.

7.2 Acid deposition

"Acid rain" became a big environmental and scientific issue during the 1970s. Forests across northern Europe and north east North America began dying, and fish were being killed in lakes across Scandinavia. This environmental destruction was blamed by environmentalists on excessively high levels of acidity in rainfall, due to the burning of fossil fuels. The increasing acidity of rainfall near industrial areas had in fact been noticed in parts of Europe since the mid-nineteenth century, but its effects were not recognised. The suspicion in the 1980s that acid rain might be leading to severe environmental damage over large areas triggered a major research effort (e.g. the Anglo-Scandinavian Surface Water Acidification Project: Box 7.1) which concluded that acid "rain" was a misnomer (dry and cloud deposition were as damaging as acidic rainfall), acidification *was* leading to significant degradation of lakes and rivers, and that the processes by which surface water become acidified were very complicated. Acidity levels in lakes and rivers were not just rising because acidic water was finding its way into rivers: chemical transformations in the catchment buffered or exaggerated the acidity in the incoming water. Figure 7.2 shows trends in pH in a set of Scandinavian lakes and rivers. An estimated 12 000 km of Welsh streams is affected by acidification (Environment Agency, 1998a). Forest dieback – attributed by environmental campaigners to acid rain – however, is not simply due to acid deposition, and factors such as drought may also be important.

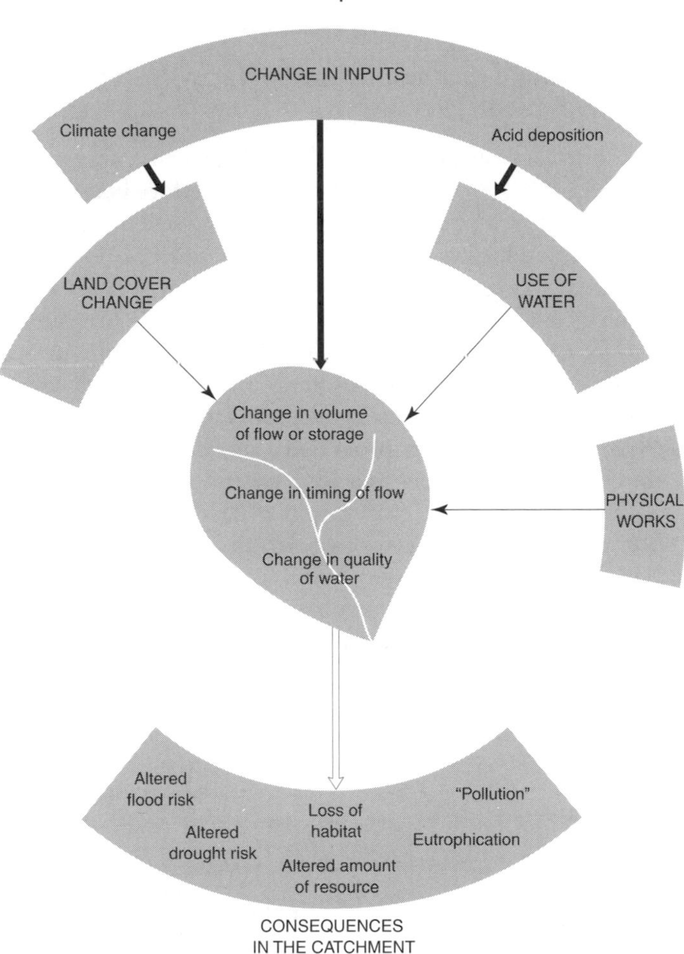

Figure 7.1 Changes to catchment inputs

There are three components to the acidification of surface waters: deposition on the catchment, chemical transformations within the soil, and hydrological processes moving acid-enriched water to the river to be mixed with other sources of water. Before looking at these three components in turn, it is necessary to summarise briefly where the acid material comes from.

Sources of material

Acid deposition is caused by two principal pollutant gases in the atmosphere, sulphur oxides (SO_x) – particularly sulphur dioxide (SO_2) – and nitrogen oxides (NO_x). These gases react chemically with moisture (H_2O), the hydroxyl radical (OH) and sunlight to form droplets of sulphuric acid (H_2SO_4) and nitric acid (HNO_3). The sulphur dioxide is formed by the oxidation of sulphur during the smelting of sulphide-based ores and the combustion of fossil fuels

Box 7.1 The Anglo-Scandinavian Surface Water Acidification Project (SWAP)

Several major research efforts into acid rain began in Europe and North America during the 1980s. SWAP was a joint effort between the UK, Norway and Sweden, aiming to determine the effects of acid deposition on fish in rivers and lakes in Norway and Sweden, and estimate the effects of defined changes in acid deposition (Mason, 1992). The project lasted for five years from 1984, involving more than 30 research groups working at six research locations in Scandinavia and four in Scotland: hydrochemical research was concentrated at four sites. The main conclusions of SWAP were that many lakes in southern Norway, Sweden and the UK have undergone progressive acidification since the mid-eighteenth century, due largely to acid deposition although afforestation enhances acidification. For a given input of acid deposition, the degree of acidification of rivers and lakes depends on the characteristics of the soil and hydrological pathways. SWAP is a good example of inter-disciplinary research, bringing hydrologists, chemists and biologists together to tackle a scientific problem.

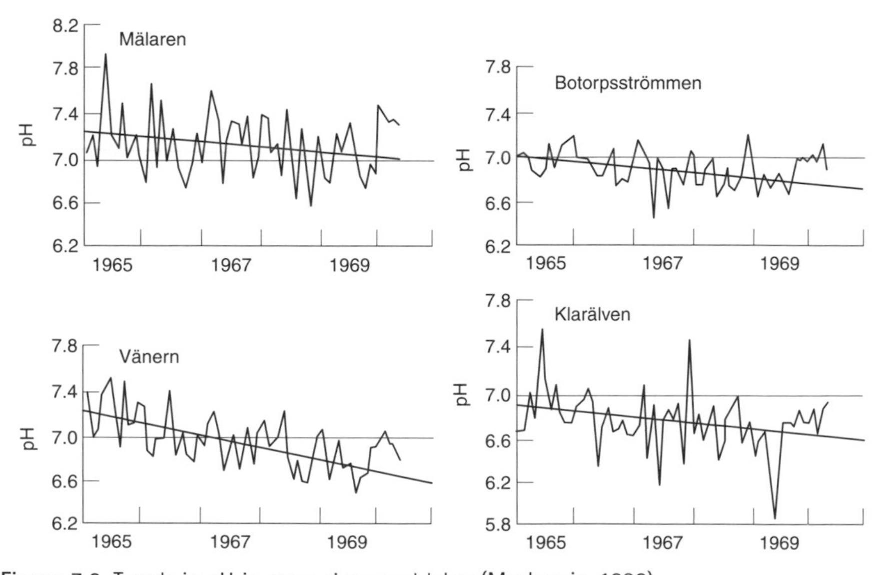

Figure 7.2 Trends in pH in some rivers and lakes (Mackenzie, 1998)

(which all contain sulphur). Nitrogen oxides, in contrast, are formed from the oxidation of nitrogen in the air at the high temperatures involved in the combustion of fossil fuels: only a small proportion of the nitrogen comes from organic nitrogen in the fuel source itself. Nitric acid can react with ammonia, producing ammonium nitrate (NH_4NO_3).

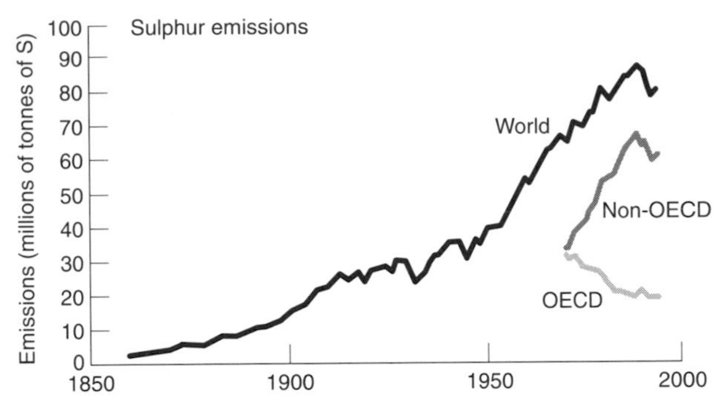

Figure 7.3 Global emissions of sulphur dioxide (as sulphur): 1860 to 1994 (data from Stern & Kaufmann, 1996, available at http://cres.anu.edu.au/~dstern/anzsee/datasite.html)

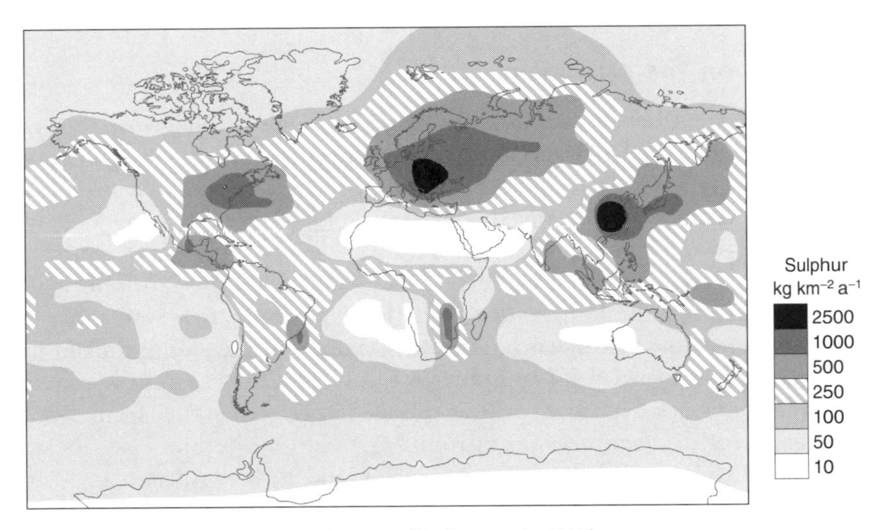

Figure 7.4 Deposition of sulphur in 1990 (Rodhe *et al.*, 1995)

Between 80 and 90 million tonnes of sulphur dioxide were released to the atmosphere each year in the 1990s from smelting, combustion of fossil fuels and biomass burning (Figure 7.3). This represents around 40% of the 210 million tonnes of oxidised sulphur (as sulphur dioxide SO_2 and sulphate SO_4) annually falling on the earth's surface (Mackenzie, 1998). Over 50% of the 56 million tonnes of oxidised nitrogen deposited each year derives from smelting, combustion and biomass burning. Sulphur dioxide, sulphate and nitrogen oxides have short residence times in the atmosphere – just a few days – so sulphur and nitrogen oxides are deposited close to areas of emission. At a regional scale, therefore, anthropogenic sources may provide the overwhelmingly dominant source of sulphur and nitrogen oxides. The vast proportion of emissions come from Europe, eastern North America and, increasingly, east Asia, and deposition of sulphur and nitrogen oxides is therefore greatest in these areas (Figure 7.4).

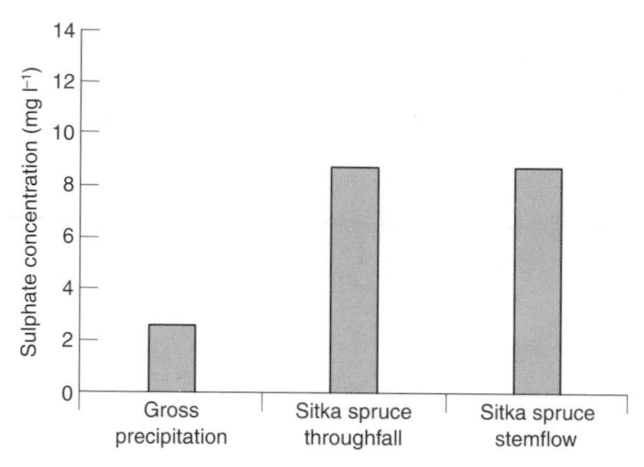

Figure 7.5 Sulphate concentrations in gross precipitation, throughfall and stemflow, for a Sitka spruce plantation in mid-Wales (adapted from Soulsby, 1997)

Deposition on the catchment

Rain water contains dissolved CO_2, so is naturally slightly acidic with an average pH of 5.7. The actual pH varies with the concentrations of other dissolved materials, and rain is only said to be "acid" if pH falls below 5. When sulphuric acid and nitric acid are dissolved in water vapour, they are separated into hydrogen (H^+), sulphate (SO_4^{2-}) and nitrate (NO_3^-) ions, and it is of course the hydrogen ions which influence the acidity of the rain water. Sulphuric and nitric acids are not the only source of hydrogen ions in rain water, but the correlation between the concentrations of sulphate, nitrate and hydrogen ions is typically greater than 0.8 (vanLoon & Duffy, 2000). Hydrogen, sulphate and nitrate ions are deposited on the surface as *wet deposition*, from rain or snow, or *occult* or *cloud deposition* from fog or mist (Section 5.3). As with other solutes, concentrations are higher in fog and mist droplets than in rain or snow, and the concentration of hydrogen, sulphate and nitrate ions in meltwater will be higher than in the newly-fallen snow. *Dry deposition* is the absorption of gases (including nitric acid, sulphur dioxide and nitrite) by vegetation. The relative importance of these different mechanisms of deposition varies geographically and with the vegetation at the surface. For a given vegetation type, the proportion of material that falls in dry deposition decreases with distance from the source of the atmospheric pollution. For Britain as a whole, dry deposition of acidity exceeds that of wet deposition, but in Sweden, downwind from pollution sources, wet deposition dominates (Mason, 1992). In a given location, the relative contribution of wet, occult and dry deposition varies with vegetation. Occult deposition is significantly greater over forests than short vegetation, typically by a factor of three to five (Table 6.4).

A proportion of intercepted precipitation evaporates, and this increases the concentration of ions in the remaining water. Figure 7.5 shows the concentration of sulphate (and implicitly hydrogen ions) in gross precipitation, together with throughfall and stemflow from a Sitka spruce plantation (Soulsby, 1997). In fact, not only is the ionic concentration increased, but throughfall and stemflow concentrate the water into particular places, leading to substantial inputs of acidic precipitation into restricted locations.

Catchment processes: from acid deposition to acidified rivers and lakes

The actual processes governing the details of soil and freshwater acidification are complex, however, and hydrogen ions are not simply "washed through" the soil into the river (Meybeck *et al.*, 1989). The impact of acid deposition on the acidity of soil and water depends largely on the buffering capacity of the soil, which is a measure of the change in acidity for a given input of hydrogen ions. It is often indexed by the acid neutralisation capacity (ANC), defined as the difference in concentration between strong bases and strong acids. Table 7.1 summarises the main processes altering the concentration of hydrogen ions in soil water (Mason, 1992). The relative importance of these processes varies significantly from location to location.

The chemical buffering mechanisms vary between soils. Carbonate-rich soils have the highest ANC and hence buffering capacity, and acid deposition therefore has minimal effect on the acidity of soil water or streams. Hydrogen ions are consumed by the dissolution of calcium carbonate. Many upland areas, however, are characterised by rock (and hence soils) with low bicarbonate or calcium carbonate concentrations, and therefore have a low buffering capacity and high sensitivity to acidification. In general, the neutralising or buffering of soil water proceeds much more rapidly by cation exchange than by mineral weathering. Hydrogen ions in the soil water are exchanged for cations (primarily Ca^{2+}, Mg^{2+}, K^+ and Na^+) held on the outside surfaces of soil particles, lowering the acidity of the soil water but increasing the acidity of the soil particles. The rate of operation of this process is determined by the cation exchange capacity, which tends to be substantially lower in thin, upland soils: again, these soils therefore have a high sensitivity to acidification. "Mobile anions" moving through soil water strip positively-charged cations from the cation exchange: sulphate is a mobile anion, and increasing concentrations of sulphate mean that more cations are leached from the soil. In well-buffered areas these cations will be primarily calcium and magnesium, but in poorly-buffered soils H^+ will be leached out. Increased sulphate concentrations therefore lead to increased leaching of soil cations.

Figure 7.6 shows the areas of the UK, Europe and North America identified as being sensitive to acidification: soil water in these areas will have increased acidity.

Biological processes significantly affect acidification of soils and soil water. Plants take up major cations as nutrients, replacing them with hydrogen ions. Respiration increases the concentration of hydrogen ions in soil water, and release of organic acids lowers pH, affecting the operation of the chemical processes neutralising acidity. Coniferous afforestation, for example, tends to result in lower ANC and increase acidity (Sullivan *et al.*, 1996).

One consequence of acidification is an increasing concentration of aluminium in soil water. Aluminium is insoluble if pH is between 5 and 9, but outside

Table 7.1 Main mechanisms affecting the concentration of hydrogen ions in soil water (adapted from Mason, 1992)

Mechanism	Most effective pH range	Effect on H^+ ions
Chemical buffering mechanisms		
Carbonate dissolution	6.2–8.0	Consumed by dissolution of $CaCO_3$
Aluminosilicate weathering	5.0–6.2	Consumed in weathering process
Cation exchange	4.2–5.0	Exchanged for base cations (Ca^{2+}, Mg^{2+}, K^+, Na^+)
Aluminium-exchanged acidity	3.0–4.2	Exchanged for Al^{3+} ions
Iron-exchanged acidity	<3.5	Exchanged for Fe^{2+} ions
Chemical processes enhancing acidification		
Oxidation of sulphur compounds in dry soils	–	Hydrogen sulphide oxidised to sulphate and hydrogen
Biological processes enhancing acidification		
Plant and microbe respiration	>4.5	Carbon dioxide dissolves to form carbonic acid, which partially dissociates into H^+ and bicarbonate
Formation of organic acids		Lower soil pH, thereby affecting other chemical processes
Plant uptake of base cations		Roots take up base cations, lowering ANC, and release H^+ into soil
Biological processes neutralising soil acidification		
Microbial reduction of nitrate (denitrification) and sulphate		These microbial processes use H^+ ions

this range it can be dissolved in water. Soil water with a pH below 5 – due to acidification – may therefore contain dissolved aluminium, which may be leached out with the aid of sulphate. Aluminium is harmful to freshwater biota. Other trace metals too may be mobilised by increasing acidity (Meybeck *et al.*, 1989).

"Acid rain", however, does not just result in changes in soil water acidity. Increasing deposition of sulphur compounds obviously increases the amount of sulphur available in a catchment. Some of this sulphur is absorbed by plants and returned by decay, but much – often up to 50% (Howells, 1990) – is stored in the soil in non-soluble forms. Nitrogen loads in a catchment are also increased, but unlike sulphur, the nitrogen cycle is very significantly affected by biological processes. Boreal forests tend to be nitrogen limited, so an increase in nitrogen deposition will increase productivity: this has been observed, for example, in Europe (Kauppi *et al.*, 1992). Excess nitrogen, however, will be

Figure 7.6 Areas sensitive to acidification (a) UK (adapted from Howells, 1990), (b) the world (GEMS-Water, 1999)

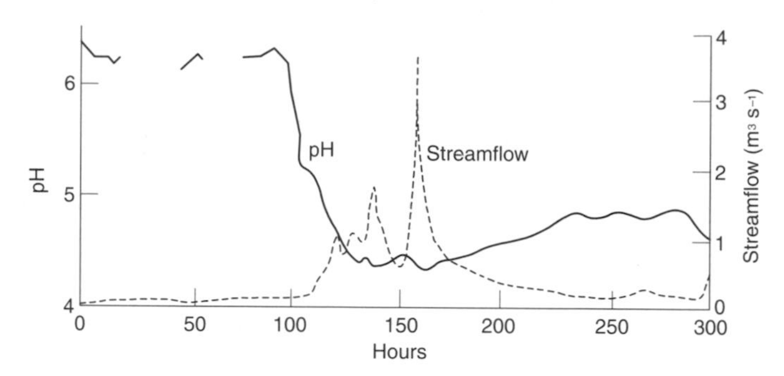

Figure 7.7 Variation in pH through a large flow event, and relationship between pH and streamflow: Plynlimon (Hill & Neal, 1997)

leached from soil and leaf litter (as shown by Wright *et al.*, 1998, for example), leading to increased concentrations of nitrogen (in various forms) in soil water.

River acidification: hydrological dynamics

River water acidified by acid deposition tends to have a pH of between 4.5 and 5.5 and high levels of dissolved aluminium, and sulphate – rather than bicarbonate – is generally the dominant anion (Meybeck *et al.*, 1989). Acid and aluminium-enriched soil water is flushed into streams during rainfall events. pH typically falls rapidly during an event (e.g. Hill & Neal, 1997; Figure 7.7), and aluminium concentrations increase by factors of between two and three (Neal *et al.*, 1997b). However, different amounts of precipitation (and hence deposition), different origins of precipitation (relative to emissions sources), different patterns of occult and dry deposition over the days before the rainfall, and different antecedent soil moisture conditions (and hence flow pathways) mean that different rainfall events can produce very different chemical responses (Soulsby, 1995). Spring snowmelt can produce large peaks of acidity. In contrast, pH tends generally to be higher when baseflows dominate and aluminium concentrations lower. Acidity at any point in time in a river therefore depends on the relative contributions of shallow flow from the surface layers of the catchment and groundwater: high groundwater contributions can reduce the effect of increasing soil acidity on river acidity.

Permanently-acidified rivers and lakes have a pH continuously below about 5, and are found in south west Sweden, southern Norway, western Britain and parts of eastern North America. Many more rivers over a larger area, however, are "episodically acidified" during high flow events.

Acid deposition also involves the deposition of sulphur and nitrogen compounds. Although not the only source, atmospheric pollution is the dominant source of anthropogenic sulphur, and it is estimated that the total transport of sulphur from land to sea in rivers has doubled from "pristine" conditions, from 60 to 122 million tonnes per year (Slaymaker & Spencer, 1998). Total nitrogen transport from land to sea has more than doubled, from 14 to 35 million tonnes per year (Mackenzie, 1998), although the relative contribution of acid deposition and other sources is not known. Acid deposition increases sulphur (primarily sulphate) and nitrogen loads in all rivers exposed to deposition, not just those sensitive to acidification.

The indirect effects of acid deposition

Although the links between acid deposition and forest dieback are not entirely understood, large areas of forest in Germany and eastern Europe have been severely damaged by acid rain in association with drought, other climatic extremes and other air pollutants (Howells, 1990). This reduction in forest cover – an indirect effect of air pollution – has had two implications for rivers. First, reduced forest cover results in the effects of deforestation reviewed in Chapter 6: higher streamflows, particularly during small floods. Dubicki (1994), for example, found that forest dieback in mountainous western Poland led to a shortening of the spring thaw period by between 15 and 25 days, leading to increased concentration of streamflow in early spring and lower flows during summer. Flood durations decreased, and peak flows increased by between 30 and 95%.

The second effect is on river water quality. Forest dieback means less material is intercepted, and therefore chemical inputs to the catchment are reduced. Forest dieback in the Krusne Hory mountains in the Czech Republic, for example, had resulted in substantially lower sulphate (and hence implicitly hydrogen ion) deposition (Havel *et al.*, 1999). The decomposition of vegetation killed by air pollution also releases materials into the soil and watercourse.

The effectiveness of mitigation

There are two basic ways of reversing the acidification of lakes and rivers: reducing acid deposition and adding lime to a catchment or watercourse to lower its acidity.

Acid deposition can be reduced by removing forest cover, but this is not really a practical or sustainable response, and the most effective way of reducing deposition is by reducing emissions of sulphur dioxide and nitrous oxides. In northern Europe and North America sulphur dioxide emissions have been reducing since the 1980s, largely in response to air pollution concerns (as shown in Figure 7.3). The Second Sulphur Protocol, signed in 1994, commits 28 European countries to reduce sulphur emissions, with the UK target set at a 70% reduction on 1980 levels by 2005. Two types of study have explored the effect of reducing sulphur and nitrogen emissions. The first has looked for observational evidence of reductions in acidity, sulphate and nitrogen compound loads. Many studies have found lower concentrations of sulphates (and implicitly hydrogen ions) in precipitation, and in some lakes sulphate concentrations have been reduced: Wright *et al.* (1994), for example, found a 42% reduction in sulphate concentrations in lochs in south western Scotland between 1979 and 1988, and sulphate concentrations have reduced in many Scandinavian

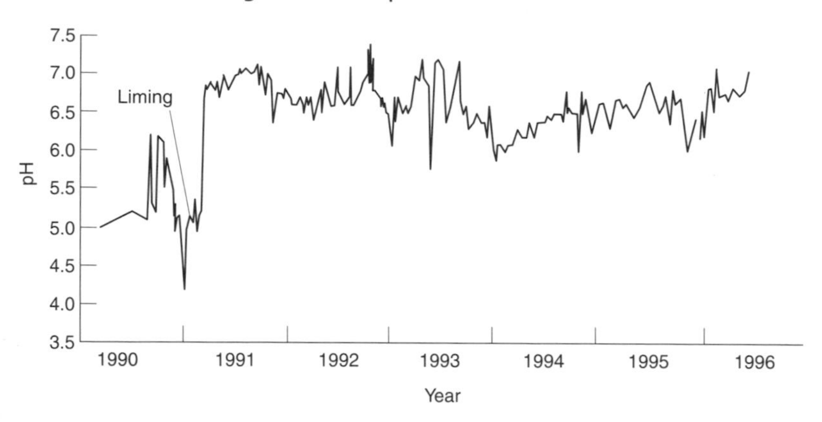

Figure 7.8 Effect of liming on the pH of Llyn Brianne, Wales (Environment Agency, 1998a)

lakes (Howells, 1990). However, whilst there have been some observed reductions in river and lake pH, the acidity of acidified rivers and lakes has generally decreased only slightly. pH in the Scottish lochs studied by Wright *et al.* hardly changed, although the Environment Agency report a widespread increase in pH in Welsh streams of around 0.2 pH units during the 1980s. This small reduction in acidity despite large reductions in sulphur emissions reflects a number of factors, including the legacy of previous acid depositions, continuing increases in nitrogen deposition, and in some cases continued catchment afforestation.

The second group of studies has used simulation models to estimate the effects of different emissions projections on future levels of river and lake acidity. The MAGIC model, for example, simulates the major chemical and hydrological processes associated with acid deposition, and has been widely applied. Cosby *et al.* (1985), for example, explored different emissions strategies in a forested catchment in Virginia, USA, and Whitehead & Neal (1987) compared emissions strategies and afforestation strategies in upland Scotland. As a general rule, sulphur emissions of at least 50% are needed in order to begin to reverse river and lake acidification, and Evans *et al.* (1998) estimated using MAGIC that the sulphur emissions required by the Second Sulphur Protocol will reduce acidity in acid-sensitive catchments across the UK, but even by 2050 only just over a quarter of the decline between 1850 and the 1990s would be counteracted. Jenkins *et al.* (1997) extended MAGIC to incorporate nitrogen deposition, showing that decreases in nitrogen deposition of 50% were necessary in addition to the already agreed reductions in sulphur emissions in order to reverse acidification in a number of acid-sensitive catchments in south west Scotland.

For many years, farmers have applied lime to reduce the acidity of acid soils. In a similar way, lime can be used to restore an acidified water body. Single applications of more than 5 tonnes of calcium carbonate ($CaCO_3$) per hectare can have substantial effects on river and lake pH (e.g. Newton *et al.*, 1996; Figure 7.8), raising it by between 2 and 3 pH units, and these effects may be sustained for at least a decade (Traaen *et al.*, 1997). Lime can be added to a lake directly or to the catchment. Experiments have shown that liming is most effective when the calcium carbonate is spread on saturated parts of the catchment (including beaver ponds: Newton *et al.*, 1996), as most is washed into the river rather than being stored in soil. Liming is, however, tackling the symptoms of acid deposition rather than the cause, and whilst it may lead to the restoration of permanently acidified watercourses, it is not necessarily a sustainable response and may itself have ecological side-effects. Also, the amount of lime needed and its effectiveness will vary from catchment to catchment.

Acid deposition and the hydrological system: an overview

"Acid rain" (or more properly acid deposition) has significant effects on water quality in acid-sensitive areas. The key controls on whether a river becomes acidified are the soil characteristics (upland acidic soils are sensitive, whilst carbonate-rich soils are not), catchment vegetation (afforested catchments scavenge more materials from the air) and hydrological processes. These hydrological processes determine the pathways water follows through the soil, and hence the dynamics of the "acid pulse" that is widely observed in acid-sensitive areas. The relative importance of acidic quickflow and more alkaline baseflow affects the overall level of acidity in rivers and lakes: the higher the groundwater contribution, the more the catchment is buffered against the effects of acidic inputs.

However, acid deposition does not just affect the acidity of river water. Toxic aluminium ions are frequently flushed into the river after rainfall or snowmelt. Both sulphate and nitrogen compound concentrations increase, and this increase is found wherever acid deposition occurs – not just in acid-sensitive catchments.

Both acidification and the recovery from acidification are related to catchment land cover, and land management practices: the effects of changes to catchment inputs are therefore influenced by changes in the catchment (Chapter 6).

7.3 Climate change due to global warming

Eight of the ten warmest years since 1850 occurred during the 1990s (Figure 7.9), the average global temperature during the 1990s was 0.31 °C higher than the average between 1961 and 1990, and the 1990s were the warmest decade for at least a millennium (Figure 4.12b; Mann *et al.*, 1999). This warming has raised concerns that human activities are affecting the global climate by increasing the concentration of certain trace gases that trap outgoing long-wave radiation from the earth's surface and warm the lower atmosphere. These concerns are not new. As far back as 1827, Jean Baptiste Fourier hypothesised that the atmosphere acted like the transparent cover of a box (hence "the greenhouse effect"), and in 1861 John Tyndall identified CO_2 and water vapour as the trace gases which absorbed long-wave radiation. In 1896 Svante Arrhenius estimated that industrial pollution might over centuries double the amount of CO_2 in the atmosphere and result in a 5 °C rise in temperature. During the early part of the twentieth century, however, it was believed that CO_2 actually had little effect on outgoing radiation because water vapour acted in the same wavelength (Drake, 2000). Callender (1938)

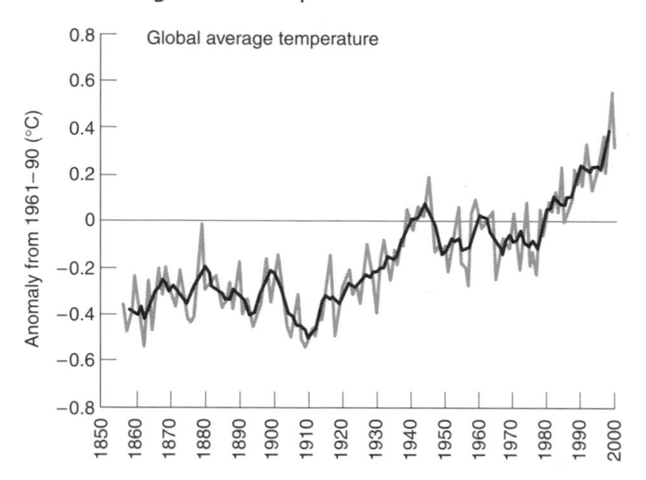

Figure 7.9 Global average temperatures, 1850 to 1999: data taken from the Climatic Research Unit's web-site (http://www.cru.uea.ac.uk)

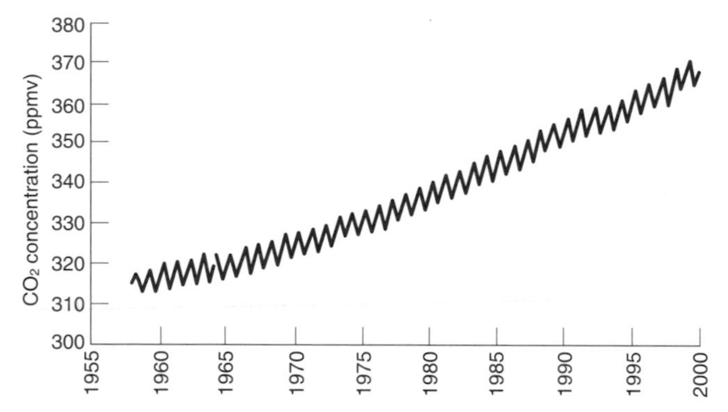

Figure 7.10 Atmospheric CO_2 levels at Mauna Loa, Hawaii: data from the CDIAC web-site (http://cdiac.esd.ornl.gov)

challenged this view, linking an apparent early twentieth century temperature increase to carbon emissions – but his view too was challenged on the grounds that any extra carbon would be absorbed by the oceans. Revelle & Suess (1957) argued that this was not the case, and the following year atmospheric CO_2 levels began to be monitored regularly at Mauna Loa in Hawaii (Figure 7.10).

It was not until the 1980s, however, that scientific and political concern over potential global warming really accelerated, and this is partly because in the previous decades global temperatures appeared to be stable or even declining. In the early 1980s, however, computer simulation models began to show how increasing CO_2 concentrations could lead to global warming (Manabe & Bryan, 1985), and in 1988 the Intergovernmental Panel on Climate Change (IPCC) was established by the United Nations. The First Assessment Report of the IPCC summarised the state of science in 1990 (IPCC, 1990), drawing attention to all the trace gases – CO_2, methane (CH_4), nitrous oxide (N_2O) and

CFCs – involved in global warming, and showing the important role of positive feedbacks in raising global temperatures. Water vapour traps outgoing long-wave radiation: water vapour concentrations increase as temperature rises, because there is more evaporation, and therefore temperature rises still further. The Second Assessment Report of the IPCC was completed six years later (IPCC, 1996). This reported major advances in the understanding of the processes of climate change, and drew attention to the effect of sulphate aerosols – the same sulphate as involved in acid rain – in *reducing* global warming. Most significantly, however, the Second Assessment Report concluded that human activities were probably having a discernible effect on climate, and it is this conclusion that (apart from being the most controversial) has encouraged countries and organisations to think seriously about responding to and attempting to prevent global warming. The Third Assessment Report was published in 2001 (IPCC, 2001).

This is not the place to review in detail the scientific aspects of global warming (see Drake (2000) and Harvey (2000) for up-to-date reviews), and the focus of this section is on the hydrological effects of increasing concentrations of the trace gases that are responsible for global warming. This section first examines whether the effects of global warming can be seen in the observed hydrological record, before looking at how the future effects of global warming can be estimated. The remaining parts of the section consider potential changes in precipitation, evaporation, streamflow and groundwater recharge, and water quality. First, however, it is important to clarify a few terms. "Global warming" is an increase in global temperature, which may or may not be due to increasing concentrations of trace gases: global warming occurs as ice ages finish, for example. "Climate change" is a change in long-term climate, which may or may not be due to increasing concentrations of trace gases: climate changed during ice ages, for example. "The greenhouse effect" is a term frequently used in the popular literature to describe the effect of radiatively-important trace gases (which are therefore often called greenhouse gases[1]). This is in itself a natural phenomenon and is what maintains temperature at the earth's surface at levels that can support life, and the term really should be "*enhanced* greenhouse effect". Clearly, these terms are linked but not synonymous. The phrase "climate change" will be used in this section as a shorthand for the correct, but cumbersome, "climate change due to global warming due to an enhanced greenhouse effect".

Can the effects of climate change be detected in hydrological data?

Global average temperatures have increased in recent decades, as shown in Figure 7.9. Temperatures have generally increased in high latitude land areas

[1] In practice, however, temperatures in a real greenhouse are usually elevated largely because the panes of glass provide shelter from wind: the term "greenhouse effect" is therefore strictly inaccurate.

in the Northern Hemisphere more rapidly than elsewhere, although there are many local departures from the general pattern, and some parts of the world saw reductions in temperature in the latter part of the twentieth century (Nicholls *et al.*, 1996). Precipitation over land has generally increased outside the tropics, with the biggest percentage increases in the north. Land precipitation has declined in large parts of the tropics, include the Sahel, the Middle East and southern Africa, and around the Mediterranean Sea (Nicholls *et al.*, 1996). Changes in precipitation, however, are much less coherent and strong than changes in temperature, partly because precipitation varies much more from year to year, and partly because precipitation is particularly influenced by patterns of natural climatic variability such as ENSO. There is evidence from the United States (Karl & Knight, 1998) and the UK (Osborn *et al.*, 2000) that "extreme" precipitation events have become more frequent, at least in some regions. Snow cover in the Northern Hemisphere has declined substantially at the end of the twentieth century, with a 10% reduction in the annual mean snow extent between 1972 and 1994 (Nicholls *et al.*, 1996): this reduction has mostly occurred during spring.

The effects of these changes should in principle be seen in evaporation, streamflow and groundwater recharge. In practice, however, trends in hydrological characteristics are less clear than trends in the climatic drivers, for a number of reasons. First, hydrological records tend to be relatively short. Second, streamflow and groundwater recharge in particular are subject to many human influences other than climate change, as examined in Chapter 6. This makes it hard to attribute an observed change in hydrological data to climate change. Third, streamflow and groundwater recharge are an integration of several aspects of climate over both time and space. Trends in one driver – such as temperature – may be offset by trends in another – such as precipitation – and trends in different parts of a large catchment may be moving in different directions. Fourth, streamflow and recharge are, even more so than precipitation, influenced by natural climatic variability (Section 4.4). Finally, whilst it is possible to take a global perspective with precipitation and temperature data, this is much more difficult with streamflow and recharge. Global databases tend to contain data from the largest catchments, and it is likely that trends will be first apparent in smaller catchments.

There have, however, been many studies into trends in streamflow during the twentieth century. A global perspective was taken by Yoshino (1999) – but only to 1990 – who found a weak increasing trend in runoff across much of western Europe and North America, decreasing runoff in the Sahel region, and an increase in relative variability from year to year in several semi-arid regions. Lins & Slack (1999) examined trends in various flow indices in 395 catchments across the United States. Over the 50-year period 1944 to 1993 no significant trend could be detected in more than half of the catchments (rising to 90% when looking for trend in maximum runoff), and in the remainder the dominant trend was towards an increase in streamflow – particularly at low flows (Figure 7.11). Decreasing trends were found in the Pacific North West, northern California and parts of the south east. Lins & Slack

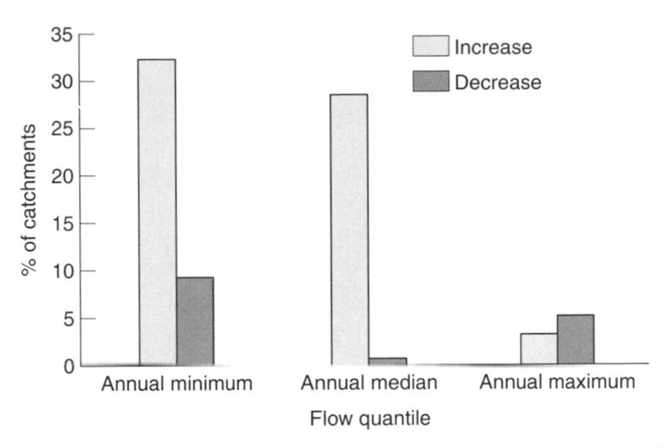

Figure 7.11 Percentage of catchments (out of 395) with statistically significant ($p < 0.05$) trends over the 50-year period 1944 to 1993 (adapted from Lins & Slack, 1999). Trends estimated using the Mann–Kendall test

observed that these trends were consistent with both trends in precipitation and the possible signal of climate change, but could also be explained by the recent persistent high phase of the North Atlantic Oscillation and patterns in the North Pacific (Section 4.4). A decrease in streamflow in south eastern South America at the end of the twentieth century is also attributed to long-term climatic variability (Genta *et al.*, 1998). Olsen *et al.* (1999) found a statistically significant increase in flood frequency in the Upper Mississippi and Upper Missouri basins, but concluded that climate change was the *least* likely cause of the trend because the direction of change is different from that projected by climate models.

Within the UK, average temperatures during the 1990s have been 0.57 °C above the 1961–90 mean (Figure 7.12a), and whilst patterns in precipitation are far less clear (Figure 7.12b), there is a tendency towards an increasing proportion of precipitation falling in winter. This is particularly so in Scotland and western Britain, and in these areas – along with much of Norway – streamflow has increased substantially during the last few years of the twentieth century (Marsh *et al.*, 2000). Meanwhile, many catchments in southern England have seen a succession of low flows, with notable droughts between 1988 and 1992, and in 1994. This tendency is illustrated in Figure 7.13, which shows streamflows in the River Leven, draining part of the Scottish Highlands, and the River Great Stour in Kent (but note the very large runoff in the Great Stour in 2000).

The late 1990s saw a number of notable floods in Britain, including the "Easter floods" in the English Midlands in April 1998 and a succession of large floods in Scotland. Whilst there is evidence that flood frequency has increased in Scotland since the late 1980s (Black, 1996: Figure 7.14), there is no evidence that the number or magnitude of floods across the UK as a whole has been affected by climate change (Robson *et al.*, 1998). There are substantial variations from one period to another, but these are attributed to climatic *variability* rather than change.

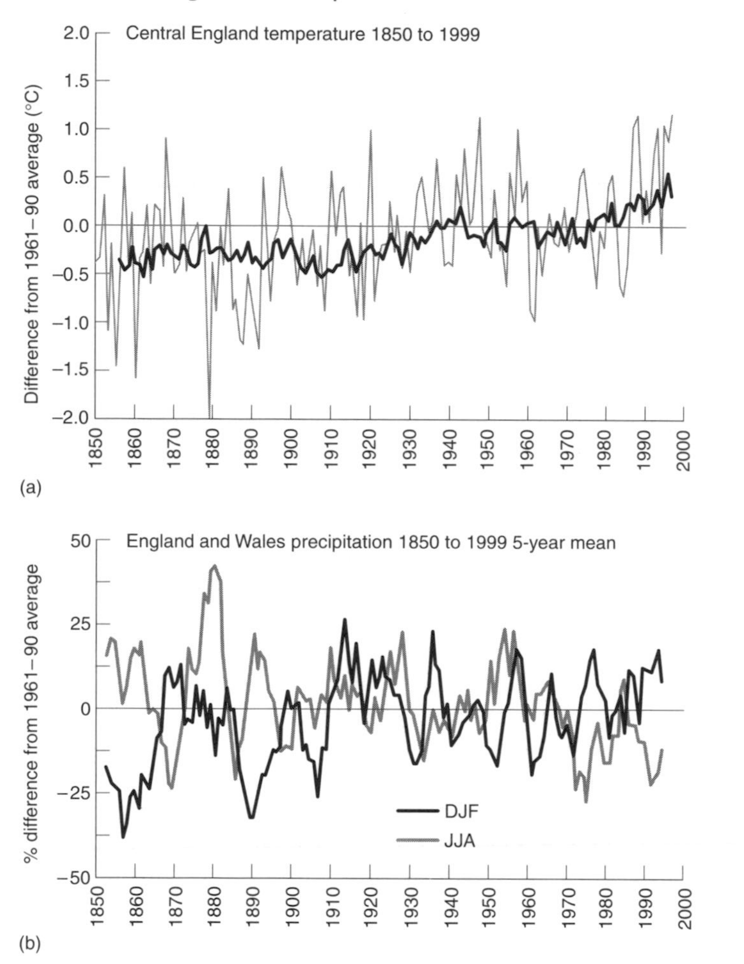

Figure 7.12 (a) Central England average annual temperature, 1850 to 1999; (b) England and Wales total rainfall (winter and summer), 1850 to 1999: data from the Climatic Research Unit's web-site (http://www.cru.uea.ac.uk)

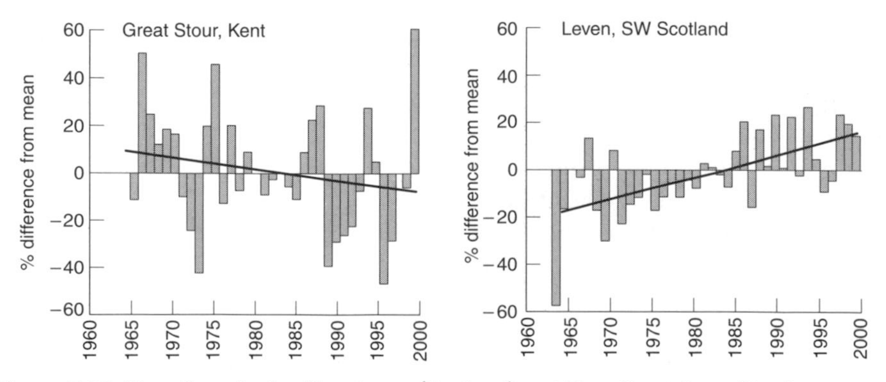

Figure 7.13 River flows in the River Leven (Scotland) and River Great Stour (Kent)

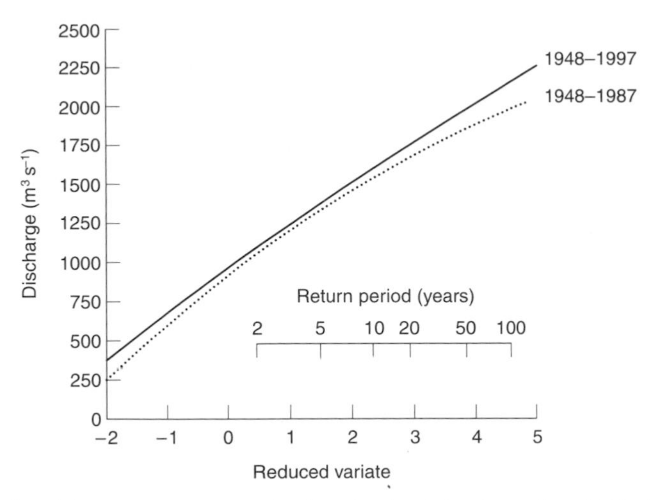

Figure 7.14 Flood frequency curves for the Tay, Scotland, for the periods 1948–87 and 1948–97 (adapted from Marsh *et al.*, 2000)

In general, therefore, it is difficult to detect widespread changes in streamflow (or indeed groundwater), and even harder to attribute these changes to climate change: most can be explained in terms of natural climatic variability. Aspects of the hydrological system more directly dependent on temperature than precipitation, however, do show clearer trends. Potential evaporation at Oxford, for example, was 3.3%, or nearly 16 mm, higher between 1991 and 1996 than between 1961 and 1990 (Burt & Shahgedanova, 1998). More generally, several studies have shown a declining contribution of snowmelt to streamflow. Bergstrom & Carlsson (1994) found reduced spring flows and increased winter flows in Sweden and the Baltic States, and Westmacott & Burn (1997) observed earlier spring snowmelt peaks in west-central Canada (see also Figure 4.21).

Estimating future hydrological characteristics

Whilst it is difficult to detect changes in hydrological characteristics in recent years, and even harder to attribute changes to climate change, systematic and consistent changes in climate are likely to affect hydrological behaviour in the future. Not only are the inputs to the catchment likely to change, but catchment land cover might change too and a changing climate might result in some cases in changes in the relative importance of different processes.

Unfortunately, it is not possible to predict precisely the future climate, so it is necessary to construct "scenarios" of change, defined as feasible, self-consistent possible futures. Climate scenarios are generally based on simulations made using general circulation models (GCMs: Box 7.2), but as Figure 7.15 shows, there are several stages in the process of estimating potential changes in hydrological behaviour. In practice, hydrological impact studies focus on the parts shown within the dotted box in Figure 7.15, using the results of studies in the

Box 7.2 General circulation models

A climate model is a mathematical representation of the climate system. A general circulation model (GCM) is a type of climate model which calculates the fundamental physical equations governing the movement of energy and water in three dimensions (vertical and horizontal). A GCM simulates the spatial and vertical variation in weather at time steps as fine as 10 minutes, and this "weather" is averaged to calculate climate. GCMs typically have a spatial resolution of several degrees of longitude and latitude, with the finest global scale model currently having a resolution of 3.75° longitude by 2.5° latitude (or approximately 80 000 km^2 in mid-latitudes). Spatial resolution is limited by computer power, and processes that operate at scales finer than the model resolution are represented by empirical "parameterisations". These processes include cloud formation and the development of small-scale weather systems – and also the land surface processes which control evaporation and the energy balance (see Chapter 8). Atmospheric GCMs (AGCMs) and ocean GCMs (OGCMs) simulate respectively the atmosphere and the oceans, using broadly the same physical procedures. Coupled AOGCMs simulate the two together, allowing for feedbacks between ocean and atmosphere. GCMs are not calibrated to reproduce current climate, but may be tuned to fit observed data as closely as possible, and future climate is estimated "simply" by changing the composition of the model atmosphere. Early GCM climate experiments used AGCMs with no dynamic interaction with the ocean, and were used to simulate *equilibrium* climate change, defined as the change between the current climate and the climate which would arise when atmospheric CO_2 concentrations had doubled and the ocean and atmosphere were in equilibrium with each other. In practice, of course, CO_2 and the other radiatively-important trace gases are increasing gradually, and the rate of change in climate will depend on the rate of change in the ocean. Coupled AOGCMs have therefore been used since the mid-1990s to simulate *transient* climate change, and these have the significant advantage that they can be used to define climate scenarios for defined time horizons. Unfortunately, their increased realism is bought at the cost of significantly increased computer time, and it may take months to simulate a hundred years of climate. GCMs are being continually revised and enhanced, and leading climate modelling centres currently include the Hadley Centre (HadCM) in the UK, the Max Planck Institute for Meteorology (MPI) in Germany, the Bureau of Meteorology (BMRC) and the Commonwealth Scientific and Industrial Research Organisation (CSIRO) in Australia, the Canadian Climate Centre (CCC), and the Geophysical Fluid Dynamics Laboratory (GFDL), the National Center for Atmospheric Research (NCAR) and Goddard Institute for Space Studies (GISS) in the USA: their climate models are generally known by the acronyms in parentheses. Harvey (2000) provides a good review of climate models.

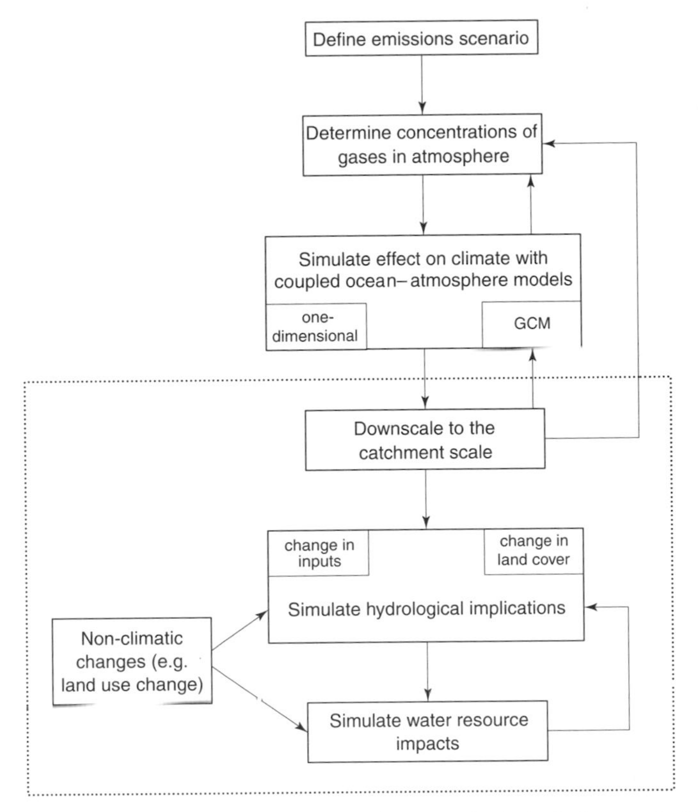

Figure 7.15 Estimating the impacts of climate change on hydrological characteristics

top part of the diagram as inputs. This book concentrates on *hydrological* changes, rather than impacts on water resources as used by people and the environment, so will not consider in detail here the bottom box in Figure 7.15.

The first stage in the process is the estimation of future emissions of greenhouse gases. Projections must be based on assumptions about the rate of population growth and economic development, which are inherently uncertain. In 1992 the IPCC published a set of emissions scenarios making different assumptions about rates of economic development, known as IS92a to IS92f. IS92a was regarded as the "business as usual" scenario, and assumed no particular policy responses to climate change. A new set of scenarios was produced in 2000 (IPCC, 2000), following a rather different approach. First, a number of different feasible "storylines" was constructed, representing different possible ways in which the world might develop (more or less globalisation, greater or lesser market orientation, and so on). Sets of future emissions of greenhouse gases were estimated for each of these storylines, producing around 30 different possible future emissions profiles: six of these profiles have been selected as "marker" scenarios, and assumed future CO_2 emissions for four of them are shown in Figure 7.16a (the scenarios make assumptions, of course, about the future emissions of other relevant gases).

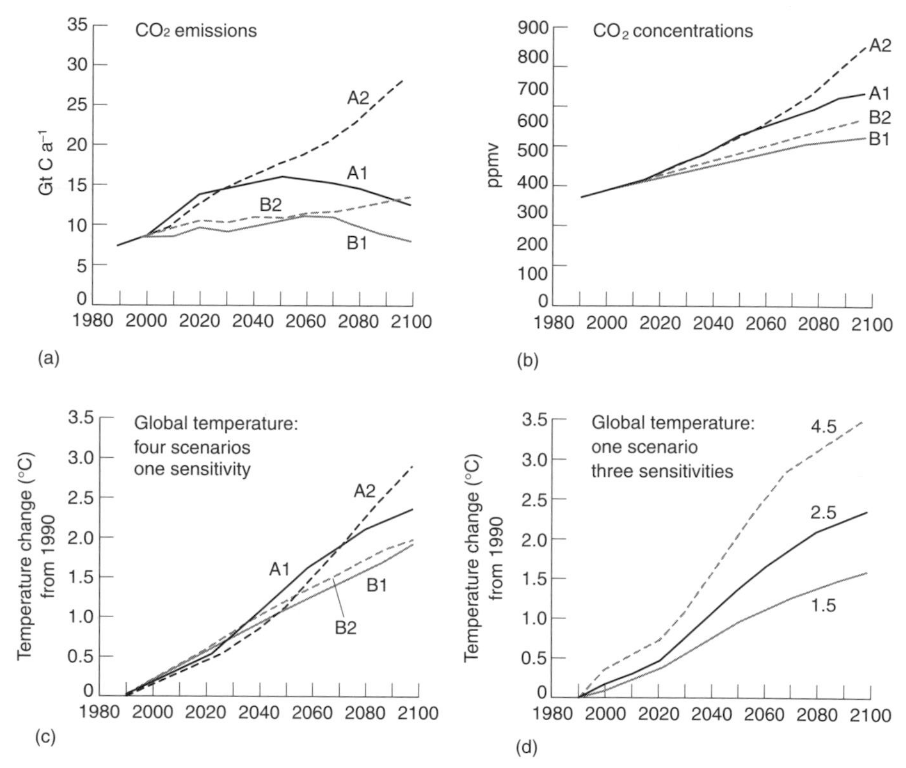

Figure 7.16 (a) Future CO_2 emissions under the Special Report on Emissions Scenarios (SRES) marker scenarios; (b) future CO_2 concentrations under the SRES marker scenarios; (c) future global average temperature change under the SRES marker scenarios; (d) future global average temperature under one emissions scenario and different assumed climate sensitivity (Smith *et al.*, 2000)

The next stage is to convert these emissions profiles into greenhouse gas concentrations (Figure 7.16b). There are three major complications and uncertainties. First, the chemical transformations which take place in the atmosphere are not fully understood. Second, and perhaps more importantly, the relative magnitudes of the sources and sinks of the various greenhouse gases – particularly CO_2 – are not known. Carbon is absorbed by both the terrestrial biosphere and the ocean, but the processes and rates are uncertain (Harvey, 2000). Third, sources, sinks and the rate of chemical transformations in the atmosphere are a function of the rate of climate change: there is a feedback between climate change and the concentration of greenhouse gases. Vegetation uptake of carbon, for example, is a function of CO_2 concentration and plant growth, which is a function of temperature and moisture availability. Changes in ocean surface temperature and the rate of operation of the "biological pump" absorbing CO_2 from the atmosphere will affect the size of the ocean sink. Methane emissions from tundra wetlands are likely to increase as temperature rises and permafrost thaws. All these uncertainties mean that it is not possible to translate a profile of emissions into greenhouse gas concentrations with precision. In practice, most transient GCM experiments assume a 1% compound increase in greenhouse gas concentrations per year, which approximates to the IS92a emissions scenario and a best guess at the processes converting emissions to greenhouse gas concentrations.

Revisiting Figure 7.15, the third box describes the translation of greenhouse gas concentrations into climate. This contains two components, both incorporated within climate models: estimation of the effect of changes in greenhouse gas concentrations on radiative forcing (i.e. the extra energy available at the earth's surface), and the estimation of the effects of this extra radiative forcing on weather and climate. GCMs can do this, but because they take so long to run it is impractical to use them to explore the effects of different emissions profiles. Single-dimension climate models "lose" the spatial dimension and are much faster than GCMs, but simulate only global average climates. They are used to estimate the effect on global average temperature of different emissions pathways. One of the parameters used by single-dimension climate models is the "climate sensitivity", defined as the increase in temperature that would arise once climate had reached equilibrium with a CO_2 concentration twice the pre-industrial value. This figure is estimated from equilibrium GCM experiments (Box 7.2), and varies between GCMs: it ranges between 1.5° and 4.5°. Figure 7.16c shows the global temperature change under the four emissions scenarios and an average climate sensitivity (2.5°), and Figure 7.16d shows for one emissions scenario the range in global temperature change with different estimates of climate sensitivity. However, whilst single-dimension climate models can be used to explore global climate sensitivity, they do not provide any information on the spatial pattern of change in climate. This can only be provided by a GCM, and different GCMs give different estimates. The broad pattern of change is consistent between GCMs (maximum warming at high latitudes, increase in precipitation at high latitudes and in most of the tropics, reduced precipitation in the sub-tropics and mid-latitudes), but there are significant local differences: in some, for example, large parts of North America become wetter, whilst in others they become drier. One way to combine the advantages of single-dimension climate models and GCMs is to rescale the spatial patterns as simulated by a GCM to match the global average temperature change as simulated by a single-dimension climate model. This makes it possible to estimate changes in regional climate under different emissions scenarios, but involves the major assumption that the pattern of change is not significantly affected by the rate of climate change.

Coupled AOGCMs simulate the development of climate over time, and therefore simulate natural climatic variability. The effect of changing greenhouse gas concentrations is therefore superimposed on year-to-year and possibly decade-to-decade variability, and this has important implications for the estimation of the effect of climate change on climate. The difference between simulated future climate at any time period and simulated present climate reflects both climate change and climatic variability, and the shorter the averaging period used, the greater the relative effect of climatic variability. Also, the shorter the time horizon, the greater the contribution of climatic variability to any apparent difference from the current climate. There are two potential responses to this problem. The first is to estimate the change in climate at the end of the transient climate model simulation, where the climate change signal is expected to be strongest, and rescale this change to a defined global average temperature change. The second is to run an ensemble of climate change simulations with the same AOGCM with slightly different starting conditions. This of course is expensive in terms of computer time, but gives an indication of the relative effects of climatic variability and climate change on the change in climate at a particular time horizon. An ensemble of four runs has been completed with the HadCM2 model (Mitchell *et al.*, 1999).

Many GCMs are currently being used to simulate the climatic consequences of increasing greenhouse gas concentrations: which is best? It is, however, very difficult to rank the performance of each GCM, as there are many indicators and different models perform well in different geographic areas. Comparisons between GCMs are being organised through the WMO's Atmospheric Model Intercomparison Project (AMIP), which by 1995 involved 31 modelling groups (Gates *et al.*, 1999). The first phase of AMIP showed that the average large-scale seasonal distributions of pressure, temperature and atmospheric circulation were reasonably simulated, but that there were large differences between models in simulated precipitation in low latitudes. Cloudiness is generally poorly simulated. The AMIP comparison showed that no single model consistently outperformed the others in all the tests.

The first three stages in a climate impact assessment are, as indicated above, usually undertaken as climate scenario development initiatives, independent of hydrological (or indeed any other sector) impact assessments: hydrologists use the scenarios as input, and whilst it is not important for the hydrologists to understand the details of the construction of these scenarios, it is essential that they understand the process followed and the uncertainties involved. Climate change scenarios derived from seven coupled AOGCMS are available on the IPCC's Data Distribution Centre web-site (http:\\www.ipcc-ddc.cru.uea.ac.uk). Because GCMs do not necessarily simulate regional climate very well in absolute terms, climate change scenarios derived from GCMs show the *change* in climate from the simulated present to the simulated future. This change must then be applied to *observed* local climate data. GCMs simulate river runoff, essentially as what is left over from precipitation after evaporation has taken place, but because of their coarse spatial scale and uncertainties in the way runoff is simulated (Chapter 8) these direct estimates of runoff are not in practice used in impact assessments.

Climate scenarios at a spatial resolution of several tens of thousands of square kilometres and showing changes in monthly mean climate are not exactly what is needed for hydrological impact studies. Catchment-scale assessments require scenarios with a spatial resolution of at most a few thousand square kilometres, and at daily or shorter time steps. The process of moving from the scale of scenarios produced by GCMs to finer space and time scales is known as *downscaling* (note that whilst GCMs operate at fine time scales results are often not output at less than monthly resolutions and in any case would reflect averages over a very large geographical area). There are in general terms three different ways of downscaling GCM scenarios to the catchment scale. The simplest is to interpolate the GCM-scale simulations down to a finer spatial scale (such as $0.5 \times 0.5°$). This has been done for the scenarios produced for the UK Climate Impacts Programme (the UKCIP98 scenarios: Hulme & Jenkins, 1998), for example, and also for the scenarios provided to the US national assessment of the potential consequences of climate change (Felzer & Heard, 1999). Changes in sub-monthly climate can be estimated either by applying the interpolated monthly changes to observed daily time series, or by changing the parameters of a stochastic weather generation model which produces simulated daily data from monthly-scale inputs (Wilks, 1989).

A second approach uses a variety of statistical relationships between large-scale climate features such as regional average precipitation or pressure fields and local climate (Wilby & Wigley, 1997). All these techniques assume that present relationships between scales continue to hold in the future, and the parameters can vary considerably with the period of record used. Most importantly, however, the downscaled climate is driven by the change in climate indices as simulated by the driving GCM. Schnurr & Lettenmaier (1998) concluded, following a study in Australia, that little reliance could be placed on the results of statistical downscaling for regional hydrological modelling, partly because the downscaling methods considered did not reproduce very accurately the observed precipitation characteristics, but largely because of errors in the driving GCM simulations: in their case studies, the differences between the GCM simulation of current large-scale circulation patterns and observations were greater than the effect of climate change as simulated by the GCM. On the other hand, Wilby *et al.* (1999) showed in mountainous Colorado that statistically-downscaled GCM data describing current climate gave more accurate simulations of streamflow than could be generated from the small number of observed climate stations, and inferred that in this case statistical downscaling was necessary to produce more "reliable" projections of the impact of climate change. Scenarios at sub-monthly time steps can be created using the same techniques as with the first approach, and additionally by using the daily large-scale features simulated by the GCM.

The third approach uses a regional climate model (RCM) nested within a GCM to give finer resolution simulations (typically of the order of $0.5 \times 0.5°$, or approximately 2500 km^2) over a restricted area (typically a continent or sub-continent). The GCM provides the boundary conditions, and the RCM uses these to simulate local climates within its domain more accurately. However, RCMs are computationally-intensive, and are generally run only for short periods and for a small number of GCMs, and whilst the absolute values of climate within an area may be different from those obtained by interpolating a GCM, the patterns of change in climate may not be very different. The anticipated advantages of using an RCM to create catchment-scale scenarios are therefore not necessarily realised, and RCMs have rarely been used so far to construct scenarios for hydrological impact studies (although Leung &

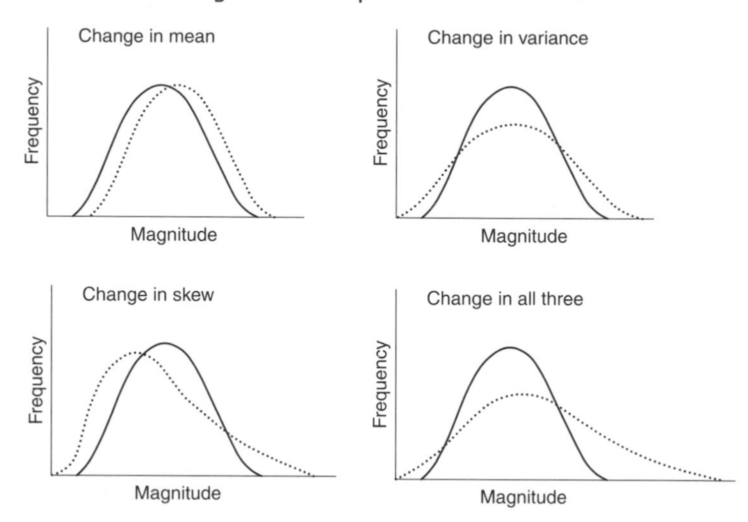

Figure 7.17 Changes in the mean, variability and shape of the frequency distribution due to climate change

Wigmosta (1999) used an RCM to create scenarios when assessing possible future hydrological changes in mountainous regions of western North America). If the RCM is used to estimate changes in monthly climate, then the same techniques as used for the other approaches could be used to create sub-monthly-scale scenarios. Additionally, RCM simulations of current and future climate at sub-monthly scales could be used directly, as long as the RCM accurately simulates current climate (and this was done by Leung & Wigmosta).

In practice, virtually all published hydrological climate change impact studies have used simple interpolation of GCM-simulated changes in monthly mean climate, applied to observed monthly or daily climate time series: only a small minority have used statistical downscaling techniques, and virtually all of the papers that describe statistical downscaling have been concerned with the down-scaling technique itself. Virtually all published studies have also concentrated on the effects of changes in *mean* climate. Global warming may also, of course, lead to changes in relative climatic variability and possibly the shape of frequency distributions (Figure 7.17) or the frequency of anomalous events (such as El Niño). However, it is difficult to create scenarios for such possible changes, as climate models do not necessarily reproduce the observed year-to-year variability in climate very well even though they reproduce average climate. The UKCIP98 scenarios (Hulme & Jenkins, 1998) define possible changes in the relative variability of monthly precipitation across the UK, but it is difficult actually to include this information in practice.

The fifth box in Figure 7.15 is at the heart of a hydrological climate change impact study: the application of a hydrological model with altered climate inputs to simulate possible effects on streamflow, recharge and water quality. Hydrologists have long experience in developing simulation models, and as a general rule little extra uncertainty is added when climate is translated into streamflow using a conceptually-sound hydrological model. However, if the hydrological model does not represent processes correctly, then it may give a misleading indication of the implications of climate change. Statistical/empirical models such as regression models may give inaccurate estimates of the effect of a change in input climate data, as the model parameters may internalise present relationships between different dimensions of climate and hydrological response. Estimating the impacts of climate change on water quality is rather more complicated, as water quality simulation models are less well defined than water balance models because many of the biochemical processes are not fully understood.

Figure 7.15 also emphasises that climate change is not necessarily the only change taking place in the catchment. Land use may be changing too, and this may affect the way climate change impacts upon the catchment.

It is clear from the above that there is a cascade of uncertainty in estimating the impact of climate change on hydrological characteristics. Much of this uncertainty comes from the climate scenarios. Some is due to incomplete scientific understanding, and can be expected to be reduced in the near future. Some uncertainty, however, lies in the initial assumptions about future population growth and economic change, and this will never be eliminated. Climate change impact studies therefore consider a range of feasible climate scenarios.

Precipitation in the future

Precipitation is the key input to the hydrological system: variations over space and time in hydrological behaviour are largely driven by precipitation. A warmer world means greater total evaporation, and therefore greater total precipitation, but not everywhere will see additional rainfall. As a result of the changes in circulation patterns due to increased energy availability at the earth's surface, some parts of the world can be expected to get much more of the extra precipitation, and others to get less than at present. In general terms, most models project an increase in rainfall in high latitudes and some tropical regions, and lower rainfall elsewhere. In some parts of the world – such as North America and Southern Asia – there is little agreement between models. Precipitation is generally projected to increase in northern Europe throughout the year, and decrease in southern Europe particularly during summer: the boundary between these two regions varies between models, but in many runs through the centre of Britain.

Increasing temperatures mean that in cold regions a smaller proportion of precipitation will fall as snow. The effect varies with climate regime. In milder regions where both rain and snow fall during winter, a rise in temperature will have very significant effects on the proportion falling as snow. In colder regions, the temperature rise will not be sufficient to make precipitation fall as rain, and the ratio between rain and snow will be little affected.

A more active hydrological cycle, driven by higher temperatures, should on first principles lead to an increase in the number of large precipitation events. Hennessy *et al.* (1997) examined two GCM experiments and found that the probability of heavy daily precipitation events increased by 2050 by more than 50% in many worldwide locations and also that for a given precipitation intensity, the average return period becomes shorter by a factor of between 2 and 5.

Drought frequencies, intensities and duration are also likely to change under conditions of global warming with the mid-continental regions becoming especially prone to increased summer drought. This occurs through a combination of reduced summer rainfall (many continental interiors such as southern Russia, central Europe and the American mid-west) and increased potential evaporation (nearly all regions). Such increased droughtiness is exacerbated by longer dry warm spells resulting from fewer summer wet days (Gregory *et al.*, 1997) and higher temperatures.

Effects on evaporation

The rate of evaporation from a surface is a function of the meteorological demands, the vegetation and surface properties, and of course the amount of water available (Chapter 3). Both the meteorological demands and the amount of water available will be affected by climate change, and so will vegetation properties.

The meteorological controls on the rate of evaporation from a surface with unlimited water (i.e. potential evaporation) are the amount of energy available (as measured by net radiation), the vapour deficit of the air (a function of water content and temperature), and the rate of removal of air from the evaporating surface (as indexed by windspeed). If everything else remained constant, an increase in temperature alone would lead to an increase in potential evaporation because the vapour pressure deficit would increase. The effect of this increase in temperature will depend on the current vapour pressure deficit. If this is already very high – as in an arid or semi-arid area – then an increase in temperature, and hence the vapour pressure deficit, will have little effect on evaporation. Table 7.2 shows the percentage change in potential evaporation for an increase in temperature of 1.5 °C in the New Forest, England, in July and October, and Niger in the Sahel during the dry season: the greater the humidity, the greater the effect of a given increase in temperature. However, all the other meteoro-

Table 7.2 Change in potential evaporation with a 1.5 °C increase in temperature

Location and date	Relative humidity (%)	Increase in potential evaporation (%)	Increase in vapour pressure which would offset temperature increase (%)
New Forest: July	32	4.5	35
New Forest: October	69	13.3	15
Niger: October	3	1.7	250

Potential evaporation calculated using the Penman equation, for a short well-watered grass surface

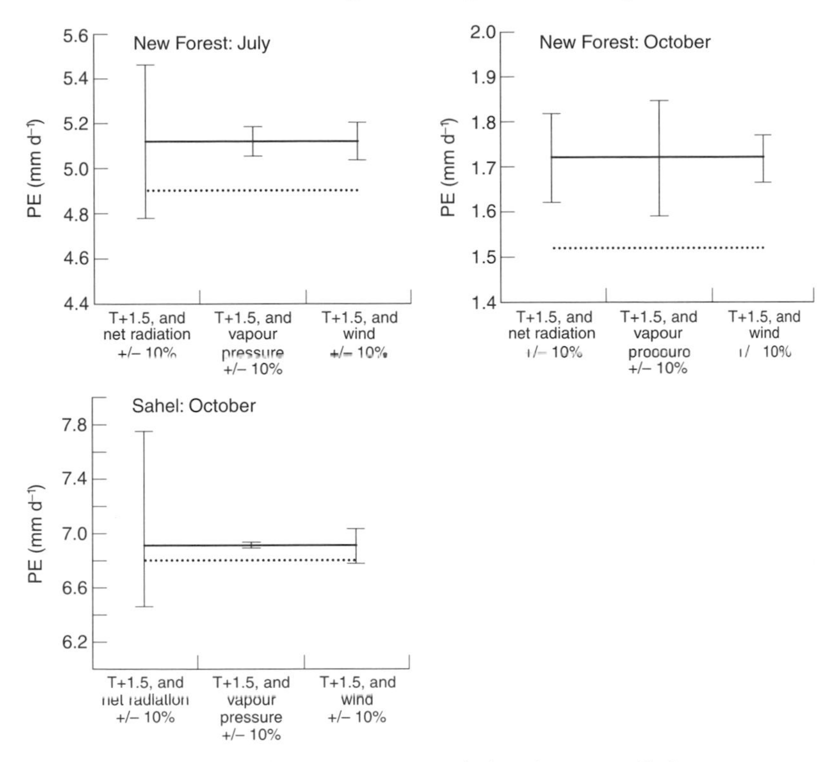

Figure 7.18 Effect of increasing temperature and changing net radiation, vapour pressure and windspeed on potential evaporation in the New Forest and Niger. The dashed line shows the "current" potential evaporation, and the solid horizontal line shows PE with a temperature increase of 1.5 °C

logical variables are not constant. Increased evaporation due to higher temperatures means a general increase in atmospheric water vapour content, which *lowers* the rate of potential evaporation (thus acting as a negative feedback). Table 7.2 shows the percentage increase in vapour pressure which would offset a 1.5 °C increase in temperature. Again, the effect varies with humidity. In areas with high humidity any temperature effect would be eliminated by a relatively modest increase in vapour pressure, but in dry areas the effect of increasing atmospheric moisture contents would be small. This is further shown in Figure 7.18, which shows the effect of a 1.5 °C increase in temperature, together with plus or minus 10% changes in vapour pressure, net radiation and windspeed. In the New Forest in summer, a change in either vapour pressure or windspeed has little effect on the increase in potential evaporation, but a change in net radiation of plus or minus 10% has a very large effect. Here, potential evaporation is energy limited. The same can be seen in the Sahel. In the New Forest in autumn, however, changes in vapour pressure and net radiation are equally important, and a combination of high vapour pressure and low net radiation (which will actually often go together) could mean a reduction in potential evaporation with increased temperature.

Vegetation affects the total amount of evaporation from a catchment through transpiration and interception (Chapter 3). Both vary with vegetation type, so a change in vegetation due to a change in climate would have an effect on catchment evaporation, with the effect of course depending on the extent of the change in vegetation. The rate of transpiration for a given plant, however, will also be influenced by climate change. Plants absorb CO_2 from the atmosphere as part of the process of photosynthesis, and this CO_2 is absorbed through stomata. An increasing concentration of CO_2 in the atmosphere has two main (hydrologically-relevant) effects on plants, which have been demonstrated through experiments at the leaf and plant scale. First, net photosynthesis increases, so plant growth increases. Photosynthesis occurs through two main pathways, known as C_3 and C_4 after the number of atoms in an intermediate carbon compound. C_3 plants include all trees and most temperate and high-latitude grasses, and these are most affected by increasing CO_2 concentrations. Growth may increase by more than 40% in some species as CO_2 concentrations are doubled, but in others may be little affected. The increase in growth of C_3 plants tends to be highest when the plant is grown under stressed conditions, and is very much affected by the availability of nutrients and also the increase in temperature associated with an increase in CO_2. Also, there is evidence that some plants acclimatise to increased CO_2 concentrations. C_4 plants include most tropical grasses, and are much less affected by increasing CO_2 concentrations.

The second effect of CO_2 enrichment is to make stomata smaller, which has the effect of increasing the stomatal resistance (or reducing stomatal conductance) to evaporation. Many studies have shown decreases of stomatal conductance with increasing atmospheric CO_2, averaging between 20 and 30%. This lowers the rate of transpiration for a given set of meteorological inputs, and the water use efficiency (WUE: ratio of carbon uptake (i.e. biomass growth) to transpiration) of plants increases. This has frequently been taken to mean that increased CO_2 concentrations will lead to a reduction in *catchment-scale* evaporation, but this is not necessarily so. In fact, many of the few studies that have looked at scales larger than the plant have shown that evaporation per unit area does not decrease as CO_2 concentrations rise, largely because the increased WUE is offset by the additional plant growth. Kruijt *et al.* (1999), for example, estimated that increasing CO_2 concentrations would have little net effect on transpiration over forest. In a study using a dynamic vegetation model coupled with a climate model to simulate the effects of both CO_2 increase and associated climate change, Betts *et al.* (2000) found that, at the global scale, the effects of CO_2-enrichment and temperature increase on plant growth together offset reduced stomatal conductance. Also, catchments contain a mix of land covers – including bare soil – which will be affected differently by increasing concentrations of CO_2. The higher the proportion of total evaporation derived from intercepted water (Section 3.4), the less the direct effect of increasing CO_2 on catchment evaporation. There is clearly considerable uncertainty over the effects of CO_2 enrichment on plant transpiration (Amthor, 1995; Jarvis *et al.*, 1999) and hence catchment evaporation (Field *et al.*, 1995), and it is not possible to draw any general conclusions.

Table 7.3 Catchment-scale assessments of the implications of climate change for streamflows (adapted from Arnell *et al.*, 2001)

Area	Reference
Europe	
Belgium	Gellens & Roulin (1998)
UK	Arnell (1996); Arnell & Reynard (1996; 2000); Pilling & Jones (1999)
Finland	Vehviläinen & Huttunen (1997)
Czech Republic	Dvorak *et al.* (1997)
Switzerland	Bultot *et al.* (1992)
Spain	Avila *et al.* (1996)
North America	
Pacific North West	Leung & Wigmosta (1999); Hamlet & Lettenmaier (1999b)
Wyoming	Stonefelt *et al.* (2000)
South west USA	Miller *et al.* (1999)
Asia	
Philippines	Jose & Cruz (1999)
Africa	
Nile basin	Conway & Hulme (1996)
Ethiopia	Hailemariam (1999)
Australasia	
Australia	Bates *et al.* (1996); Viney & Sivapalan (1996)

Note: The table is not exhaustive

The *actual* rate of evaporation from the land surface depends of course on the amount of water available, and if climate change results in less soil moisture storage at any time, then evaporation may fall even if potential evaporation increases.

Effects on streamflow regimes

Colour plate IV shows the percentage change in average annual runoff across the world under the two climate change scenarios. The pattern of change in runoff is broadly similar to that of precipitation, although increased evaporation means that runoff decreases in some parts of the world even when precipitation increases. As a general rule, the lower the current ratio of runoff to precipitation, the greater the relative effect of a given percentage change of precipitation on streamflow (Chapter 4). Different details are produced by scenarios based on different climate models, but all show a broadly similar pattern. The effects of climate change on the *timing* of streamflow through the year in a catchment depend on the catchment's climatic and hydrological regime, and Table 7.3 lists some recent catchment-scale assessments.

In temperate areas, most precipitation in winter falls as rain and the seasonal cycle in streamflow is essentially due to the changing balance between rainfall and evaporation: streamflow is lowest during summer because evaporation is

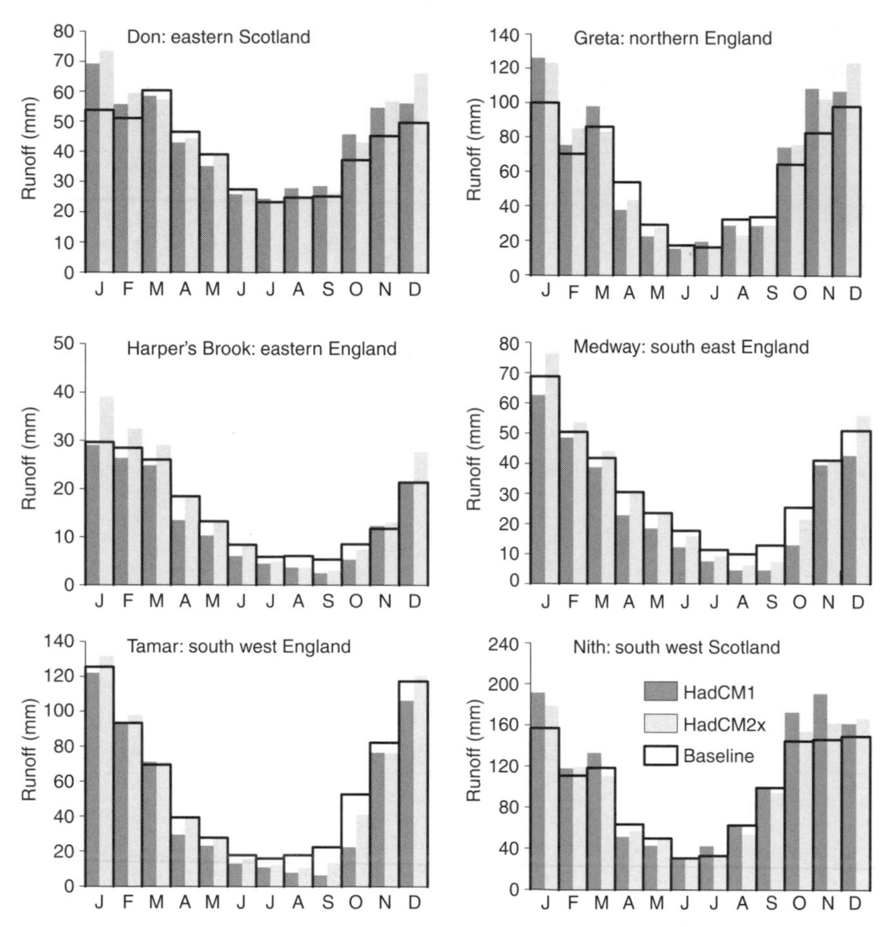

Figure 7.19 Monthly runoff by the 2050s under two scenarios for six British catchments (Arnell & Reynard, 2000)

highest. These areas include the UK and much of western Europe, together with large parts of eastern and western North America. In most of these regions climate change scenarios tend to suggest a general intensification of current weather patterns, leading to increases in rainfall during winter and either increases or decreases during summer. This, coupled with increased evaporation, results in a more extreme seasonal cycle in streamflow, but little change in the timing of flow through the year. Figure 7.19 shows the change in monthly streamflow under two scenarios for six catchments in Britain. In the catchments in northern Britain, streamflow increases in winter but changes little in summer. In southern England, lower summer rainfall coupled with increased evaporation means that streamflows decrease during summer. The effects, however, are dependent not only on climate change scenario but also on catchment geology: catchments with large amounts of storage (in an aquifer, for example), show smaller reductions in summer flow than very responsive

Table 7.4 Change in the frequency of high and low flows by the 2050s for six British catchments, under two scenarios (adapted from Arnell & Reynard, 2000)

	Low flows (*Q*95)				**High flows (*Q*5)**			
	% change in *Q*95		**Average annual number of days with flow <*Q*95**		**% change in *Q*5**		**Average annual number of days with flow >*Q*5**	
	a	**b**	**a**	**b**	**a**	**b**	**a**	**b**
Don: eastern Scotland	6	2	15	17	14	14	25	25
Greta: northern England	−5	−9	20	21	13	11	23	22
Harper's Brook: eastern England	−22	−24	34	36	13	23	22	24
Medway: south east England	−34	−30	37	34	3	8	19	20
Tamar: south west England	−39	−31	41	34	8	3	21	19
Nith: south west Scotland	7	−4	16	19	13	4	24	20

Scenarios a and b are both based on Hadley Centre climate models

catchments (Arnell, 1996; Gellens & Roulin, 1998). The changes in seasonal flows feed through to changes in low flows, and Table 7.4 summarises the change in both the magnitude of the flow exceeded 95% of the time (*Q*95) and the number of days that flows would be below the current *Q*95 in the future, under two scenarios. Increased winter flows mean increased flood frequency in temperate areas, and Table 7.4 shows the increase in magnitude of the flow exceeded 5% of the time. In temperate areas, climate change may therefore increase both the frequency of floods and low flows, but will probably have little effect on the timing of flows through the year.

In colder regions all or much of precipitation during winter falls as snow, and is stored on the catchment to be released during spring. Here, rising temperatures may have a very significant effect on the timing of flow through the year. At the mild extreme, even a modest increase in temperature might mean that snow becomes virtually unknown. In this case, the spring snowmelt peak would be eliminated, and replaced by higher flows during winter: these winter peaks would all be smaller than the previous spring peak. At the cold extreme, all winter precipitation falls as snow even with a rise in temperature, and the spring snowmelt peak would therefore still occur, although it may be brought forward. Many catchments will be somewhere between these two extremes, and an increase in temperature would mean a shift in the relative importance of winter and spring flows. This effect has been seen in a great many climate change impact assessments in snow-affected areas (including many of the studies listed in Table 7.3), and is illustrated for some European examples under one scenario in Figure 7.20. In the Polish example, the spring snowmelt peak is eliminated by the 2050s, in the Ukrainian example it is brought forward, and in western Russia it is little affected. There is therefore a clear climate change threshold in catchments with snow-dominated regimes

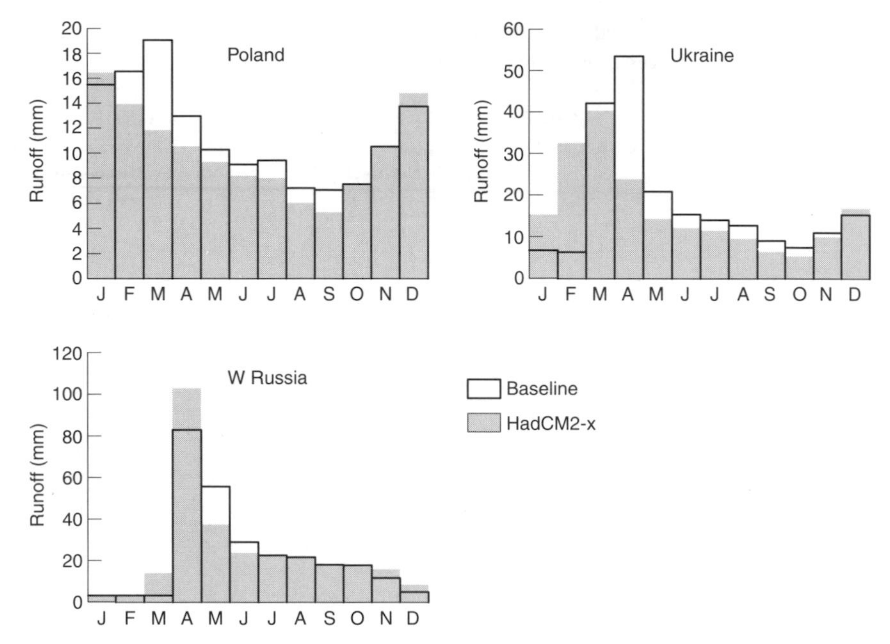

Figure 7.20 Effect of a climate change scenario on streamflow in three snow-affected catchments (Arnell, 1999c)

– the temperature increase which causes significant reductions in the snowmelt peak. Figure 7.21 illustrates the effect of gradually increasing temperature on catchment behaviour in two snow-affected catchments (Arnell, 2000). In western Norway snow still falls as temperatures rise to 2.5 °C above present values. The snowmelt peak in May continues, and even increases slightly up to an increase of 1.5 °C, because there is more precipitation to be stored as snow. Beyond an increase of 1.5 °C, more of this extra precipitation falls in autumn and spring as rain, so the size of the winter snowpack begins to reduce: nevertheless, the hydrological regime is still dominated by the May snowmelt peak. In the Bavarian example, however, a rise of between 0.75 and 1 °C would effectively eliminate snowfall. Up to this point, the winter snow melts earlier and earlier, leading to an increase in February flows and a decrease in March flows. Beyond around 0.5 °C, the amount of snow stored is diminished so much that the snowmelt peak begins to decline, and beyond 1 °C the lack of snow means that thereafter the catchment responds to climate change in a linear manner. In a snow-dominated catchment the effect of climate change is therefore primarily felt through the temperature increase, affecting the timing of flows, and there may be very significant thresholds and non-linearities in response.

There have been far fewer assessments of the potential effect of climate change on hydrological regimes in semi-arid, subtropical or tropical environments, with the notable exception of a few studies in Australia. These environments have very seasonal rainfall regimes and high potential evaporation.

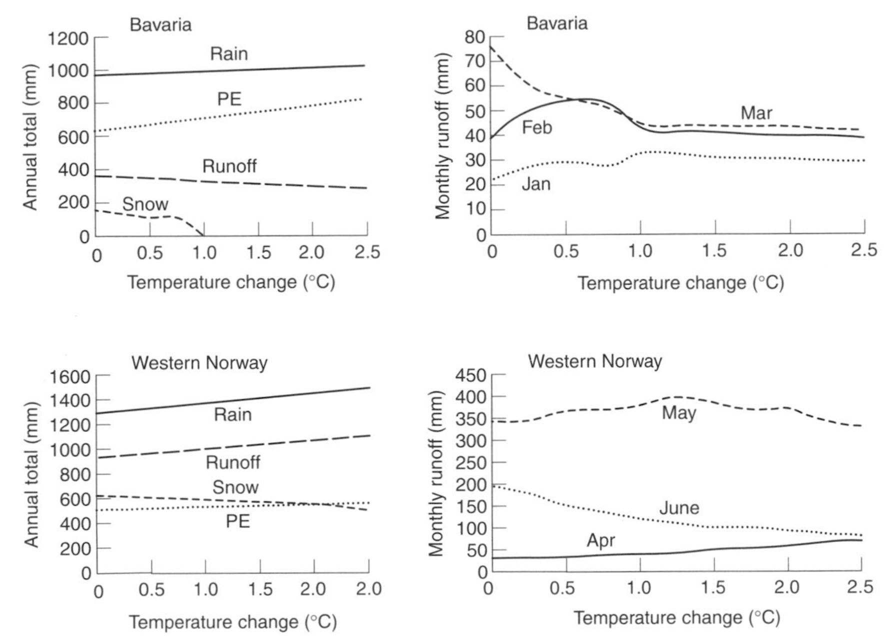

Figure 7.21 Effect of increasing temperature on snowfall and snowmelt peaks (Arnell, 2000)

In the drier regions an increase in potential evaporation is likely to have relatively little effect on the volumes of streamflow generated, as evaporation is constrained by the amount of water available on the ground rather than energy or humidity. In wetter tropical regions where evaporation proceeds at the potential rate for longer, it can be expected that increases in potential evaporation will influence flow volumes. The seasonal cycle of flows, however, will in all these regions be affected by changes in both the magnitude of rainfall during the wet season and the timing of onset and duration of the wet season.

Virtually all studies into the potential effects of climate change on streamflows have used computer simulation models of the hydrological system, and the few that have not have been based on empirical relationships between observed climatic and hydrological variability. All these studies have implicitly assumed that just the *inputs* to the catchment would change, leading to a change in outputs and perhaps a shift from a snow-dominated to a rain-dominated regime. Climate change might also affect the *processes* operating within a catchment, particularly at the margins between different hydrological regions. The vegetation in a catchment has a major influence on the water balance and the translation of rainfall into streamflow (Chapter 3), and this might be altered by climate change. A shift towards more drought-tolerant vegetation, growing at a lower density, for example, would expose more bare soil and thus alter the potential for the generation of infiltration-excess overland flow. Changes in soil characteristics, such as an increased propensity to crack in warmer, drier conditions or an increased frequency of waterlogging due to increased rainfall, may affect flow pathways. In high-latitude areas higher temperatures might lead to progressive degradation of permafrost, again altering flow pathways. All these effects, however, remain as speculative hypotheses at present.

Effects on groundwater recharge

There has been far less research into the effects of climate change on groundwater recharge, but again from first principles (Chapter 3) it is possible to speculate on possible changes. These changes will depend on the groundwater recharge process.

Some aquifers, particularly those in temperate areas, are recharged by direct percolation from rain water. This occurs during the recharge season, once the soils overlying the aquifer have been filled to field capacity by rainfall, and stops when soil moisture deficits begin to appear during spring and early summer. Climate change will affect both the amount of water available for recharge and the duration of the recharge season. In most temperate areas, most scenarios suggest an increase in winter rainfall. This will lead to increased recharge rates, but perhaps over a shorter recharge season – possibly leading to lower *total* recharge. In drier areas, direct percolation only occurs when rainfall exceeds some threshold. Sandstrom (1995b), for example, showed that a 15% reduction in rainfall in Tanzania would lead to significant reductions in the number of recharging events, and a 40 to 50% reduction in recharge.

Indirect recharge derives from overland flow running across the surface of the catchment (in arid and semi-arid areas) and also from infiltration from river beds. This recharge will obviously be affected by changes in the number of overland flow events and flow in the river. Changes in recharge to aquifers recharged by river flows may be determined by the changes in climate occurring in the headwaters of the catchment, which may be several hundred kilometres upstream.

The effects of changes in recharge on discharge from groundwater to streams depend on aquifer properties: the faster the rate of water movement through the aquifer, the more rapid the response. Cooper *et al.* (1995), for example, showed how changes in recharge in chalk aquifers in England would begin to be seen in streamflow after just a few recharge seasons, whereas it might take decades for changes to filter through the less permeable Permo-Triassic sandstone aquifers.

Effects on water quality

Climate change would impact on water quality in four main ways: by changing the chemical and biological processes operating in the soil and bedrock, by changing flow pathways, by increasing water temperature and by altering the volume and timing of streamflow.

Many biochemical processes in the catchment are climate-related. Avila *et al.* (1996) showed how chemical weathering rates, and hence supply of base cations, in a Spanish catchment were enhanced when temperature rose and moisture availability increased, although weathering reduced in drier conditions. The mineralisation of organic nitrogen by soil organic matter decomposition, for example, is increased by higher temperatures (Murdoch *et al.*, 2000), and Wright *et al.* (1998) simulated very large percentage increases in nitrogen exports from a small catchment in Norway following an increase in the rate of decomposition of only 5%. More generally, most microbial nutrient cycling processes are enhanced in warmer soils.

Flow pathways and the means by which precipitation gets to the river channel influence water quality dynamics (Chapter 5), and these may be affected by climate change. An increase in the soil water table can mobilise material, and alter the routes taken by water. Increased soil cracking due to higher temperatures might increase the frequency of preferential flow, encouraging the flushing of stored material. A reduction in the amount of precipitation stored as snow, and hence the operation of processes within the snowpack over winter and the release of material during snowmelt, will also affect water quality during spring.

River water temperature is directly affected by air temperature (Chapter 5), and an increase in air temperature will generally produce a slightly smaller increase in water temperature (Pilgrim *et al.*, 1998): the effect will be smallest in groundwater-dominated rivers. Water temperature has a significant effect on chemical and biological processes in the river channel. Most importantly, dissolved oxygen concentrations are lower at higher temperature, falling by around 0.2 mg l^{-1} for every 1 °C increase in water temperature (Figure 5.3). Higher temperatures also encourage increased growth of algal blooms (as long as there are enough nutrients available), which consume oxygen on decomposition. Chemical processes within a river are also temperature-dependent. Both the rate of nitrification and denitrification increase with temperature, but the rate of denitrification increases more rapidly so, other things being equal, nitrate concentrations would reduce in warmer water. The duration of river ice cover also reduces with temperature, although the extent of effect is unknown.

Perhaps most significantly, however, water quality and the load of material are most affected by changes in the volume of streamflow. Lower flows will reduce dilution, whilst higher flows will increase it. In Finland, simulated changes in nutrient transport are very dependent on the simulated changes in streamflow (Kallio *et al.*, 1997). Hanratty & Stefan (1998) simulated reductions in total transport of nutrients from a small Minnesota catchment, even though peak concentrations increased, because the total volume of flow declined. The effect of natural climatic variability on annual fluxes of material was shown in Chapter 5. Suspended sediment transport is particularly affected by changes in streamflows. Favis-Mortlock & Guerra (1999) found in a Brazilian catchment that two climate model scenarios led to a reduction in soil erosion by water, because rainfall decreased, but another scenario resulted in an increase in mean annual sediment yield by 27% – and proportionately more in wet years.

The effects of climate change on physical and chemical aspects of water quality are therefore currently highly uncertain – and are likely to be very significantly affected by changes in catchment land use. There has been considerably more research into the potential effects of climate change on lake water quality. Lake water temperature will increase with climate change, depending on the thermal stratification processes. Hostetler & Small (1999) simulated potential impacts on hypothetical shallow and deep lakes across North America, showing widespread increases in lake water temperature slightly below the increase in air temperature in the scenarios used. The greatest increases were found in lakes which were simulated to experience substantial reductions in the duration of ice cover: the boundary of ice-free conditions shifted northwards by 10° of latitude or more (1000 km). Such increases in lake water temperature tend to reduce lake mixing, particularly in deep lakes (Hassan *et al.*, 1998). Reduced mixing, together with warmer temperatures generally, result in lower dissolved oxygen levels. If these are coupled with increased algal growth, then there is an even greater risk of damaging oxygen depletion.

Climate change and hydrological regimes: an overview

This section has reviewed the implications of human-induced climate change for hydrological systems and regimes. There are clearly many uncertainties in the estimation of impacts, some because of uncertainties in the driving scenarios, but many – particularly concerning possible changes in process or water quality – due to incomplete understanding of the way the hydrological system works. It is clear, however, that the precise implications of a given climate change scenario for the volume, timing and quality of groundwater and streamflows in a catchment will depend very much on the physical and biological characteristics of the catchment: it is difficult to generalise in quantitative terms.

Future climate change must, however, be placed in its proper context. First, a changing climate is only one of the pressures on a catchment. The others were reviewed in Chapter 6, and may overwhelm, mitigate or exacerbate the effects of climate change. Second, Chapter 4 showed how climate cannot be assumed to be stable even in the absence of human-induced climate change. This variability from decade to decade will continue, and over the next few decades may mask many of the effects of climate change (Hulme *et al.*, 1999). The temperature-related signal, however, in potential evaporation and particularly snowfall, will become increasingly strong.

Finally, virtually all the assessments of the potential effect of climate change on hydrological systems have considered the effects just of a change in mean climate. As noted above, climate change is likely to affect relative variability too, at all time steps, and this variability is a key characteristic of hydrological behaviour in a catchment. Unfortunately, scenarios for possible changes in climatic variability are difficult to define, and the potential effects of climate change on the key modes of climatic variability, such as ENSO and the North Atlantic Oscillation, are currently very uncertain.

7.4 Overview

This chapter has explored the implications of changes in the inputs to a catchment on hydrological behaviour. These changes to the inputs affect the energy and water available within the catchment, and can also affect the catchment land cover and its effect on the hydrological system. Changes to the inputs are being superimposed on completely separate changes to catchment land cover, as described in Chapter 6. It may therefore be very difficult to detect, for example, a climate change trend, or indeed separate out the effects of the many changes affecting a catchment. Acid deposition and climate change are also related, in two main ways. First, the sulphate that is formed in the atmosphere from sulphur emissions as one of the components of "acid rain" also acts as an aerosol, counteracting to a certain extent the greenhouse effect. Cutting down sulphur emissions to alleviate acid rain therefore exaggerates further the effects of increasing concentrations of CO_2 and other greenhouse gases. Second, a changing climate alters the way materials put into the catchment by acid deposition are transformed and transported through the catchment. The nitrogen deposited as nitrates, for example, is cycled through the catchment, with the rates of biological decomposition largely dependent on temperature: higher temperatures releases more of the nitrogen to be leached into rivers, and may slow down the rate of recovery of catchments damaged in the past by excessive deposition of sulphur and nitrogen compounds.

Finally, the chapter has focused on the hydrological changes triggered by changes in catchment inputs, as indeed Chapter 6 concentrated on the hydrological changes due to changes in catchment land cover and water use. The *impacts* of these changes on human use of the water environment depends on how the water environment is managed, and cannot necessarily be inferred directly from changes in streamflow and groundwater. A small percentage change in streamflow in one catchment may seriously threaten the reliability of a water supply reservoir, for example, whilst a substantially larger percentage change in another may have very little effect on a second reservoir. Similarly, a given reduction in dissolved oxygen concentrations may have a greater polluting impact in one river than another. There are many published examples of the implications of changes in streamflow due to climate change for the reliability of managed water systems (e.g. Lettenmaier *et al.*, 1999), and all show the importance of the system characteristics and operating rules. Similarly, the impacts of hydrological changes on the health of the in-stream and riverine environment depend on the nature of the ecological system. Understanding the way the catchment is managed, the pressures on the catchment and the vulnerabilities and sensitivities of the water environment is therefore crucial in understanding the impact of hydrological change.

Chapter 8

Hydrological processes and the earth system

8.1 Introduction

This book so far has seen the hydrological system as receiving inputs from "above" – energy and precipitation – and sending outputs of water laterally (as streamflow) or vertically (as evaporation). The focus has been on the transformation of input to output, with little attention given either to where the inputs come from or what effects the outputs have on the receiving systems. Indeed, this has been the traditional focus of much hydrological research. This chapter looks beyond the catchment system, and considers how the outputs from the hydrological system impact upon the atmosphere and the ocean. Land surface processes (essentially the partitioning of energy into sensible and latent heat fluxes) affect atmospheric processes (Section 8.2), and the rate of inflows of water, sediment and nutrients to the coastal zone affect ocean processes (Section 8.3). Human impacts on land surface processes and hydrological fluxes (Chapters 6 and 7) affect the relationships between hydrosphere, atmosphere and ocean (Figure 8.1).

The final part of this chapter explores how hydrological processes are incorporated into weather and climate models.

8.2 Hydrological processes and the atmosphere

Processes in the atmosphere, and hence weather and climate, operate on four broad scales. Planetary-scale features include trade winds and westerlies. Synoptic-scale features include depressions and anticyclones, with a spatial extent of several hundred kilometres. Meso-scale features have a spatial scale of between 1 and 50 km, and include convective storms, small depressions and tornadoes. The fourth scale – micro-scale – covers gusts and very localised features, and is not very important hydrologically. Planetary and synoptic-scale patterns are determined by large-scale variations in energy availability (between equator and pole) and broad contrasts between land and sea, so are not themselves directly affected by the details of what is happening at the land surface. Meso-scale features, however, may be very much determined by interactions between the land surface and the atmosphere, and these may characterise weather and climate in a region when synoptic flow is at its lowest. This happens when high pressure dominates, particularly during summer. Meso-scale processes can, however, substantially modify the way synoptic and planetary-scale features

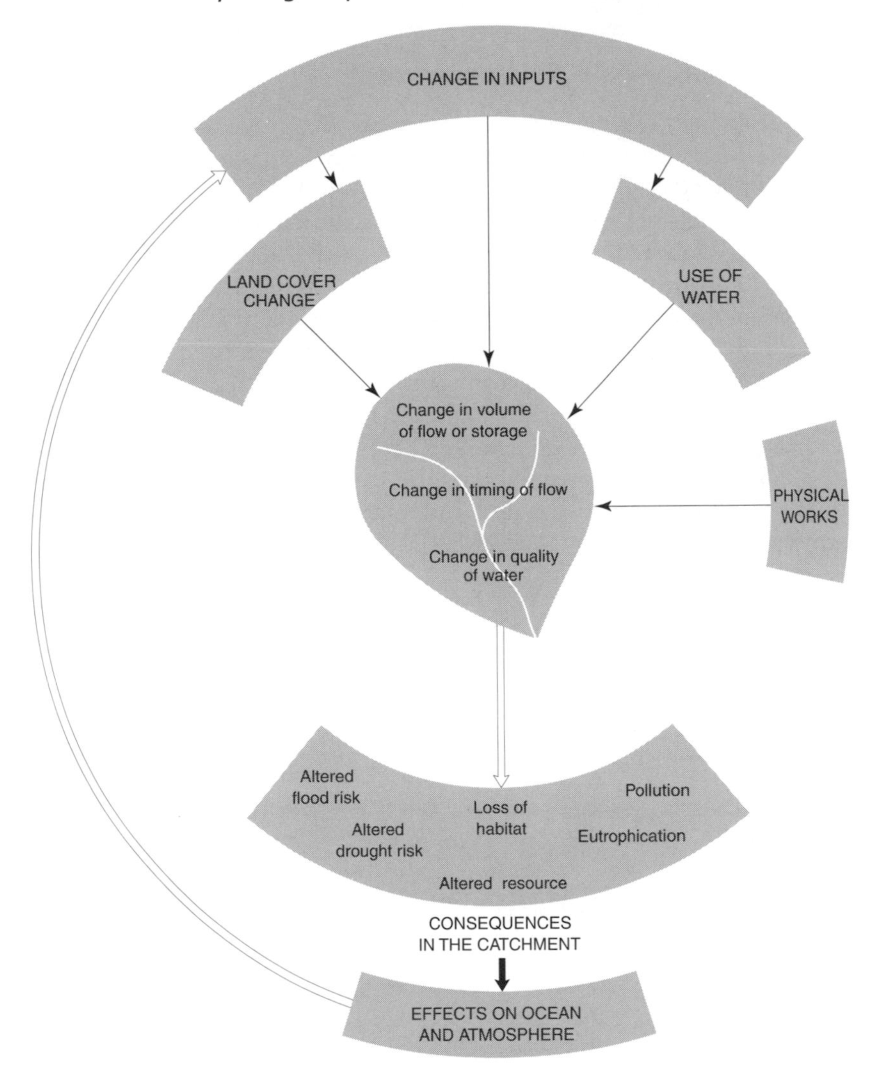

Figure 8.1 Beyond the hydrological system: links with atmosphere and ocean

translate into local weather and climate. Local rainfall along the Mai-yu front in China, associated with the Asian monsoon, for example, is influenced by meso-scale convective storm development, as are squall lines along cold-fronts in North America.

Since the mid-1980s an increasing desire by atmospheric scientists to understand and model atmospheric processes has led to a greater interest in how the land surface affects the atmosphere. A series of *meso-scale* and *continental-scale experiments* has therefore been established, in which hydrologists are playing an increasing role. These experiments, which are different in scope and concept from conventional hydrological catchment experiments (Chapter 6), are described in Boxes 8.1 and 8.2.

Box 8.1 Meso-scale field experiments

Meso-scale field experiments are concerned with measuring and modelling the fluxes of energy and water between land surface and atmosphere. They typically cover an area of 50 × 50 km, with measurements taken at a range of spatial scales from the individual leaf to the entire study area (Figure 8.2). Remote sensing is used to define spatial patterns of fluxes and to extrapolate from point measurements. "Routine" field measurements of point fluxes are typically taken for at least two to three years, and are supplemented during short-lived field campaigns by detailed point measurements and airborne remote sensing from many platforms. These short field campaigns are termed "intensive observation periods" and individual "golden days" are selected for detailed analysis. The field and remotely-sensed data are used to characterise fluxes and their variability, and to develop and validate models of land surface–atmosphere interactions (Section 8.4). Meso-scale field experiments are

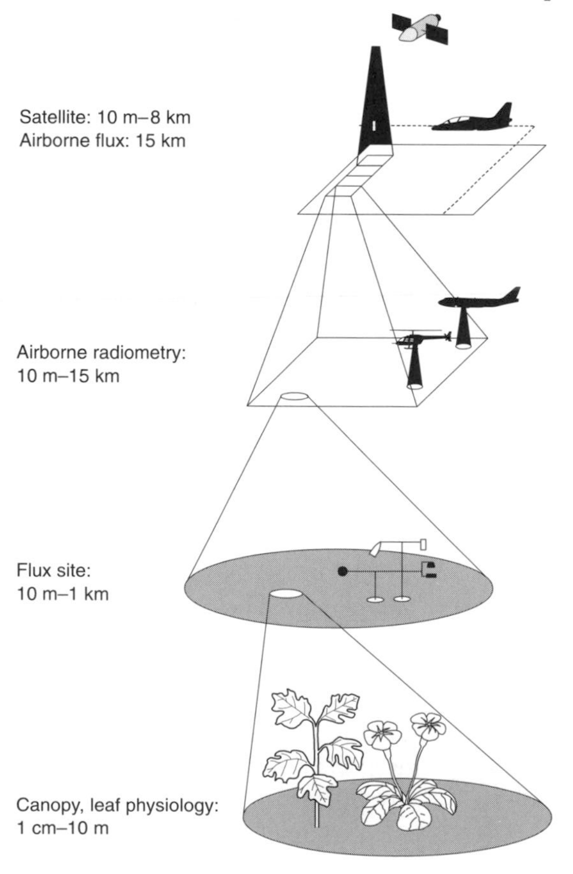

Figure 8.2 Scales addressed in meso-scale field experiments (Sellers *et al.*, 1992)

(continued)

(continued)

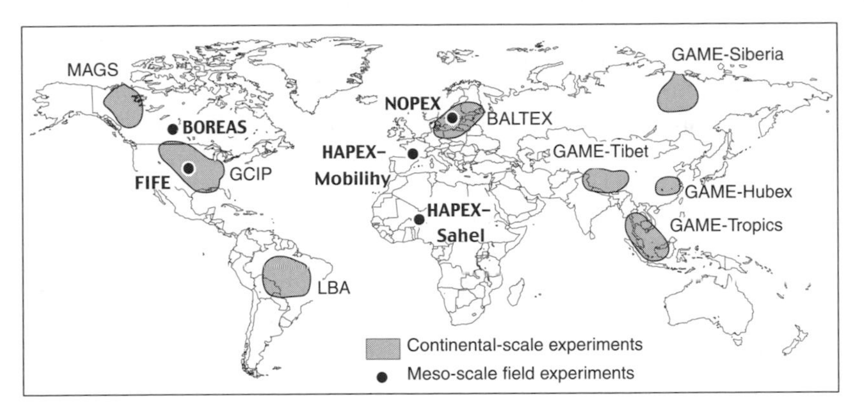

Figure 8.3 Meso-scale field experiments and continental-scale experiments

Table 8.1 Meso-scale field experiments

Experiment	Location	Vegetation type	Reference
FIFE	Kansas, USA	Prairie grassland	Sellers *et al.* (1992)
EFEDA	La Mancha, Spain	Semi-arid	Bolle *et al.* (1993)
HAPEX-MOBILHY	SW France	Largely coniferous forest	Andre *et al.* (1988)
HAPEX-Sahel	Niger, west Africa	Semi-arid savannah	Goutourbe *et al.* (1997)
ABRACOS	Amazon	Tropical rain forest	Gash & Nobre (1997)
NOPEX	Sweden	Mixed forest, lake and farmland	Halldin *et al.* (1998)
BOREAS	Northern Canada	Boreal forest	Hall (1999)

expensive, and involve collaboration between scientists as diverse as meteorologists, hydrologists and plant physiologists. Figure 8.3 maps the major meso-scale field experiments, which are tabulated in Table 8.1. All are known by their acronyms (Appendix 1).

The part of the atmosphere closest to the surface of the earth is termed the planetary boundary layer (PBL). Movement of air within the PBL is affected by friction, creating turbulence, and during daytime thermal convection causes more vigorous stirring. The height of the PBL varies over time, but is typically between 500 m and 2 km. All the physical processes that control the exchange of water and energy between atmosphere and land surface occur within the PBL. The land surface affects these processes primarily through the effects of moisture availability at the surface, the albedo of the surface and, to a lesser extent, through the effects of the turbulence set up by surface types with different degrees of roughness.

Box 8.2 Continental-scale experiments

Continental-scale experiments (CSEs) are, like meso-scale experiments, concerned with the links between the land surface and the atmosphere, but cover a far larger spatial domain and concentrate on the use of routine meteorological and hydrological observations, supplemented by remotely-sensed data. These experiments are run under the auspices of the World Meteorological Organization's World Climate Research Programme (WCRP), and are summarised in Figure 8.3 and Table 8.2. CSEs involve the calculation of water and energy fluxes over large geographical areas. These observations are used to validate and refine regional climate models (Section 8.4).

Table 8.2 Continental-scale experiments

Experiment	Location	Reference
GCIP	Mississippi Basin, USA and Canada	Coughlan & Avissar (1996)
BALTEX	Baltic Sea catchment, Europe	www.gewex.com
GAME	East Asia	www.gewex.com
LBA	Amazonia	www.gewex.com
MAGS	Mackenzie Basin, northern Canada	Stewart *et al.* (1998)

The key hydrological processes and features which affect the interaction between land surface and atmosphere are therefore soil moisture, interception and transpiration, and the transport of water along river networks to lakes and wetlands from which it may be evaporated. The next section reviews the effect of variations over space and time in moisture in soil and vegetation, and the following section explores the specific effects of wetlands and lakes. Hydrological processes and regimes can of course be affected by human activities, and the consequences of human interventions for the atmosphere are examined in the third section.

Effects on the atmosphere of variations over space and time in surface moisture

The amount of moisture available at the surface on plants and in the soil, and its variability over space and time, is very important for the atmosphere and hence subsequent downwind precipitation. A number of early climate model experiments (e.g. Shukla & Mintz, 1982; Delworth & Manabe, 1989) showed that if the surface of the land was continuously dry, there would be little or no precipitation over continental regions in some seasons.

Most obviously, surface moisture provides a source of water for the atmosphere. However, the most significant effects of surface moisture are in fact on the energy balance: both the amount of energy at the available at the surface

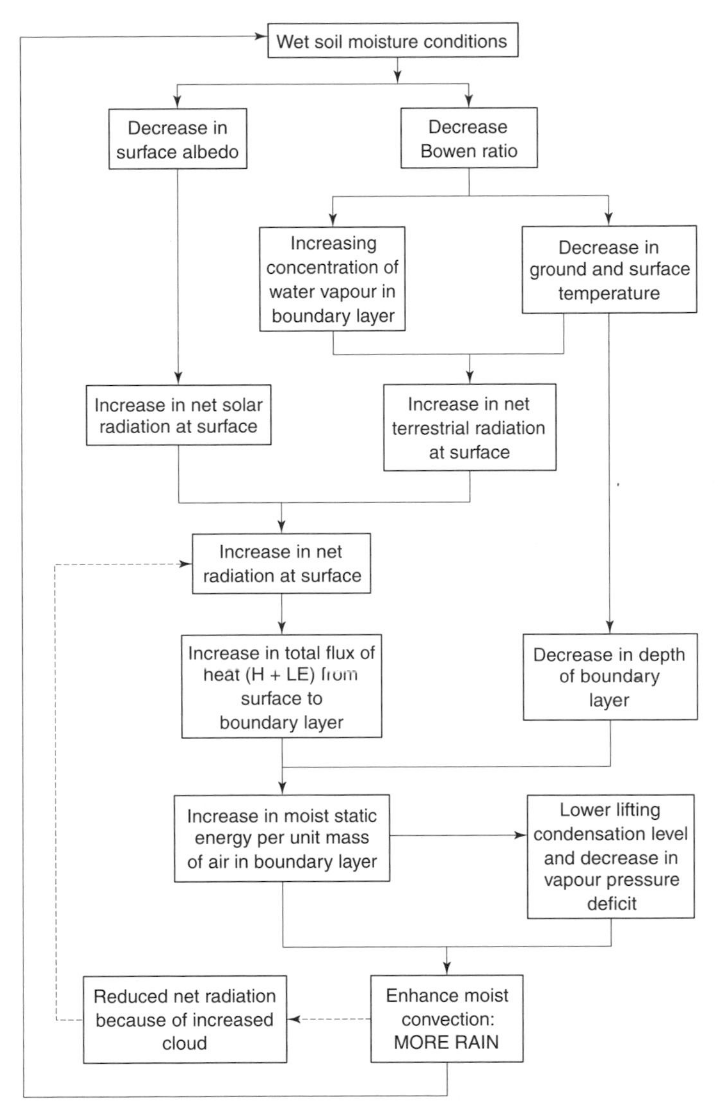

Figure 8.4 The positive feedback between a wet surface and precipitation (adapted from Eltahir, 1998)

and the partitioning of that energy into latent and sensible heat fluxes are affected by soil moisture and its controls on evaporation from soils and transpiration from plants. Figure 8.4 shows a conceptual model (Eltahir, 1998) of the positive feedback between a wet surface and rainfall over or downstream of that surface. The two primary effects, shown at the top of the figure, are on albedo and energy partitioning (as characterised by the Bowen ratio (Chapter 3), which is the ratio of sensible to latent heat flux).

Surface albedo determines the proportion of the incoming short-wave radiation that is reflected back into the atmosphere. Albedo varies with vegetation type, but the albedo of soil is affected by soil moisture content (Chapter 3) and snow cover has a very significant effect on albedo. Long-wave emissions from the surface are a function of temperature (Chapter 3). Wet, dark and cool surfaces therefore not only absorb more incoming radiation, but emit less long-wave radiation, increasing the available energy still further.

The smaller the proportion of the available energy used for evaporation, the warmer the surface. This partitioning of energy affects the stability and height of the PBL: the warmer the surface, the deeper the PBL. Evaporation also puts water into the PBL, so the PBL will be shallower and wetter over a moist surface. A shallow, wet PBL is more likely to generate moist convection, and hence rainfall. There is a negative feedback, because extra moisture leads to greater cloudiness, which leads to less short-wave radiation reaching the surface, but this is generally small. The amount of water available for evaporation at a given time varies with antecedent rainfall, but also with soil type and vegetation type, as shown in Chapter 3. Deep-rooted vegetation can access soil moisture at depth, and can continue to transpire during long dry spells. An area of forest will also hold more intercepted water in the few hours after rainfall than an adjacent area of grassland, for example, and the subsequent evaporation from the forest would therefore be greater. More generally, the greater the opportunities for evaporation from a particular type of vegetation – because it has intercepted water or its roots are allowing transpiration to continue during dry periods – the greater the proportion of the incoming energy used for evaporation and the smaller the proportion of energy used to heat the air.

Over a given area moisture availability varies from day to day, month to month and year to year. Interactions with the atmosphere therefore also vary over time, and there is generally a positive feedback between soil moisture content and subsequent precipitation. Taylor (2000) showed, for example, how convective precipitation in the Sahel tended to occur in the same places for several days in succession. At a larger scale, Schar *et al.* (1999) demonstrated the presence of the surface–atmosphere feedback across much of continental Europe during summer. Unusually high soil moisture contents, following a wet winter and spring, contributed to the high rainfall which generated the Mississippi floods of 1993 (Beljaars *et al.*, 1996; Trenberth & Guillemot, 1996).

Soil moisture is not the only "wet" feature of the land surface that affects the energy balance at the land surface. Snow cover has a very significant effect on albedo. At a given time, the spatial extent of snow cover determines the amount of surface heating over a region, affecting large-scale surface temperatures. As temperature falls in early winter and an increasing proportion of precipitation falls as snow, albedo rises: this lowers the temperature still further, in a positive feedback. Positive feedback occurs during the melt season too, as the reducing snow cover lowers albedo and increases the fraction of incoming energy absorbed at the surface.

The importance of the surface and its moisture content varies with atmospheric conditions. The greater the synoptic flow across a region – depressions or westerlies, for example – the less the effect of the land surface on weather and climate, and the greater the effect of upwind, usually ocean, controls. For this

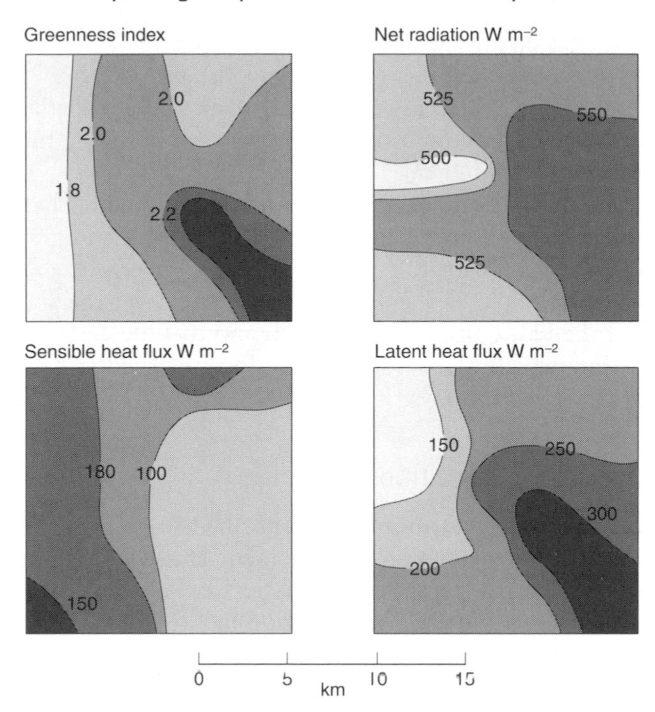

Figure 8.5 Spatial variation in latent and sensible heat fluxes over four hours on one day in summer 1989: Kansas (Desjardins *et al.*, 1992)

reason, the effects of the land surface tend to be greatest in mid-continents during the warm season, and in arid and semi-arid areas.

It is not just the absolute amount of moisture available at the surface that is important, however. The variation in moisture availability across space (due to variations in soil type or vegetation) means that albedo and energy-partitioning vary, and therefore surface temperature varies across space. Temperature gradients, even at scales as small as a few hundred metres, can set up meso-scale disturbances. This has been shown, for example, over the boreal forest of Canada (during the BOREAS experiment: Vidale *et al.*, 1997) and over the prairie grassland in Kansas (during FIFE: Wai & Smith, 1998): these disturbances may be superimposed onto synoptic features such as weather fronts. Variations over space in soil moisture at a given time occur because of differences in soil texture, soil depth and the history of rainfall over the preceding few days or weeks. The top panels of Figure 8.5 show the variation in grass "greenness" (an indirect measure of soil moisture content), and net radiation during the middle of a summer's day across the 15 × 15 km area of Kansas prairie grassland in the FIFE meso-scale field experiment (Desjardins *et al.*, 1992). The grass is greener in the south east part of the square, because here the soil moisture contents are highest because the soil is shallowest. Net radiation is relatively constant across the square (which has gentle topography), with a slight increase in the greener, darker south east because the albedo here is lowest. The bottom two panels show the partitioning of the net radiation into sensible heat and latent heat fluxes: the latent heat flux is considerably greater over the green, wet grass, and is here far greater than the sensible heat flux. Similar results were found in the HAPEX-Sahel study over a larger spatial area, where much of the heterogeneity in soil moisture at any point in time is due to variability in rainfall (Said *et al.*, 1997).

Variations in soil moisture over several hundred kilometres in the Sahel set up secondary circulation patterns over similar scales which migrate north and south through the year with the Inter-Tropical Convergence Zone (ITCZ) and its associated rainfall (Wai *et al.*, 1997). Air rises over the warm dry areas to the north of the ITCZ, moves southwards and then descends over the cool, wetter areas, bringing additional rainfall. This positive feedback delays the end of the wet season by a few days.

Wetlands, lakes and the atmosphere

Evaporation occurs not only from soil and through vegetation, but also from open water. This water may exist in permanent lakes or in seasonal lakes or wetlands, and these expanses of water may be fed from a large distance upstream. Water may therefore be imported into a region by the river network, and made available for evaporation. If this water is imported into a dry region, it may have very significant effects on local and regional energy balances, moisture recycling and climate.

Several significant seasonal wetlands in arid and semi-arid areas are fed by rivers bringing water from more humid headwaters. The Sudd Marshes in Sudan, for example, are fed by the Nile, which obtains most of its water from tropical east Africa. Table 8.3 gives some estimates of the proportion of the water entering some major floodplain wetlands that is "lost" in evaporation: half the water in the Nile, for example, evaporates as the Nile inundates and passes through the Sudd Marshes. Energy is therefore used in evaporation rather than to warm the surface, altering the local and regional energy balance. The evaporated water also falls as precipitation downwind. Figure 8.6 shows a simulation of where the water evaporated in July from the Sudd Marshes falls as precipitation (Koster *et al.*, 1988). Most falls just to the east of the marshes in Ethiopia (and thus supplied the Blue Nile and some of the tributaries of the

Table 8.3 Evaporation from some major wetlands

Wetland	Inflows ($km^3\ a^{-1}$)	Evaporation ($km^3\ a^{-1}$)	Percentage evaporation
Sudd Marshes (Nile)	26.5	13.3	50
Tigris and Euphrates	39.0	32.1	82
Niger internal delta	46.9	22.5	48
Banguela, Zambia	20.7	12.5	60
Okavango, Botswana	10.5	9.9	94

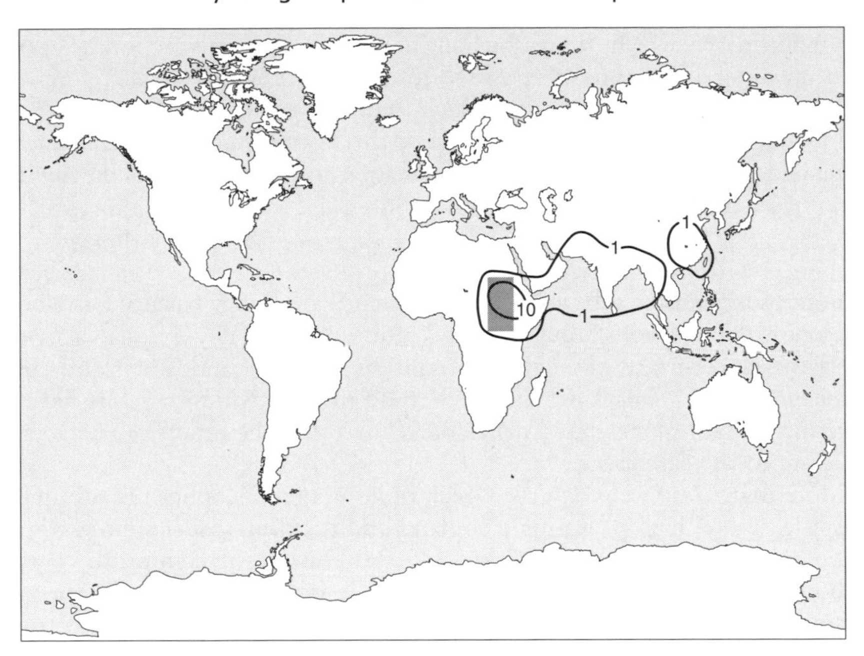

Figure 8.6 Precipitation (in millimetres) originating from the Sudd Marshes over 30 days in July (adapted from Koster *et al.*, 1988)

Nile joining upstream of the Sudd Marshes). The extent and timing of inundation in these seasonal wetlands, and hence the total evaporation and energy partitioning, vary from year to year with the climate in the headwaters of the feeding rivers.

The climatic effects of *permanent* lakes, which have a reasonably consistent area throughout the year, are much less affected by the timing and magnitude of river inflows, and hence by hydrological transport processes. Their effects on energy partitioning and moisture recycling are more affected by their size, depth and thermal characteristics, and vary little from year to year because of variations in inflows (although variation in temperature from year to year can lead to variations in effects). Effects do, however, vary seasonally, depending on the variation through the year in the contrasts between lake and "dry" land. The effect of the North American Great Lakes on downwind precipitation, for example, is greatest in winter when the open lake water provides not only a source of moisture but also a source of heat, which produces a low-pressure trough. This trough causes atmospheric convergence, which increases precipitation to the east of the lakes still further (Barry & Chorley, 1998).

The importance of lakes and wetlands for local and regional climate is further illustrated by model simulations of the mid-Holocene climate (6000 years BP) in northern and western Africa (Coe & Bonan, 1997; Brostrom *et al.*, 1998). At this time, palaeoecological evidence shows that climate of the region was wetter than at present, due to a combination of global-scale orbital changes and different regional vegetation (the wetter conditions favoured increased vegetation cover, which, for reasons outlined in the previous section, triggered even greater rainfall through positive feedbacks). There were therefore also more wetlands and lakes – for which there is geological evidence – and simulations with climate models show that this open water both altered energy fluxes and provided a source of water, increasing still further rainfall downwind.

Effects of human interventions

The land surface affects the atmosphere, through the energy balance (via albedo and long-wave emissions), roughness and moisture availability. Human activities affect the land surface, through deforestation, drainage and other land cover and hydrological changes (Chapter 6). Could these activities therefore have an effect on the atmosphere, and hence on climate and the inputs of energy and moisture to the catchment?

More than 2000 years ago the Greek philosopher Theophrastus, a pupil of Aristotle, wrote that by clearing woodland and draining marshes farmers were altering their local climates (Glacken, 1967). During the eighteenth century in both Europe and eastern North America it was widely accepted that the removal of trees warmed the land, and indeed proposals were made to try to manage land clearance in such a way as to "enhance" the local climate for the benefit of health and crops (Glacken, 1967). These ideas continued to be expressed through the nineteenth century (Meyer, 1996), culminating in the early twentieth century with the belief that in the Great Plains of North America "rain follows the plough" (Riebsame, 1990): by planting trees and breaking up the soil to allow infiltration, farmers would cause rain to re-evaporate locally and generate rainfall, rather than run off to the Gulf of Mexico. This did not happen and in fact the Great Plains suffered in the 1930s from a prolonged and intense drought. During the middle part of the twentieth century, the dominant scientific view was therefore that human activities had at most only very local effects on climate (Meyer, 1996). Urbanisation, for example, has had a long-recognised effect on weather, with the absorbent, relatively dry surfaces triggering convective precipitation downwind (Barry & Chorley, 1998). Other localised effects may arise at the boundaries between different land uses (as found by Brown & Arnold (1998) in Illinois) or close to irrigated land (Barnston & Shickedanz (1984) found increased precipitation in summer downwind of irrigation in the southern Great Plains, for example).

In the 1970s, however, scientific interest in the effect of the land surface on *regional* weather and climate increased significantly, first with Charney's (1975) hypothesis that land clearance in the Sahel was accelerating the rate of desertification, and second with the suggestion that deforestation of tropical rainforests – particularly in the Amazon – was reducing the rate of precipitation recycling. During the 1980s and 1990s there were therefore many empirical and modelling studies exploring the possible effect of land surface changes on the atmosphere, and indeed several of the major field experiments – including HAPEX-Sahel – were primarily established in order to provide data to test hypotheses of effects and responses.

From the 1970s, rainfall in the Sahel region of west Africa was persistently below earlier values, resulting in a long-duration, severely damaging drought. Charney suggested that this persistent rainfall anomaly was due to land clearance, which both increased the albedo and reduced evaporation, leading to a reduction in rainfall: the drought was therefore locally-induced. An alternative explanation, however, was that the drought reflected persistent large-scale climate anomalies, associated for example with persistent sea surface temperature anomalies (Folland *et al.*, 1986). Analysis of observations and modelling studies have shown that in fact rainfall anomalies in the Sahel region are driven by climatic anomalies, but enhanced by feedback from the land surface. During dry years there is less plant growth and less evaporation, leading to further reductions in rainfall. In other words, the land surface reinforces atmospheric anomalies and "locks" the weather into drought mode (Nicholson, 2000). The hypothesis that desertification causes drought is now regarded as untenable (Nicholson, 2000), although land clearance does exacerbate the land–atmosphere feedback. Zheng & Eltahir (1997) used a climate model to explore the effects of land degradation on rainfall in the Sahel, and found little impact. However, removal of the tropical forest along the coast of west Africa would have a far greater impact on rainfall across the whole of the region.

The presence of tropical forest affects significantly local and regional climates. Intercepted water is evaporated and deep roots (extending as far as 8 m) sustain transpiration through the dry season: an estimated 50% of the rainfall in the Amazon basin is generated from evaporation within the basin (Chapter 3). The availability of moisture affects the partitioning of energy, and the dark surface of the forest lowers albedo. Removal of this forest cover will therefore alter both the energy and water balance, and several studies have used climate models to estimate the effects of future deforestation. Although different models vary, a complete removal of the Amazonian rainforest could reduce evaporation by up to 20%, and decrease rainfall by up to 30% across the basin as a whole (Nobre *et al.* (1991): Lean *et al.* (1996) estimated smaller changes). Atmospheric circulation patterns would be weakened, leading to potential changes in climate beyond the Amazon. Deforestation of the other major areas of tropical forest in Africa and south east Asia would have less effect, because these forests recycle less water, but complete replacement of all tropical forest with scrubland would reduce the Asian monsoon and have small, but noticeable, effects on climate in middle and high latitudes (McGuffie *et al.*, 1995). Sud *et al.* (1996) simulated the effects of *past* deforestation in the Amazon. By 1988 around 14% of the tropical forest had been removed, and whilst evaporation was estimated to have fallen by 33% in the deforested area, the total basin-scale reduction was 8% (Figure 8.7 shows the change in evaporation due to deforestation, together with the deforested area). The reduction in rainfall in the deforested area was less than the reduction in evaporation, because of increased convective activity in the warmer boundary layer.

During the Roman Classical Period, about 2000 years ago, the land around the Mediterranean Sea was much more densely forested than at present. Glacken (1967) describes the concerns expressed at the time over the consequences of

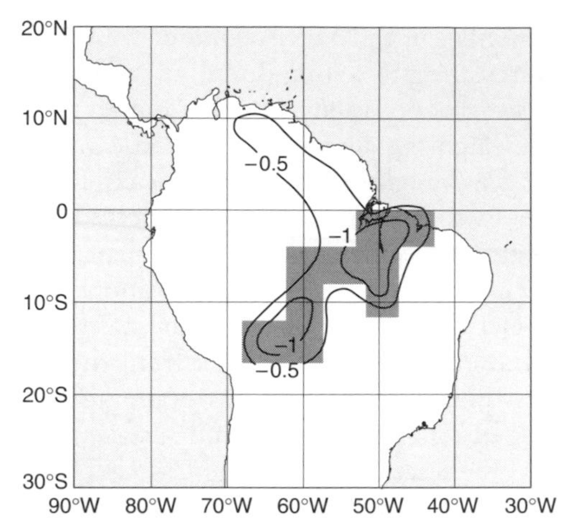

Figure 8.7 Effect of past deforestation on regional evaporation in the Amazon basin (Sud *et al.*, 1996). The shaded area was deforested, and the contours are in mm d^{-1}

deforestation, and these concerns have actually been borne out by recent modelling studies. Reale & Shukla (2000) showed that the ITCZ would have extended further north in Africa and circulation over the Mediterranean would have been enhanced with the original vegetation, and concluded that widespread deforestation initiated during the Roman Classical Period has contributed to the dryness of the current climate around the Mediterranean.

It has also been suggested that large-scale agricultural development in the Great Plains has affected regional climate, and hence inputs to the catchment, albeit in ways not anticipated by those who in the early twentieth century thought that rain would follow the plough. Simulation experiments with climate models suggest that replacing the natural permanent grasslands with a mixture of dry and irrigated farmland affects the partitioning of energy at the surface, leading to increased evaporation and, most significantly for local and regional climate, lower surface temperatures during the summer (Stohlgren *et al.*, 1998; Chase *et al.*, 1999) – which have been observed in parts of the Rocky Mountains. Replacing forest with grassland in the same part of the United States alters winter snow cover extent, and this affects not only the local energy balance but also the timing of streamflows and consequently downstream hydrological regimes (Greene *et al.*, 1999).

Repeated land cover change in many locations affects not only local and regional climate, but also impacts upon global-scale climate. Chase *et al.* (2000) conducted a climate model experiment simulating average January climate with both the "natural" vegetation and with current human-influenced vegetation. Although vegetation was changed in only around 15% of the land surface, and the global-average effect was small, there were clear and systematic regional impacts of land cover change: much of North America and Europe, parts of Siberia and south east Asia, were warmer under the current climate than the "natural" vegetation (by up to 1.5 °C). These effects were not limited to the areas of land cover change, and indeed some of the largest changes in precipitation were found over tropical oceans. The two main mechanisms influencing these large-scale patterns of change were a reduction in the amount of tropical convection and hence reduced high-level outflow from the tropics (resulting in smaller temperature gradients between tropics and high latitudes), and increased albedo in both tropical and mid-latitudes. The implications of these changes in climate for hydrological regimes has not yet been explored – but may regionally be substantial. Higher winter temperatures in mid-latitudes, for example, can be expected to lead to a greater proportion of winter precipitation falling as rain, and running off the catchment in winter rather than as snowmelt in spring. These climatic effects of land use change will be combined with the effects of increasing greenhouse gas concentrations. Pitman & Zhao (2000) used a global climate model to show that whilst historical land use change did not affect the globally-averaged increase in temperature due to increasing CO_2 concentrations during the twentieth century, in some seasons and regions the effects of land use change on temperature were similar to or greater than the effects of climate change. These regions were not necessarily those experiencing land use change. Above 50°N, for example, the effects of land use change were on average as great over the twentieth century as those of increasing CO_2 concentrations, and in some areas twice as large. In tropical areas, in contrast, cooling due to land use change has offset to a small extent the effects of climate change.

Human activities not only affect land cover, but also alter hydrological regimes. Changes in hydrological regimes can be expected to have the greatest effect in arid and semi-arid areas. The construction of a reservoir in a dry area, for example, will introduce a source of moisture which would not previously have existed: local evaporation will be enhanced and weather affected. The impoundment may, however, prevent water reaching wetlands downstream, and thus lead to a *reduction* in evaporation downstream. Abstractions of water (primarily for irrigation) can also lead to reductions in inflows to wetlands or lakes, as shown in Chapter 6. The decline in size of the Aral Sea has changed its effect on the local and regional climate, leading to higher local temperatures and less precipitation (Small *et al.*, 1999). Draining of large wetlands in semi-arid areas would also reduce local evaporation (the Jonglei Canal, draining the Sudd Marshes, for example, would increase flow through the marshes to Egypt, but reduce local evaporation).

8.3 Hydrological processes and the oceans

Water moves from land to sea, along rivers and through groundwater. This transports material from the land to the sea, and the influx of freshwater affects the water balance and circulation of the oceans – which in turn affects climate.[1] A broad distinction can be drawn between the open ocean and the

[1] It is worth noting here that when oceanographers refer to "hydrological processes" and the "hydrological cycle", they are generally referring just to precipitation and evaporation.

coastal zone, which includes the relatively shallow seas of the continental shelf. The coastal zone accounts for approximately 8% of the total area of the ocean, but only 0.5% of its volume (Mackenzie, 1998). However, it is the most productive part of the ocean, contributing between 10 and 30% of the ocean's total organic production (Mackenzie, 1998). This high productivity is due to both the upwelling of ocean waters rich in nutrients and the input of nutrients through rivers. Human activities can impact significantly upon these interactions. Research into the links between river and coastal systems, and the effects of global change, has been stimulated by the IGBP (International Biosphere–Geosphere Programme) LOICZ (Land–Ocean Interactions in the Coastal Zone) project (IGBP, 1995). In the UK, the LOIS (Land–Ocean Interaction Study) investigated the fluxes of water and materials in and out of the North Sea, in rivers, groundwater, sea and atmosphere (Leeks & Jarvie, 1998).

Fluxes of water from land to sea

Oceans play an important role in global, regional and local climate, in two linked ways: through their surface temperature, and through the transport of heat from equator to pole (see Wells (1997) for a broad introduction to physical oceanography). The temperature of sea water at a particular location is partly affected by currents of water and partly by the atmosphere above, in a complex and interactive way. Shallow currents at the surface of the ocean are driven by surface winds, but salinity and temperature gradients also set up vertical circulations (known as thermohaline circulations). Whilst many of these are of local significance only, the oceanic "conveyor belt" which connects all the oceans and transports large amounts of water and heat is a thermohaline circulation. This conveyor belt starts in two places. Low temperatures and the formation of sea ice create cold, saline water in the North Atlantic. This water sinks (forming North Atlantic Deep Water) and flows southwards through the Atlantic Ocean, and joins cold water created off Antarctica (Antarctic Bottom Water). This cold, saline water flows eastwards through the Southern Ocean to the Pacific Ocean. As this deep current flows northwards through the Pacific Ocean it warms, and water reaches the surface in the North Pacific. This warmer water then flows southwards, passes through the Indian Ocean, flows past southern Africa and flows northwards, close to the surface, through the Atlantic Ocean, where it cools and sinks (Broecker, 1987). The north east movement of warm surface water from the tropical Atlantic towards Europe is the Gulf Stream, which makes the climate of western Europe warmer than it otherwise would be. Changes in the intensity of this thermohaline circulation affect climate, particularly in Europe, and it too may be affected by climatic fluctuations. This global thermohaline circulation has weakened substantially in the past – leading to rapid cooling in Europe – and it has been hypothesised that it may do again in the future (see below).

The variations in water temperature and salinity that generate both the global and more local-scale thermohaline circulations are triggered by air temperature and the ocean water balance. The greater the volume of freshwater inputs to a

Table 8.4 Precipitation, evaporation and runoff over ocean basins (10^3 km^3 a^{-1})

	Precipitation	Evaporation	Precipitation – evaporation	Runoff	Runoff as % of total inflows
Atlantic	74.6	111.1	−36.5	20.8	22
Pacific	228.5	212.6	15.9	14.8	6
Indian	81.0	100.5	−19.5	8.2	7
Arctic	0.8	0.4	0.4	5.2	87
Mediterranean	0.9	3	−2.1	0.45	33
Baltic	0.237	0.184	0.053	0.428	64

Sources: Baumgartner & Reichel (1975); Gleick (1993); Piexoto *et al.* (1982); Schinke & Matthaus (1998)

part of the ocean, and the lower the rate of evaporation from the sea surface, the lower its salinity. The total inputs to a part of the ocean comprise rainfall, river runoff, groundwater discharge and inflows from other oceans, and these are balanced by evaporation and outflows to other oceans. They must balance, because a consistent sea surface level must be maintained. The inflows and outflows to and from other oceans are saline, and the other inputs and outputs are freshwater. Freshwater is also created when sea ice melts, because salt is not incorporated into ice when sea water freezes. Table 8.4 gives a set of estimates of precipitation and evaporation from oceans and inputs from rivers. The Atlantic and Indian Oceans "lose" more water in evaporation than they receive in precipitation, and runoff from the land does not make up the difference. Precipitation exceeds evaporation over the Pacific Ocean, and runoff doubles the surplus; runoff provides close to 90% of the inputs to the Arctic Ocean. The "surplus" water in the Pacific flows out into the Indian Ocean and through the Bering Straits to the Arctic Ocean, where it joins the water generated in the Arctic Ocean and flows into the Atlantic. This movement of water is not the same as the conveyor belt, and it occurs at shallower depths. The numbers in Table 8.4 are subject to considerable error – especially the estimates of evaporation and precipitation (Chahine, 1992) – and different assessments give different answers (as shown by Oki (1999) and Chen *et al.* (1994)): estimates of inputs and outputs are also difficult to reconcile with estimates of freshwater fluxes from one ocean to another (Wijffels *et al.*, 1992).

The vast majority (around 95%) of discharge from land to sea is through rivers, although in some regions groundwater contributions may be greater (Moore (1996) showed that direct groundwater discharge to the sea in parts of South Carolina was 40% of that of rivers during summer). Figure 8.8 shows the parts of the world draining into five "oceans", together with areas of endorheic drainage which does not reach the open ocean: Table 8.5 presents some summary statistics. Whilst there are 6152 separate river basins discharging either into a sea or an internal "sink" (at a spatial resolution of 0.5 × 0.5°), 65% of the total land mass is drained by around 100 major river basins. Relatively little of the inputs to the Pacific Ocean come from the land (Table 8.5), but a substantial proportion of the inputs to both the Arctic and Atlantic

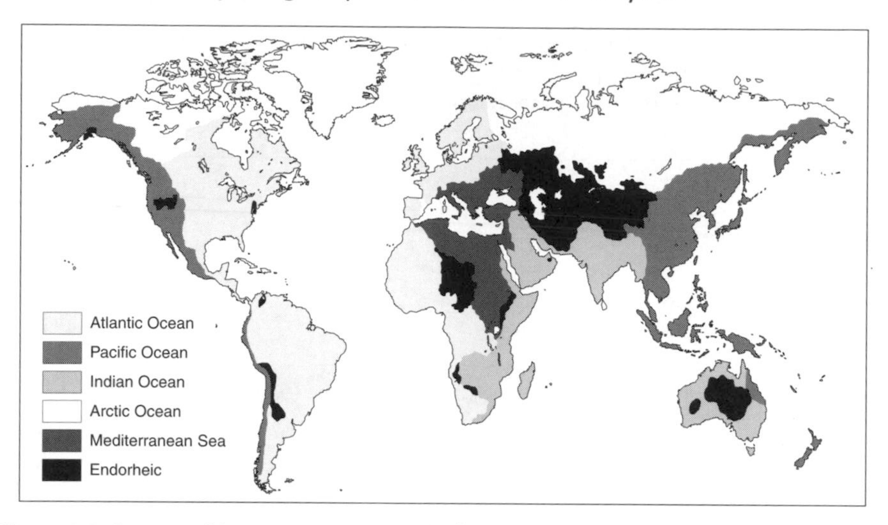

Figure 8.8 Ocean and internal drainage basins (Vörösmarty *et al.*, 2000)

Table 8.5 Characteristics of land areas draining to oceans (adapted from Vörösmarty *et al.*, 2000)

Ocean/receiving body	Land area (10^6 km^3)	Percentage of ice-free land	Ratio of land area to ocean area	Number of separate basins
Arctic	17.0	13	1.21	1644
Atlantic	45.6	34	0.55	1387
Indian	21.1	16	0.29	811
Mediterranean and Black Sea	10.9	8	4.36	305
Pacific	21.1	16	0.13	1555
Internal drainage basins	17.4	13	–	450

Oceans come from land. Greater proportions of the inflows to enclosed seas, such as the Mediterranean, Baltic and North Sea, come from river runoff.

The pattern of circulation within an enclosed sea (i.e. one with a relatively narrow outlet to the ocean) is determined by the relative balance of inputs of runoff and rainfall and output of evaporation. If runoff plus rainfall is greater than evaporation, then relatively fresh water flows out into the adjoining ocean, and a lesser amount of saline water enters the enclosed sea at depth: this happens in the Baltic Sea. If evaporation is greater than runoff plus rainfall, however, then more water comes into the sea than leaves it, and the outflow tends to be more saline than the inflow: this happens in the Mediterranean Sea.

The location and timing of inflows to the open ocean can also influence circulation patterns and other important aspects of the marine environment. In volumetric terms, the greatest influx of water from the land to the oceans is in the tropical Atlantic, particularly from the River Amazon. The Amazon discharge forms a plume of low-salinity water that is typically between 3 and 10 m thick, between 80 and 200 km wide, and that extends beyond the continental shelf (Lentz & Limeburner, 1995). This plume entrains twice as much water again as it moves north westward up the Brazilian coast. Ternon *et al.* (2000) found that the low-salinity Amazon plume water acted as a sink for atmospheric CO_2, unlike the rest of the equatorial Atlantic Ocean: this finding links the land surface to the atmosphere via the ocean. Low-salinity plumes have been recorded off many major rivers, and there is a significant statistical correlation between salinity in the Atlantic Ocean near Bermuda and discharge from the Mississippi River into the Gulf of Mexico (Atkinson & Grosch, 1999).

The most significant effect of river flows, in climatic terms, however, is perhaps through their contribution to the freezing and thawing of sea ice in the Arctic Ocean. The extent of ice cover on the Arctic Ocean varies through the year (ranging from typically 8.5 million km^3 in September to 15 million km^3 in March), and from year to year. This variability is important, because it affects the energy balance at the earth's surface (ice has a high albedo and prevents the flux of heat between sea and air (Curry *et al.*, 1995), and because ice melt produces freshwater which lowers salinity. The onset of the freezing of sea ice depends on the energy balance and windspeed (higher windspeeds delay freezing) at the surface and also the salinity of the water. The higher the salinity, the lower is the freezing point, and there is a tendency for an increased rate of formation of sea ice around river plumes in autumn (Manak & Mysak, 1989). The onset and duration of thaw in early summer are also functions of both the atmosphere and the flux of freshwater from rivers. A number of the world's largest rivers drain into the Arctic Ocean, including the Mackenzie (Canada), Ob, Yenisei and Lena (all in Russia). These rivers are fed by the melting of snow in the Rocky Mountains and the northern Himalaya, and this snowmelt is transported to the Arctic Ocean before the sea ice has begun to melt. This influx of water brings a flux of sensible heat to the sea ice: approximately half of the heat that melts coastal ice near the mouth of the Mackenzie River comes from the river (Searcy *et al.*, 1996). Figure 8.9 shows Mackenzie River flows and the thickness of ice in the coastal zone (Macdonald *et al.*, 1999), together with a schematic showing how the coastal ice breaks up as river flows increase (Searcy *et al.*, ibid.). Whilst some of the river discharge overflows onto the ice, most flows under the floating sea ice and melts it from below. This melting of sea ice increases the formation of freshwater, and alters locally the energy balance by increasing the amount of energy absorbed by the surface, although the effect of this on local climate is not clear (Searcy *et al.*, 1996). River inflows also generate some buoyancy-driven currents in parts of the Arctic Ocean. The Kolyma River, for example, sets up a westward-flowing surface current in the East Siberian Sea (Munchow *et al.*, 1999).

The extent of sea ice in both the Arctic and Antarctic Ocean varies from year to year. These variations feed through to atmospheric or ocean anomalies elsewhere. Both the analysis of observations and modelling studies have shown that they are primarily due to variations in atmospheric circulation, and

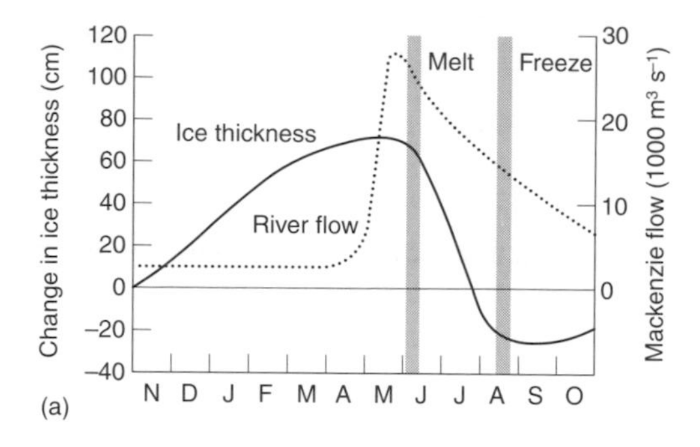

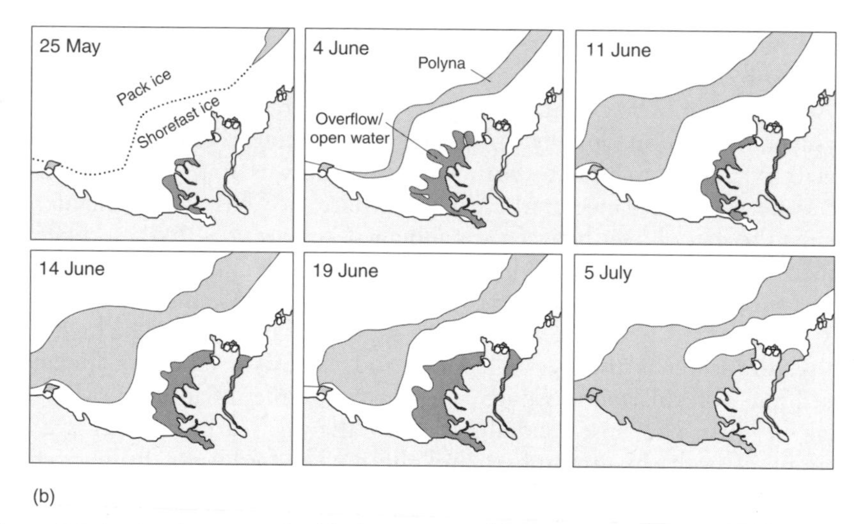

Figure 8.9 Ice break up near the Mackenzie River: (a) inflows and ice thickness (Macdonald *et al.*, 1999), and (b) a schematic of the spatial pattern of break up (Searcy *et al.*, 1996)

particularly in windspeed and temperature (Tremblay & Mysak, 1998). Year-to-year variations in river flows have relatively little effect on sea ice anomalies in the Arctic Ocean at the ocean-scale, but do affect the timing and extent of both thaw and break up around river mouths.

Rivers flow into oceans through estuaries and deltas, the size and characteristics of which depend on local topography. Estuaries and deltas are the interface between land and ocean, and this interface is dynamic, depending on the state of the tide and the volume of river inflows. The lower the freshwater inflows, the further up river saline water (the "salt wedge") will penetrate at high tide. The detailed flow patterns, and hence the patterns of erosion and deposition, within an estuary or delta are a function of the dynamic balance between river flows and sea water. Ibanez *et al.* (1997), for example, showed how river discharge was the major factor controlling the movement of the salt wedge in both the Rhône and Ebro estuaries, both of which have a low tidal range. In both estuaries, depositional conditions prevail when river flow is below average, and erosion occurs when flows are above average.

Table 8.6 Effect of human interventions on sea level during the twentieth century (Gornitz *et al.*, 1997), as presented by Harvey (2000): parentheses show ranges of possible effects

	Effect on global sea level over the twentieth century (cm)	Effect on global sea level in the 1990s (mm a^{-1})
Increase runoff and sea level		
Urbanisation	1.51	0.38 (0.35–0.41)
Aquifer mining	0.98	0.30 (0.1–0.4)
Deforestation	0.67	0.13 (0.12–0.14)
Destruction of wetlands	<0.01	0.0015 (0.001–0.002)
Total increase	*3.17*	*0.81 (0.57–0.95)*
Decrease runoff and sea level		
Storage behind dams	−0.38	−0.34 (−0.38−−0.3)
Deep infiltration behind dams	−1.71	−0.68 (−0.75−−0.61)
Evaporation from reservoirs and irrigation	−0.54	−0.16 (−0.16−−0.15)
Deep infiltration of irrigation water	−1.82	−0.44 (−0.48−−0.44)
Total decrease	*−4.45*	*−1.62 (−1.77−−1.5)*
Net change	*−1.28*	*−0.81*

Human impacts on the flux of water from land to sea

Human activities that affect the volume of runoff in the catchment can affect sea level, the circulation of water in estuaries, deltas and the ocean, and the thawing of sea ice. Table 8.6 gives estimates of the effect of the key human interventions that cause increasing runoff (and hence increasing sea level) and decreasing runoff (and hence lowering sea level): the estimates are very uncertain. The net effect of these human interventions is to have lowered the global average sea level by just over 1 cm during the twentieth century, although the rate has increased substantially over time. Most of the net decrease has occurred during the last 30 years of the twentieth century, and over that period the rate of decrease is only slightly smaller than the 1–2 mm a^{-1} increase in sea level due to glacier melt and the thermal expansion of sea water. Human interventions in the catchment are therefore possibly slowing substantially the rate of sea level rise.

Human interventions in the catchment have affected flow patterns in many individual estuaries and deltas. A reduction of flows by around 28% in the Ebro River, for example, due to irrigation and evaporation from reservoirs, has led to an increase in both the penetration of the salt wedge into the estuary of the Ebro and the duration of saline conditions (Ibanez *et al.*, 1996). Reservoirs in the catchment regulate river flows, and the removal of high peak flows has further altered saline penetration dynamics. Similar examples can be found in many other estuaries, particularly those in which river flows are affected by upstream reservoirs. As a general rule, land cover change tends not to affect estuaries and deltas of *large* rivers because only rarely does land cover change affect a large enough part of the catchment. Impoundment, however, does influence the flow regimes even of very large rivers (Chapter 6).

Changes to river flows entering the sea can in some circumstances affect not only the estuary or delta, but also larger-scale ocean circulation patterns. Before the construction of the Aswan Dam in the 1960s, for example, the Nile was the biggest single source of streamflow to the Mediterranean. Evaporation losses from Lake Nasser behind the dam together with abstractions for irrigation have reduced significantly the volume of Nile flows reaching the Mediterranean. This has resulted in an increase in inflows of water from the Atlantic, and also an increase in the salinity of water moving from the Mediterranean to the Atlantic (Turley, 1999). Regulation of some of the major rivers supplying the Baltic Sea has also altered the timing of flows – increasing flows during winter and reducing them during summer – which has led to changes in circulation patterns within the Baltic Sea and a significant reduction in the frequency of inflows of saline oxygenated water from the North Sea (Schinke & Matthaus, 1998).

During the 1960s engineers in the Soviet Union planned large-scale diversions of major Siberian rivers, directing them to flow southwards towards arid central Asia rather than northwards into the Arctic Ocean. It was suggested that these diversions would significantly weaken the cold, stable halocline which separates the surface mixed layer from the deeper, warmer Atlantic water, enhancing mixing and therefore reducing sea ice formation over an area large enough to impact upon climate (Aargard & Coachman, 1975). Semtner's (1984) model experiments implied that reducing runoff would in fact have little effect on Arctic stability and hence ice formation, but the numerical experiments of Weatherly & Walsh (1996) showed that the effects could be substantial. Fortunately, however, the proposed river diversions were abandoned during the early 1990s, but the studies triggered by them have nevertheless indicated how potentially sensitive the Arctic Ocean is to changes in river inputs.

Climate change may also, of course, affect the volume and timing of inflows of water to oceans, with subsequent impacts on ocean processes. As indicated in Section 8.4, current climate models do not generally account explicitly for the movement of water from land to sea (effectively ignoring the river-supplied freshwater inputs to oceans), so potential effects of changes in streamflow due to climate change on oceans are not well studied. Model calculations suggest, however, that increasing river flows into the Arctic Ocean (due to increased precipitation on land at high latitudes) would increase the export of freshwater from the Arctic to the North Atlantic (Miller & Russell, 2000), and there are indications that the relative magnitudes of freshwater fluxes through different straits would alter (Steele *et al.*, 1996), potentially affecting the preferred locations for the formation of deep water in the North Atlantic Ocean. It has been hypothesised that increasing inputs of freshwater into the North Atlantic could reduce the rate of formation of North Atlantic Deep Water (NADW),

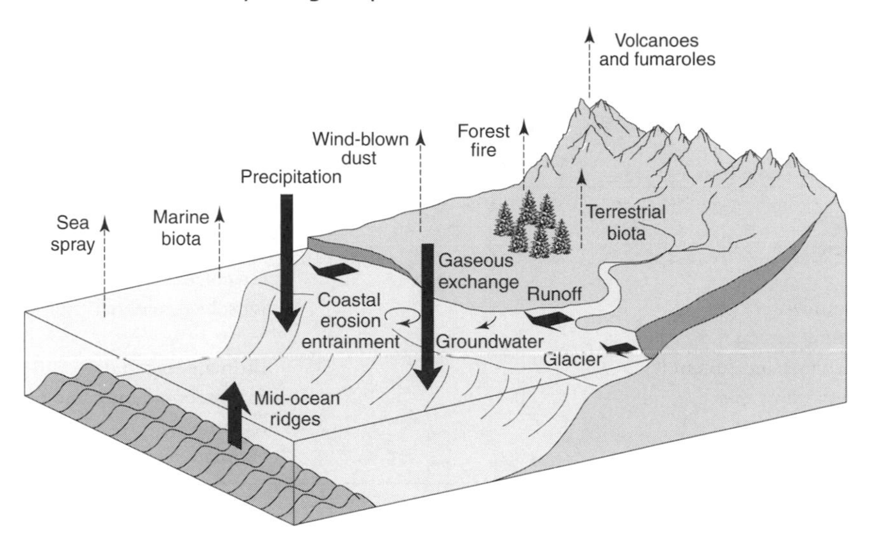

Figure 8.10 Sources of material for the ocean. The dashed lines show the sources of material for the atmosphere, which subsequently enters the ocean

slowing down the "conveyor belt" circulating water around the world, and reducing the amount of heat transported towards Europe (Rahmstorf, 1995). There is evidence that this has happened in the past – during the Younger Dryas, around 10 000 BP – rapidly lowering temperatures across much of Europe. This has been attributed to a sudden large influx of cold freshwater from the decaying Laurentide Ice Sheet, draining through the St Lawrence River (Broecker, 1987). Rahmstorf (1995) suggested that relatively small changes in freshwater inputs – from precipitation onto the Arctic Ocean or in the catchments draining into the Arctic – could change dramatically the thermohaline circulation, and this possibility has triggered considerable scientific and indeed public interest. Whilst most studies have found at most a weakening of the thermohaline circulation (e.g. Mitchell *et al.*, 1999), there remains the possibility that changes in climate in high latitudes might trigger abrupt changes in the thermohaline circulation because the sensitivity of the thermohaline circulation to changes in inputs is very uncertain.

Material fluxes to the sea

Although the dominant chemical constituents of sea water are sodium and chloride (which together make up close to 90% of the material held in sea water), sea water contains virtually every stable element, mostly at very low levels. These materials come from three directions (Figure 8.10): within the ocean, from the atmosphere and from the land. Material produced within the ocean largely comes from mid-ocean ridges, where cold sea water percolates into fissures in the hot basaltic material being forced upwards and becomes enriched in certain dissolved metals (including potassium, calcium and silicon). Some material is mobilised from sediments on the coastal shelf which were

Table 8.7 Contribution of different mechanisms to the transport of material to the oceans (from Garrels *et al.* (1975), with additions)

Mechanism	**Percentage of total transport**	**Comments**
Rivers	89	72% as suspended load
Glaciers	7	Mainly from Antarctica and Greenland
Groundwater	2	Dissolved material
Coastal erosion	1	
Volcanoes and fumaroles	0.3	Estimate uncertain
Wind-blown dust	0.2	Major source for some ocean regions
Mid-ocean ridges	<<0.1	Estimate uncertain

Note: The total does not include fluxes of carbon dioxide

initially deposited on dry land and are now underwater due to sea level rise since the last glacial period. Material that enters the ocean through precipitation derives from many sources, including wind-blown dust, volcanoes and fumaroles, forest fires and emissions from terrestrial and oceanic biota. The biggest source of material in the atmosphere, however, is through sea spray (the source of around 64% of total particulate material in the atmosphere: Andrews *et al.*, 1996) which puts salt particles into the atmosphere. This material can be thought of as being recycled, and so does not really represent a net input to the ocean. Some of the material from the atmosphere reaches the ocean in precipitation, but much enters the ocean through gaseous exchange. Most of the carbon that gets into the ocean from the atmosphere, for example, enters as carbon dioxide. There are four ways in which material moves from land to the ocean: rivers, groundwater, glaciers and coastal erosion. Rivers are by far the most significant (Table 8.7), with around 72% of the material carried as suspended load and just 28% in dissolved form, and in fact most of the constituents of sea water ultimately derive from river inflows (with the notable exception of chloride, which was degassed from the mantle early in the earth's history and has been continuously recycled since). The vast majority of the trace metals in the western Pacific Ocean, for example, come from river inflows and particularly from the islands of the East Indies (Milliman *et al.*, 1999; Sholkovitz *et al.*, 1999). Table 8.8 gives some approximate estimates of the volumes of material transported from land to oceans by rivers. In some areas and for some chemical species, groundwater may make a greater contribution than river flow. Most of the barium load in the coastal waters off the south eastern USA, for example, comes from groundwater rather than rivers (Shaw *et al.*, 1998).

There is considerable spatial variation in riverine inputs of material to the coastal zone and ocean, with most coming from south east Asia (Figure 8.11). Sediment yields tend to be greater for smaller coastal catchments than for large ones, because gradients are steeper, and small mountainous catchments therefore

Table 8.8 Estimated total volumes of material transported from land to sea through rivers

	Before major human interference Million tonnes per year	**Late twentieth century Million tonnes per year**	**Reference**
Particulate sediment	10 000	20 000	Milliman & Syvitski (1992)
Dissolved material	5800		
Total loads (particulate and dissolved)			
Carbon	800	1230	Mackenzie (1998)
Nitrogen	36.5	49–62	Berner & Berner (1996)
Phosphorus	12	23	Berner & Berner (1996)
Dissolved load			
Chloride	217	310	Berner & Berner (1996)
Sodium	195	270	"
Magnesium	127	138	"
Sulphate	198	430	"
Potassium	49	52	"
Calcium	501	550	"
Bicarbonate	1945	1982	"
Silica	389	389	"
	Thousand tonnes per year	**Thousand tonnes per year**	
Cadmium	n.k.	0.3	Andrews *et al.* (1996)
Copper	n.k.	10	"
Nickel	n.k.	11	"
Iron	n.k.	1100	"
Lead	n.k.	2	"
Zinc	n.k.	6	"

n.k. not known. See Table 5.2 for a summary of the effects of human intervention on loads

may contribute a greater proportion of sediment to a coastal zone than their area would imply (Milliman & Syvitski, 1992). For example, small mountainous streams in south west France drain less than half of the land draining to the southern Bay of Biscay, but contribute 52% of the sediment input (Maneux *et al.*, 1999), and mountainous Greek rivers transport around 14% of the sediment inputs to the Mediterranean, despite draining a much smaller proportion of the total catchment (Poulos *et al.*, 1996). In many environments there is considerable variation from year to year reflecting variations in streamflow (Inman & Jenkins, 1999; Restrepo & Kjerfve, 2000: Chapter 5), and sediment transport may be concentrated into just a few events.

Between 80 and 90% of the sediment that reaches the sea is deposited in the nearshore coastal zone, so most material that is carried as suspended sediment does not reach the open ocean. Much deposition occurs in estuaries

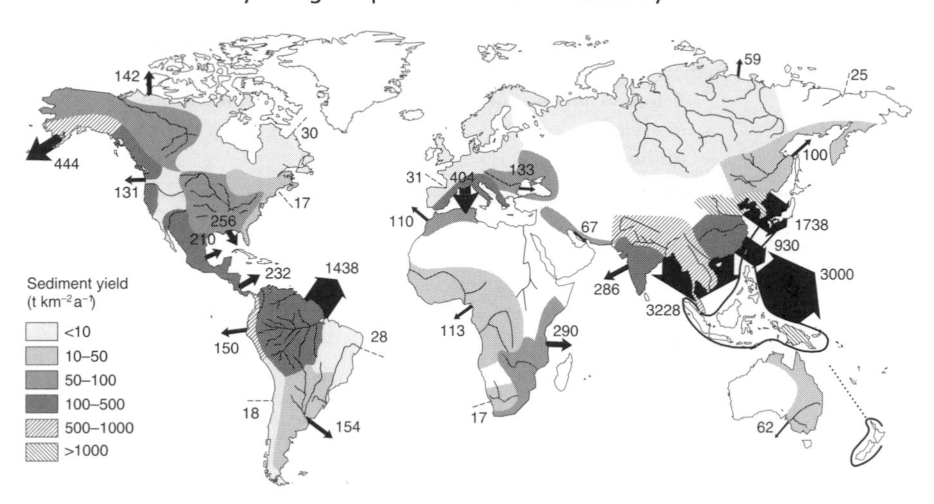

Figure 8.11 Annual discharge of suspended sediment to the ocean, and sediment yield: the width of the arrows corresponds to the relative sediment discharge (Milliman & Meade, 1983; Milliman & Syvitski, 1992)

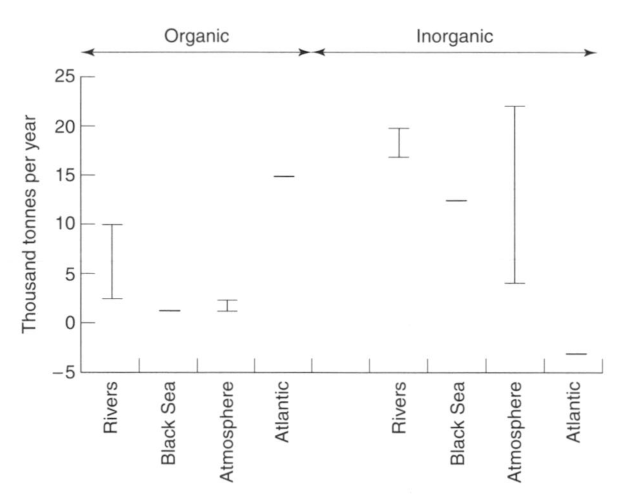

Figure 8.12 External sources of carbon for the Mediterranean Sea (adapted from Sempere *et al.*, 2000): the Black Sea source derives from rivers draining into the Black Sea

and deltas, partly because of the reduction in river water velocity as it meets salt water, and partly because of chemical reactions which occur when fresh-water mixes with saline sea water.

It is not only the volume of material that is important, however, but also the composition of that material. As indicated above, rivers are the dominant source for most trace metals in the ocean, and rivers also represent major sources for nutrients and minerals. An estimated 60% of the dissolved organic carbon in the Laptev Sea (part of the Arctic Ocean) comes from the Lena River in Siberia (Kattner *et al.*, 1999), mostly following the annual inundation of the floodplain in the lower reaches of the river. Figure 8.12 shows estimates

Table 8.9 Effect of reservoir impoundment on sediment discharge to the ocean (Milliman & Meade, 1983; Palanques *et al.*, 1990; Milliman & Syvitski, 1992; Algan *et al.*, 1999)

River or region	Million tonnes per year		
	Sediment discharge before dam construction	Sediment discharge after dam construction	Current discharge as percentage of earlier discharge
Mississippi	500	210	42
Rio Grande, USA	20	0.8	4
Colorado, USA and Mexico	125–150	20	13–16
Zambesi	50–75	20	27–40
Nile	125	3	2.4
Niger	40–65	5	8–13
Indus	250	50	20
Anatolian rivers	53	24	45
Ebro	18	1.5	8
Rhône	12	4–5	33–42

of the relative magnitudes of different external sources of organic and inorganic carbon for the Mediterranean Sea: rivers are clearly an important source. Rivers are particularly important in the transport of nutrients to the coastal zone and open ocean, including approximately half of all nitrogen inputs to the ocean (Cornell *et al.*, 1995). Human activities, however, have severely affected this transport in many catchments.

Human impacts on material fluxes

The human interventions in the hydrological system summarised in Chapters 6 and 7 impact not only on the flux of water from land to sea, but also on the flux of sediment, nutrients, trace metals and other materials (Table 8.8).

These human activities have two contrasting effects on sediment transport to the ocean (Walling, 1997). Land cover change, and particularly deforestation, tends to increase sediment supply and transport (Chapter 6). Impoundment, however, *reduces* sediment transport by trapping sediment behind dams: some of this sediment may be generated by upstream land cover change, so deforestation does not necessarily result in increased sediment fluxes to the ocean. Table 8.9 summarises the effect of dams on sediment transport to the ocean along some rivers. The net effect of the two activities has been to increase total sediment supply to the ocean, from perhaps less than 10 billion tonnes per year before widespread farming and deforestation, to perhaps 20 billion tonnes by the middle of the twentieth century (Milliman & Syvitski, 1992), but this increase is largely concentrated in small, undammed rivers draining the East Indies. In such areas increased sedimentation has encouraged the development of mudflats and deltas, and has also been blamed for the smothering of coral reefs (Hunte & Wittenberg, 1992).

Reductions in sediment discharge to the ocean are geographically more widespread, and this starvation of sediment has in many instances led to coastal erosion. Beach erosion along the Gulf Coast of the United States, for example, is partly attributed to the impoundment of sediment behind dams (Davis, 1997), and the reduction in sediment transport in the Colorado River has led to enhanced erosion and deposition in its delta at the head of the Gulf of California (Carriquiry & Sanchez, 1999). Erosion in the Nile delta – up to 40 m a year – is due to the virtual cessation of sediment transport since the closure of the Aswan Dam in 1964 (Frihy *et al.*, 1998): there are many other examples.

Increased nutrient fluxes to the ocean are largely due to fertiliser use (Chapter 6) and the deposition of nitrogen compounds following the combustion of fossil fuels (Chapter 7). Howarth *et al.* (1996) estimated nitrogen and phosphorus for 14 regions draining into the North Atlantic Ocean, showing that the largest fluxes per unit area came, unsurprisingly, from north west Europe and north east North America: fluxes to the ocean in these regions were over 1000 kg N km^{-2} a^{-1}. In highly-populated temperate regions, nitrogen and phosphorus fluxes were generally between 2 and 20 times as high as would be expected under "natural" conditions. These high nutrient fluxes have particularly important implications for estuaries and enclosed seas, and can lead to eutrophication. An estimated 70% of all eutrophication pollution in coastal zones originates from upstream sources (Mackenzie, 1998). Examples of estuaries and enclosed seas prone to increased eutrophication include Chesapeake Bay in the eastern United States, the northern Adriatic Sea (with the highly-eutrophic Venice lagoon at its head), and the Baltic Sea. However, whilst estuaries and enclosed seas are very prone to eutrophication, the very large store of nutrients in the deep ocean, in relation to riverine inputs, means that deep ocean nutrient concentrations are likely to be little affected by human interventions (Andrews *et al.*, 1996).

Silica is also a nutrient. In the coastal environment, it is used particularly by diatoms (a type of phytoplankton), which are themselves food for many higher-order aquatic animals. Silica concentrations are not increased by human activities, but can be reduced by reservoir impoundment because much silica is transported in association with particulate sediments. Humborg *et al.* (1997) showed how the silicate load of the Danube fell by two-thirds after the construction of a series of dams in the 1970s, with a resulting similar decrease in silica concentrations in the Black Sea. This triggered a shift from diatoms towards non-siliceous phytoplankton, affecting the food web of large parts of the Black Sea.

Rivers can also transport heavy metals to the coastal zone, occasionally leading to increases in sediments to toxic levels. Since these materials are conservative, any material which finds its way into a river has the potential to be transported to the sea, although the travel time will depend on patterns of deposition along the river network. Increased metal concentrations have been observed in many estuaries downstream of industrial areas. The Humber estuary, draining the industrial north of England to the North Sea, has been well studied, for example

(Tipping *et al.*, 1998). This, like many heavily-developed catchments, also receives inputs of organic and other contaminants (e.g. Long *et al.*, 1998).

8.4 Incorporating hydrological processes into climate models

Many of the studies cited above used climate models to simulate the effects of land cover change on climate: the component of the climate model that interfaces with the land surface obviously needs to be realistic, and it was noted above that different land surface models could give different estimates of the effect of land cover change. Climate models also, of course, need accurate land-surface components in order to simulate accurately both weather and climate, and the timing of inflows into the ocean affects climate. Different land-surface components can lead to different forecasts (e.g. Beljaars *et al.*, 1996). There has therefore been much research to refine the way in which climate models simulate hydrological processes.

Modelling the partitioning of energy

Climate models (Box 7.2) simulate weather and climate by solving the fundamental equations of the conservation of mass, energy and momentum at regular intervals. The spatial resolution of climate models ranges from less than 50 × 50 km for regional climate and weather prediction models to up to 500 × 500 km for global climate models. Processes operating at finer spatial scales, such as cloud formation, are "parameterised"; estimated by empirical or conceptually-based relationships. Processes at the land surface operate at this sub-grid scale, and are therefore simulated using "land surface parameterisations". These are similar in principle to hydrological models as used by hydrologists, and a climate modeller's land surface parameterisation is a hydrologist's conceptual model.

The earliest land surface parameterisation is known as the "bucket" model (Figure 8.13a). The bucket is filled by rain, and emptied by evaporation. "Runoff" is generated when the bucket is full. The rate of actual evaporation is a function of the amount of water held in the bucket (as indeed is the flux of heat into the soil). The capacity of the bucket controls the generation of runoff and the amount of water which enters the soil and can be evaporated (Milly & Dunne, 1994). A "leaky bucket" model allows water to drain slowly out of the bottom of the bucket, further depleting soil moisture storage. Whilst this representation does allow soil moisture to vary dynamically, it is rather unrealistic for two main reasons: it does not represent hydrological processes properly, and it does not account for variability in process across a climate model grid cell. Also, the simple bucket model does not include vegetation.

The first set of refinements has therefore been to enhance the representation of process, by introducing more soil layers and adding vegetation with different interception and transpiration properties. Several such refinements have

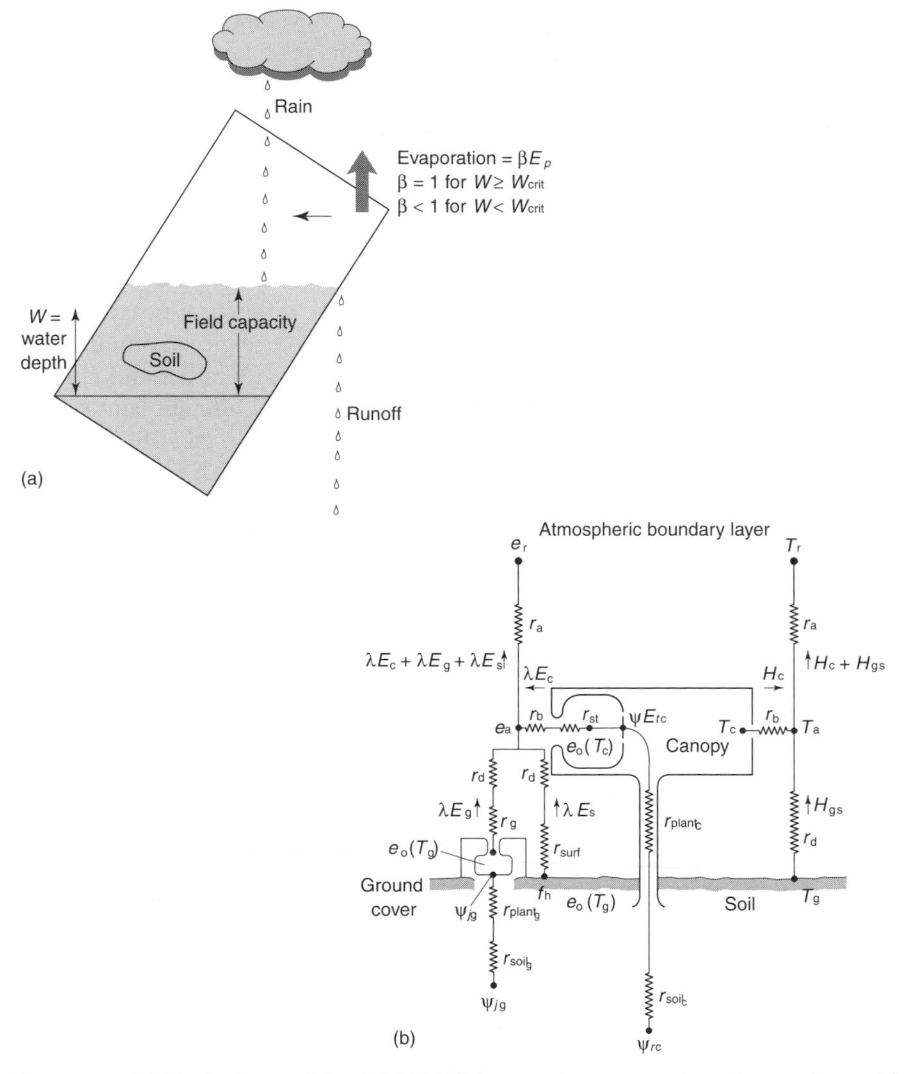

Figure 8.13 (a) The bucket model and (b) SiB (Sellers *et al.*, 1989: see the reference for model details)

been developed, the best known of which are the Simple Biosphere model (SiB: Sellers *et al.*, 1989: Figure 8.13b) and the Biosphere–Atmosphere Transfer System (BATS: e.g. Gao *et al.*, 1996). The models account for the vertical fluxes of water (and energy) through vegetation and soil, with parameters estimated from vegetation and soil properties: the models have become known as Surface–Vegetation–Atmosphere (SVAT) models. Model parameters are frequently based on the results of meso-scale field experiments (e.g. Sellers *et al.*, 1989), and indeed one of the objectives of most meso-scale experiments is to help in the refinement of SVAT models.

Treating a large area of land (even the 2500 km^2 of a high-resolution regional model) as a single column of homogeneous soil with uniform inputs of precipitation and energy is very unrealistic. If the rainfall generated within a climate model grid cell was spread evenly across the entire grid cell, then rain would effectively be falling as a light drizzle everywhere, much of the time. Virtually all of this rain would be evaporated, because the total at each point in a grid cell would be small relative to evaporative demand. However, in reality rain falls only on part of the land surface at any time (Chapter 3), and in this portion of the catchment – or grid cell – rainfall will be considerably greater than evaporative demand, and thus runoff will be generated. Climate models since the early 1990s have generally accounted for this by assuming that the rainfall simulated with a grid cell is distributed statistically across the cell, usually following an exponential distribution with parameters based on the observed spatial variability in rainfall amounts. The partitioning of precipitation into rapid runoff and soil moisture storage varies substantially depending on the assumed distribution of rainfall across the cell (Pitman *et al.*, 1990).

Even with a homogeneous input of precipitation and energy, however, this precipitation and energy is partitioned differently in different parts of the catchment or grid cell because soil and vegetation characteristics vary. It was shown in Chapter 3, for example, how rapid runoff tends to be generated from specific parts of the catchment, typically those with soils that are rapidly saturated. There are two basic ways of including this heterogeneity in a land surface parameterisation. The first is to divide a grid cell into a number of homogeneous patches, and calculate fluxes separately for each (Dolman & Blyth, 1997). The total flux over the climate model grid cell is the sum of the component fluxes. Scale is a key issue here, because the response of the average surface is not necessarily the same as the summed response of different parts of the surface (Sellers *et al.*, 1997; Wood, 1997; Song *et al.*, 1997). This occurs because of the non-linear relationship between, for example, soil moisture contents and the ratio of actual to potential evaporation. Figure 8.14 illustrates this, assuming that the ratio of actual to potential evaporation declines in a negative exponential function as soil moisture storage falls below the soil moisture storage capacity (pathway D in Figure 3.19). The soil moisture storage capacity is constant across the entire area at 150 mm, and potential evaporation is 5 mm d^{-1} everywhere. If the soil moisture content is 80 mm across the entire area, actual evaporation is 4.88 mm d^{-1}. If the soil moisture content varies over four different parts of the area, however (perhaps because of previous rainfall), then actual evaporation ranges from 5 down to 1.86 mm d^{-1}, with an aggregated flux from the whole area of 4.15 mm d^{-1}, around 15% less than if soil moisture contents were assumed constant. The greater the variability in soil moisture content and the more highly non-linear the relationship between evaporation and soil moisture content, the greater the difference between the aggregated and the average fluxes, and the greater the overestimation of evaporative flux that arises if spatially-averaged values are used (this is also shown using real data for the FIFE study area by Song *et al.*, 1997).

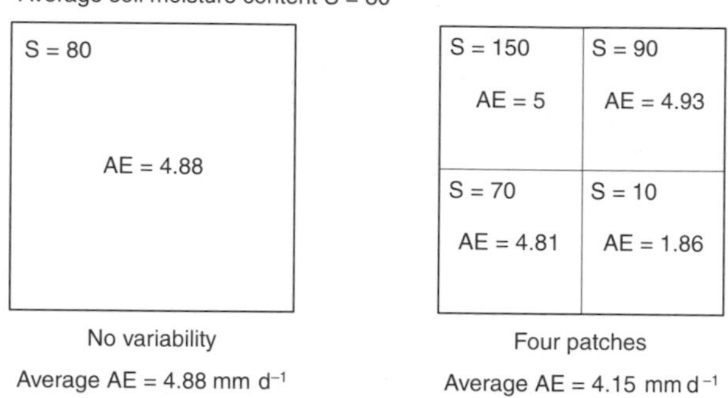

Figure 8.14 Effect of variability in soil moisture contents on aggregated evaporation flux

The primary advantage of an explicitly geographical approach, dividing the grid cell into patches, is that it is possible in principle to estimate model parameters from geographical databases, perhaps calculating *effective* parameters which compensate for the scale bias noted above. The main disadvantage is that it is necessary to disaggregate the incoming fluxes of energy and water to each of the patches. The simplest response to this is to assume that each patch has the same inputs, perhaps statistically disaggregated across each patch. A more complicated response would use physically-based downscaling techniques using some understanding of the space–time structure of input data. This is, of course, more computationally and mathematically demanding.

The second way of coping with heterogeneity in the land surface takes a catchment, rather than a column, perspective, and assumes that key surface properties are distributed across the catchment or cell following some frequency distribution. As long as the parameters of this distribution can be estimated, then it is not necessary to know exactly which parts of the grid cell have which properties. The VIC (Variable Infiltration Capacity) model, for example (Wood *et al.*, 1992), assumes that soil moisture storage capacity varies across space, with the effect that rapid runoff can be generated more quickly from some parts of the catchments than others. This model, and other models with a similar conceptual structure (Moore, 1985; Todini, 1996; Arnell, 1999a), works well at the catchment scale, and also simulates well the partitioning of precipitation into rapid runoff and soil moisture storage, and hence evaporation, in climate models. Stamm *et al.* (1994) showed how the VIC model produced less evaporation and more streamflow across Africa when applied in a climate model than a bucket representation (Figure 8.15), and generally produced a more accurate simulation of evaporation. The variation in *average* soil moisture over time is not very different between the two land surface representations, but because VIC assumes greater spatial variability in soil moisture, evaporative fluxes are greater.

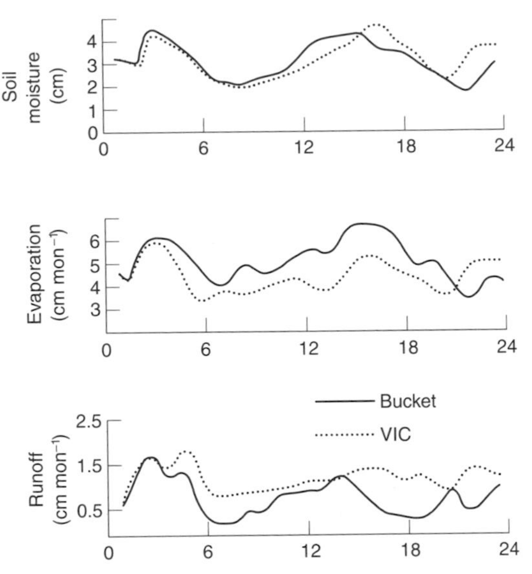

Figure 8.15 Comparison of soil moisture content, evaporation and runoff across Africa as simulated by a bucket model and VIC, incorporated in a GCM (Stamm *et al.*, 1994)

A rather more physically-based hydrological model, TOPMODEL, has also been proposed for use in climate models (Stieglitz *et al.*, 1997). This model essentially assumes that runoff generated from a part of a catchment is proportional to the slope and area draining to that part, as characterised by a topographic index. If the frequency distribution of this index is determined, using data from a digital elevation model, then it is possible to estimate the rapid runoff response from an input of precipitation: this response will vary over time, depending on the degree of saturation of the catchment or grid cell. TOPMODEL differs from VIC in that it is possible to explicitly calculate the frequency distribution of the relevant surface property from topographic data, but unlike VIC TOPMODEL was explicitly designed to simulate flows in small catchments using the physical equations of flow, so the spatial resolution of the input topographic data is of the order of metres. Moving from the small catchment scale to the GCM grid cell will involve considerable generalisation.

Different land surface parameterisations produce different estimates of the fluxes of moisture (and also energy). The international Project for the Intercomparison of Land surface Parameterisation Schemes (PILPS) tests different schemes "off-line", using observed input data and standardised sets of model parameters. Shao & Henderson-Sellers (1996) report on one phase of PILPS, showing the very considerable variability in estimated annual fluxes of moisture and energy. Figure 8.16 shows that annual evaporation estimates ranged from 550 to 775 mm (with a precipitation of 856 mm), and runoff ranged between 100 and 300, among the 14 models compared. The ratio of sensible to latent heat also varied substantially. Models are being continually refined, and ranges between estimates narrowed.

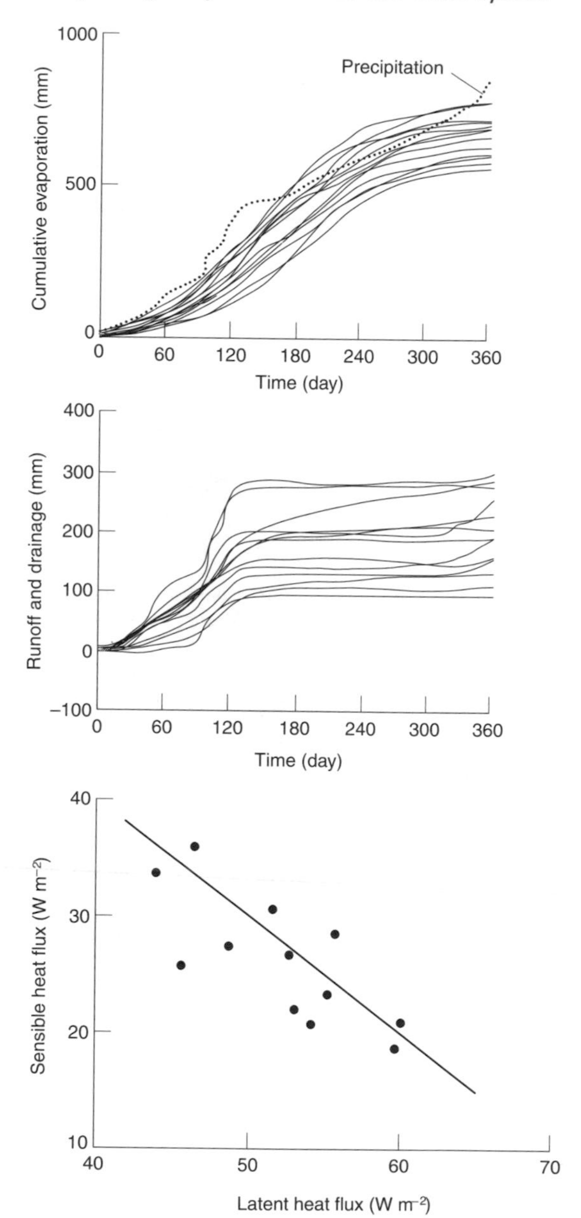

Figure 8.16 Estimates of annual evaporation, runoff and energy fluxes from 14 land surface parameterisations (Shao & Henderson-Sellers, 1996)

Modelling the transport of water along the river network

Simulation of the vertical fluxes of water and energy, however, is only one aspect of the hydrological system relevant to climate models: rivers transport water horizontally, to wetlands, lakes and the ocean, and this transport affects atmospheric and ocean processes. Permanent lakes can relatively easily be included into a climate model simply by assuming that a grid cell, or a fixed part of the grid cell, is continuously wet (Bonan, 1995). Wetlands, particularly in semi-arid and arid areas, however, may be very dynamic, depending on the variation in river flow through the year. The timing of inflows to oceans also depends on the timing of streamflow generation and the routing of flow through the channel network.

Most climate models forget about runoff as soon as it is generated from a grid cell, but a few route runoff, usually at a monthly time step, from grid cell to grid cell, and then to the sea (Sausen *et al.*, 1994; Miller *et al.*, 1994), primarily in order to simulate accurately the timing of inflows to the ocean. However, the spatial scale of global climate models is rather too large for this routing to be particularly accurate, and many significant seasonal wetlands are smaller than a GCM grid square. Miller *et al.* found that quartering the $4° \times 5°$ grid cells of their GCM and assuming that the same runoff was generated within each quarter, improved the simulation of the timing of river flows (the accuracy of the simulated volume of river flows was of course dependent on the accuracy of the simulated rainfall and evaporation). There has therefore been research into the routing of runoff along the river network across the globe at still finer spatial scales. Several global data sets showing flow pathways have been constructed automatically from digital elevation models, at scales ranging from $5' \times 5'$ (approximately 15×15 km) to $1° \times 1°$ (approximately 100×100 km) (e.g. Graham *et al.*, 1999), and several simple routing algorithms have been developed using flow direction, grid cell average slope and assumed channel lengths (e.g. Nijssen *et al.*, 1997; Coe, 2000). So far these finer-resolution runoff routeing models have not been incorporated on-line within GCMs, although there are some regional climate and weather prediction models which include river flow routeing (e.g. Yu *et al.*, 1999). However, even in these models the routed flow does not feed dynamic wetlands or lakes, and is primarily used to help validate simulated streamflow against observed data.

8.5 Overview

This chapter has explored the effect of hydrological processes on the atmosphere and the ocean. These effects are important not only for local weather, estuary currents and coastal deposition, but for regional and perhaps global climate and oceanic circulation. The amount of water at the ground surface alters energy partitioning, and river-fed wetlands and lakes can be a source of moisture in an otherwise arid environment. Inflows of freshwater affect ocean salinity, and hence currents and the rate of freeze and thaw of sea ice, potentially affecting regional climates. Fluxes of sediment and nutrients influence coastal zone erosion, deposition and productivity. Ocean productivity affects, amongst other things, the rate of uptake of carbon by the ocean, thereby affecting the radiative balance in the atmosphere.

Human activities can significantly impact upon these hydrological processes. Land cover change has the most extensive effect on the atmosphere, although draining wetlands and constructing reservoirs can alter water availability in dry areas. The volume and timing of water reaching the ocean is most affected by reservoir impoundment and diversions for irrigation, rather than land cover change itself. Sediment loads are affected by the competing effects of impoundment and land cover change, with some parts of the world experiencing increasing sediment flows to the coast, and others substantial reductions. Nutrient and heavy metal fluxes to the ocean are more affected by land use than by impoundment, but atmospheric deposition can also carry these pollutants to catchments distant from sources.

The last decade of the twentieth century saw a significant shift in focus of hydrological science, so that instead of looking down on the catchment hydrologists began to look up towards the atmosphere and, to a lesser extent, towards the sea. This has been reflected in a number of international scientific initiatives (such as GEWEX, parts of which are concerned with land–atmosphere, BAHC, which is concerned with the effects of vegetation on the movement of water between land and atmosphere, and LOICZ, which focuses on the links between land and sea) and associated field experiments and multi-disciplinary field measurement campaigns. Modelling too has become more integrated, and hydrologists have faced the challenge of developing models which can simulate the flux of energy and water between land and atmosphere, and indeed route water across the landscape, without being calibrated on observed hydrological data.

Chapter 9

Hydrology and global environmental change: an overview

9.1 Introduction

Hydrological processes are at the heart of the earth system. The hydrological cycle moves water from one store to another, and hydrological processes at the land surface strongly influence the partitioning of energy into latent and sensible heat. Water is an erosive agent, and acts as a solvent. It transports material from one place to another – it is the major source for many of the materials entering the ocean – and biogeochemical processes operating within water alter the speciation of elements. Ecosystems within a watercourse and along a floodplain are influenced by – and may be determined by – streamflow, soil moisture and groundwater regimes. People need water for drinking, washing, cooking and growing crops; industry uses water as a raw material and for cooling, power can be generated from running water, watercourses can be transport arteries, and rivers are also used to carry away waste.

The hydrological characteristics of a catchment, however, are not constant, and vary naturally over different time scales for many reasons. The physical landscape in many parts of the world reflects the legacy of different hydrological processes operating at different periods in the past (ranging from the dry valleys of the chalk of northern Europe, formed under periglacial hydrological conditions, to the massive erosional and depositional features in many rivers draining the Himalaya caused by the sudden failure of ice-dammed lakes during deglaciation). For several thousand years human activities have been affecting hydrological behaviour. These effects were initially local, but during the latter part of the twentieth century these human activities have been having effects over sufficiently large spatial scales to merit the term "global environmental change".

This final chapter draws some general conclusions about hydrology and global environmental change, before highlighting key areas for future hydrological research. Although this book has focused deliberately on *hydrology*, rather than *water resources*, the final section of this chapter explores the implications of global environmental change for water resources management in general terms.

9.2 An overall summary

The individual chapters have all drawn specific conclusions; this section highlights some general points.

- As mentioned above, hydrological behaviour varies over all time scales, due to both natural climatic variability and human-induced change. This human-induced change affects the catchment (in terms of land use and land cover, the use of water within the catchment and the development of structures to alter deliberately the timing and volume of streamflow) and the quantity and quality of inputs to the catchment. It is often difficult to separate these diverse causes of hydrological "change", and indeed to attribute an observed hydrological change to human intervention.
- Human-induced changes to the catchment and its inputs affect the volume, timing and quality of streamflow and groundwater recharge. The effects of a given intervention depend on the characteristics of the catchment and its climate. Opposing changes – such as deforestation and afforestation – do not always lead to opposite responses. Both deforestation and afforestation, for example, can lead to increases, at least in the short term, in flood magnitudes and reductions in the lag-time between rainfall and response. Individual interventions in the catchment – such as urbanisation or agricultural change – can have multiple effects; meanwhile, an individual catchment can be exposed to multiple human interventions. The implications of diverse human activities in a catchment may therefore be very difficult to ascertain.
- The relative importance of different human interventions varies with catchment scale. Land use and land cover changes tend to have less relative effect as catchment size increases, because a smaller proportion of the catchment tends to be affected. River flows in larger catchments are generally more affected by direct interventions along the river network, and specifically the construction of large reservoirs. Changes in catchment inputs can be expected to affect catchments of all sizes, although the larger the catchment the greater the likelihood that different parts of the catchment will experience different changes. Also, as catchment size increases the catchment will become more sensitive to different aspects of changes in inputs. Changes in the flood frequency curve in very small catchments, for example, will be sensitive to changes in short-duration intense rainfall; in contrast, flood frequency curves in large catchments will be much more sensitive to changes in rainfall accumulated over longer durations.
- The effect of human interventions may also vary with weather conditions. As a general rule, the effect of land cover change in a catchment on streamflow response to rainfall decreases as the magnitude of rainfall increases. Land cover changes largely affect the volume of storage available for incoming rainfall in a catchment, and these stores would generally be filled by extreme rainfall events regardless of land cover. *Catchment changes are therefore unlikely to have a discernible effect on the response of the catchment to very extreme rainfall inputs.*
- These hydrological effects of human-induced global environmental change have *impacts* on the water environment and the human use of the water resource. The translation from effect to impact, however, is not necessarily very simple, and depends on how the water resource is being used and the sensitivity of aquatic and riverine environments to change. *A similar*

hydrological change can have very different impacts in different catchments, and the greatest impact does not necessarily occur where the hydrological change is greatest.
- Pressures on both water resources and the water environment are changing in many catchments as population totals and economic activity within a catchment change. *An apparent increase in hydrological impacts – such as damaging floods or droughts – does not therefore necessarily imply a change in catchment hydrology.*
- Changes in hydrological processes and regimes have the potential to significantly affect biogeochemical cycling of many materials. These biogeochemical cycles determine the chemical composition of the atmosphere and the oceans, particularly in the coastal zone.
- Water management and land use policies in many catchments have the aim of minimising the adverse impacts of human activities on hydrological processes and regimes.
- Changes in hydrological processes and regimes can alter weather and climate, by altering the partitioning of energy, and coastal and ocean processes, by changing the flux of material to the sea. *Changes in the catchment can therefore affect global environmental change: the relationship between global environmental change and hydrological processes is not one-way.*

9.3 Hydrological science in the twenty-first century

Hydrological science developed rapidly during the last two decades of the twentieth century, partly because of the pressures of global environmental change. The focus of research by hydrologists has shifted from the development of mathematical techniques to predict hydrological characteristics for engineering design purposes to the understanding of process and the flux of mass and energy.

Hydrologists have for several decades looked at processes within the catchment, and have a good conceptual awareness of possible multiple effects of a given intervention on the volume and timing of water. They also know that it is difficult to extrapolate the quantitative effects of an intervention from one catchment to another, so there will always be a need for catchment studies in different geographical environments. However, the key *scientific* developments in hydrology – in terms of hydrology's goal of understanding the hydrological cycle – over the next couple of decades are likely to occur in seven areas of science and two linked methodological issues. These "growth areas" are (in no order):

- understanding the dynamics of river and groundwater chemistry, the fluxes of materials from land to sea, and the role of multiple human interventions;
- understanding the effects of natural climatic variability on hydrological processes and hydrological regimes;
- understanding the effect of hydrological processes at the land surface on the atmosphere, and hence the fluxes of water in all its states over time and space;

- understanding the effect of river and groundwater flows on the coastal zone and ocean processes;
- understanding the interaction between hydrological processes in the top few metres of the land surface and groundwater: how does groundwater affect river flows and water levels in the floodplain?
- understanding the role of hydrological processes in biogeochemical cycling: what are the links between carbon and water fluxes in plants, for example, and how are methane emissions from wetlands influenced by soil moisture dynamics?
- understanding the effect of hydrological processes and regimes on river and floodplain ecosystems.

The two key areas of methodological research, which will help tackle the above "science" issues, are

- the development of improved technologies to measure fluxes of water and materials held within water, using both automatic direct measurement and remote sensing, and
- the development of process-based computer simulation models that can operate over large geographical domains without the need for catchment-scale calibration, perhaps coupled to atmospheric models, which can be used to explore the potential effects of different climatic or catchment changes.

All of these potential growth areas require collaboration between hydrologists and other earth and environmental scientists.

Much hydrological research continues to be driven by the needs of water managers for information on the characteristics of hydrological regimes at sites with little or no data, and on the possible future effects of catchment and climate changes. Improved understanding of hydrological processes will underpin efforts to meet these needs, but operational hydrology (as practised day to day by water managers) needs particular progress in the following areas (again in no order):

- translating understanding of natural climatic variability into seasonal hydrological forecasts;
- understanding the statistical properties of extreme hydrological phenomena which may be generated from many different mechanisms;
- developing procedures to estimate the risk of hydrological extremes (in terms of quantity and quality) under changing catchment and climate conditions: current risk estimation procedures generally assume unchanging conditions.

As with "scientific" hydrology, progress in these areas is conditional on the development of realistic yet practical catchment simulation models and parameter estimation procedures and, most crucially, the enhanced collection of routine hydrological data. Hydrological services in many countries are cutting back on data collection, and in many areas very little information is collected. New measuring and recording technologies are lowering the relative cost of

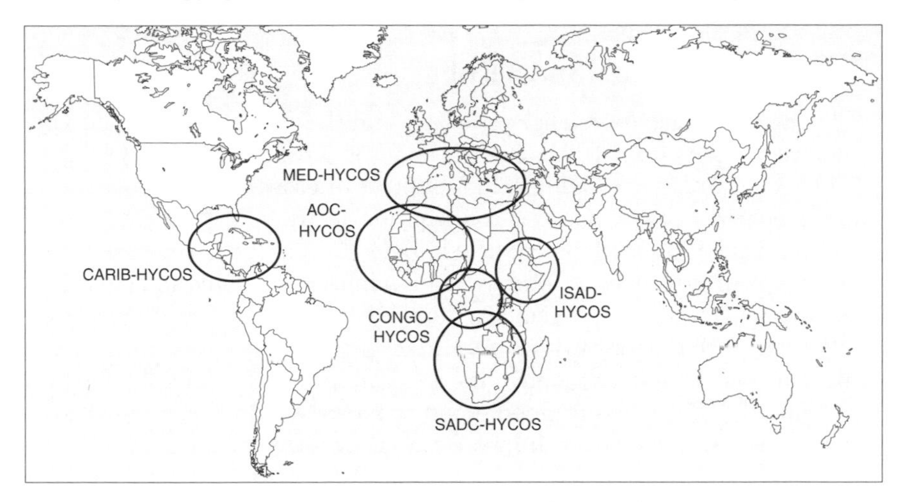

Figure 9.1 World Hydrological Cycle Observing System (WHYCOS): project locations

data collection, but at the expense of increased sophistication. The World Meteorological Organization's WHYCOS programme aims to increase the monitoring of river flows in developing countries using a combination of local skills, technical and institutional capacity-building and up-to-date satellite data collection and transmission technologies (Figure 9.1).

9.4 Hydrology, global environmental change and water management

The previous section has highlighted the areas in which hydrological research – both pure and applied – is likely to focus over the next few years. Whilst water resources management should be based on sound science, the effective management of water resources involves much more than hydrology. Water resources pressures, and pressures on the water environment, are driven by human forces and deliberate human decisions, and both minimising these pressures and reconciling conflicts over water resources between users requires an understanding of the human dimension. In practice, many water resources management problems are tackled not by managing the water but by attempting to "manage" people.

Increasing pressures on water resources and the water environment, together with global environmental change altering the resource base, have led to changes in water management philosophy in many countries and institutions (Gleick, 1998). Agenda 21, the strategy for sustainable development implemented in many countries following the 1992 "Earth Summit" (UN Conference on Environment and Development: UNCED) in Rio de Janeiro, requires the sustainable management of water resources (Kundzewicz, 1997). The "Dublin Statement", agreed at the International Conference on Water and the Environment in 1992, makes a series of specific recommendations to help water managers implement sustainable water management and implement Agenda 21 (Box 9.1).

Box 9.1 The Dublin Statement

The International Conference on Water and the Environment involved 500 participants from 100 countries, and was held in Dublin in January 1992. The conference highlighted a number of growing water concerns, and prepared a statement as a contribution to the UN Conference on Environment and Development held in the following June. The statement contains four principles – (i) freshwater is a finite, vulnerable and essential resource; (ii) water management should be based on a participatory approach; (iii) women play a central role in the provision and management of water; and (iv) water has an economic value and should be recognised as an economic good – and proposed action under ten headings. The first seven cover different aspects of water management:

Action items	Example specific recommendations
• Alleviation of poverty and disease	Provide safe water to those with inadequate sanitation
• Protection against natural disasters	Formulate disaster preparedness action plans
• Water conservation and reuse	Increase efficiency of water use
• Sustainable urban development	Encourage appropriate water charging and discharge controls
• Agricultural production and rural water supply	Increase efficiency of water use
• Protecting aquatic ecosystems	Implement integrated river basin management
• Resolving water conflicts	Integrated planning in international watersheds

The last three headings concern the investment, information needs and capacity-building necessary for sustainable water management:

Action items	Example specific recommendations
• The enabling environment	Increased investment
• The knowledge base	Research and monitoring
• Capacity-building	Identify and meet training needs

Source: http://www.wmo.ch/web/homs/icwedece.html

Although it is actually very difficult to define sustainable water management in practice, beyond stating that it aims to manage in a way that prevents progressive deterioration of the water environment, it should include four broad components: (i) the use of both supply-side techniques – managing the water – and demand-side techniques – managing the demand for water or exposure to hazard; (ii) continuous monitoring of the water environment; (iii) the adoption of flexible and adaptive techniques and strategies which are periodically reviewed and which can be easily altered; and (iv) an inclusive approach, involving all stakeholders. Underpinning all this is a sound scientific understanding of hydrological processes and the effects of human interventions in the hydrological cycle, and enhancing the "knowledge base" is one of the ten key actions identified in the Dublin Statement. The fundamental challenge for hydrological science is therefore not only to seek to understand the implications of global environmental change for hydrological processes, but also to undertake research in a way that will be directly relevant to the alleviation of global water problems. Unesco's International Hydrological Decade (IHD) ran from 1965 to 1974, and during the late 1990s calls began to be made for a second IHD aimed at tackling the big hydrological science questions. What has emerged, however, is a new joint Unesco/WMO initiative called HELP – Hydrology for Environment, Life and Policy – which seeks to "deliver social, economic and environmental benefits to stakeholders through sustainable and appropriate use of water, by deploying hydrological science in support of improved integrated catchment management" (www.unesco.org/science/help). Unlike conventional international hydrological research initiatives, HELP is problem-led rather than science-driven. Global environmental change, together with changes in demand for water resources and protection against water-related hazards, has stimulated new ways of both looking at and doing hydrology, and the twenty-first century is set to be an exciting time for hydrological science.

Appendix 1 Acronyms

4DDA	4-dimensional data assimilation
ABRACOS	Anglo-Brazilian Amazonian Climate Observational Study
AE	Actual evaporation
AMIP	Atmospheric Model Intercomparison Project
ANC	Acid neutralisation capacity
AOGCM	Atmosphere and Ocean General Circulation Model
BAHC	Biospheric Aspects of the Hydrological Cycle
BALTEX	Baltic Sea Experiment
BATS	Biosphere–Atmosphere Transfer System
BFI	Base Flow Index
BOD	Biochemical oxygen demand
BOREAS	Boreal Ecosystem Atmosphere Study
BP	Before Present (1950)
CLIVAR	WCRP Climate Variability and Predictability programme
CSE	Continental-scale experiment
CV	Coefficient of variation
DALR	Dry adiabatic lapse rate
DIC	Dissolved inorganic carbon
DMS	Dimethyl sulphide
DO	Dissolved oxygen
DOC	Dissolved organic carbon
EFEDA	European Field Experiment in a Desertification-threatened Area
ENSO	El Niño–Southern Oscillation
FAO	UN Food and Agriculture Organization
FIFE	First ISLSCP Field Experiment
FRIEND	Flow Regimes from International Experimental and Network Data
GAME	GEWEX Asian Monsoon Experiment
GCIP	GEWEX Continental-scale International Project
GCM	Global climate model
GEMS	Global Environmental Monitoring System
GEV	General extreme value distribution

GEWEX	Global Energy and Water Cycle Experiment
GPCP	Global Precipitation Climatology Project
GQA	General quality assessment
HAPEX-MOBILHY	Hydrological–Atmosphere Pilot Experiment, Modelisation du Bilan Hydrique
HAPEX-Sahel	Hydrological–Atmosphere Pilot Experiment, Sahel
HELP	Hydrology for Environment, Life and Policy
IGBP	International Geosphere–Biosphere Programme
IHD	International Hydrological Decade
IHP	International Hydrological Programme
IPCC	Intergovernmental Panel on Climate Change
ISLSCP	International Satellite Land Surface Climatology Programme
ITCZ	Inter-tropical convergence zone
LAI	Leaf area index
LBA	Large-scale Biosphere–Atmosphere Experiment in Amazonia
LEAP	Local Environment Agency Plan
LOICZ	Land–Ocean Interactions in the Coastal Zone (part of IGBP)
LOIS	Land–Ocean Interaction Study
LUCC	Land-Use and Land-Cover Change (part of IGBP)
MAGIC	Model of Acidification of Groundwater in Catchments
MAGS	Mackenzie GEWEX Study
MORECS	Meteorological Office Rainfall and Evaporation Calculation System
MWP	Medieval Warm Period
NADW	North Atlantic Deep Water
NAO	North Atlantic Oscillation
NASA	National Aeronautical and Space Agency
NOPEX	Northern Hemisphere Climate Processes Land Surface Experiment
NWP	Numerical Weather Prediction model
PAGES	Past Global Changes (part of IGBP)
PBL	Planetary Boundary Level
PDO	Pacific Decadal Oscillation
PE	Potential evaporation
PILPS	Project for the Intercomparison of Land Surface Parameterisation Schemes
PNA	Pacific/North America pattern
POC	Particulate organic carbon
POM	Particulate organic matter
*Q*95	The flow exceeded 95% of the time
RCM	Regional climate model
REA	Representative hydrological area

SALR	Saturated adiabatic lapse rate
SiB	Simple Biosphere Model
SVAT	Soil–Vegetation–Atmosphere Transfer scheme
SWAP	Surface Water Acidification Project
TDR	Time domain reflectometry
TDS	Total dissolved solids
TIGER	Terrestrial Initiative in Global Environmental Research
TIR	Thermal infrared
UKCIP	UK Climate Impacts Programme
UNCED	United Nations Conference on Environment and Development
UNEP	United Nations Environment Programme
Unesco	United Nations Educational, Scientific and Cultural Organization
VIC	Variable infiltration capacity model
WCRP	World Climate Research Programme
WHYCOS	World Hydrological Cycle Observing System
WMO	World Meteorological Organization
WUE	Water use efficiency

Appendix 2 Web-sites

The World Wide Web is increasingly being used in hydrology to exchange data sets, present new results and describe international research projects. A good starting point is the "Hyperlinks in Hydrology" web-site, located at the NERC Centre for Ecology and Hydrology in Wallingford (*http://www.nwl.ac.uk/ih/devel/wmo*) and also accessible through the home page of the WMO's Hydrology and Water Resources programme (see below). This appendix lists some of the "entry-point" web-sites, and addresses were valid in January 2001.

Web-sites for international research programmes and initiatives

The WMO's home page (*http://www.wmo.ch*) provides a gateway to the World Climate Research Programme, which is also directly accessible through *http://www.wmo.ch/web/wcrp/wcrp-home.html*). The WCRP home page links to the GEWEX home page, which is at *http://www.gewex.com*. The GEWEX home page provides links to all the continental-scale and meso-scale experiments (except NOPEX: *http://www.hyd.uu.se/nopex*).

The WMO Hydrology and Water Resources programme can be reached through the WMO home page or directly through *http://www.wmo.ch/web/homs/hwrphome.html*. This site links to many national hydrological agencies and research institutes, and also to the WHYCOS initiative.

The IGBP web site (*http://www.igbp.kva.se*) describes its research programmes, and the site *http://www.unesco.org/water/ihp* accesses the Unesco research initiatives. The new programme HELP is accessible through *http://www.nwl.ac.uk/ih/help/index.html* as well as through the Unesco site.

The GEMS global water quality programme can be accessed through *http://www.cciw.ca/gems*

Links to water management programmes and agencies can be found at *http://www.worldwater.org*

Data sets

Climate data are available through the web-sites of the Climatic Research Unit (*http://www.cru.uea.ac.uk*), the Global Precipitation Climatology Project (*http://orbit-net.nesdis.noaa.gov/arad/gpcp*), and the ISCLSCP site (*http://www.gewex.com/islscp.html*), and teleconnection indices are available through *http://www.cpc.ncep.noaa.gov/data/teledoc/telecontents.html.* Data on emissions to the atmosphere of greenhouse gases and sulphur can be found at *http://cdiac.esd.ornl.gov*

The Global Runoff Data Centre contains monthly data from many large catchments, and can be reached through *http://www.bafg.de/grdc.htm.* Many of its products are available through *http://www.grdc.sr.unh.edu*. The Dartmouth Flood Observatory (*http://www.dartmouth.edu/artsci/geog/floods*) maintains an up-to-date list of major floods worldwide.

The UK's National Water Archive is accessible through *http://www.nwl.ac.uk/ih/nwa/index.htm*, and the US national archive can be reached through *http://water.usgs.gov/nwis*: the address *http://water.usgs.gov* gives information about hydrology and water resources generally in the USA.

References

Aargaard, K. and Coachman, L.K., 1975. Towards an ice-free Arctic Ocean. *Eos* 56: 484–6.

Abbott, M.B., Bathurst, J.C., Cunge, J.A., O'Connell, P.E. and Rasmussen, J., 1986. An introduction to the European Hydrological System "SHE" (1) History and philosophy of a physically-based distributed modelling system. (2) Structure of a physically-based, distributed modelling system. *Journal of Hydrology* 87: 45–59, 61–77.

Abdulla, F.A., Lettenmaier, D.P., Wood, E.F. and Smith, J.A., 1996. Application of a macroscale hydrologic model to estimate the water balance of the Arkansas Red River basin. *Journal of Geophysical Research – Atmospheres* 101(D3): 7449–59.

Aboal, J.R., Jimenez, M.S., Morales, D. and Hernandez, J.M., 1999a. Rainfall interception in a laurel forest in the Canary Islands. *Agricultural and Forest Meteorology* 97: 73–86.

Aboal, J.R., Morales, D., Hernandez, M. and Jimenez, M.S., 1999b. The measurement and modelling of the variation in stemflow in a laurel forest in Tenerife, Canary Islands. *Journal of Hydrology* 221: 161–75.

Acreman, M.C., 1985. Predicting the mean annual flood from basin characteristics in Scotland. *Hydrological Sciences Journal* 30: 37–49.

Acreman, M.C., Adams, B., Birchall, P. and Connorton, B., 2000. Does groundwater abstraction cause degradation of rivers and wetlands? *Journal of the Chartered Institution of Water and Environmental Management* 14: 200–6.

Adams, B., Gale, I., Younger, P., Lerner, D. and Chilton, J., 2000. Groundwater. In: M. Acreman (ed.), *The Hydrology of the UK: A Study of Change*. Routledge, London, pp. 150–79.

Aguado, E., Cayan, D., Riddle, L. and Roos, M., 1992. Climate fluctuations and the timing of West Coast streamflow. *Journal of Climate* 5: 1468–83.

Alexander, R.B., Murdoch, P.S. and Smith, R.A., 1996. Streamflow-induced variations in nitrate flux in tributaries to the Atlantic Coastal Zone. *Biogeochemistry* 33: 149–77.

Alexander, R.B., Slack, J.R., Ludtke, A.S., Fitzgerald, K.K. and Schertz, T.L., 1998. Data from selected U.S. Geological Survey national stream water quality monitoring networks. *Water Resources Research* 34: 2401–5.

Algan, O., Gazioglu, C., Cagatay, N., Yucel, Z. and Gonencgil, B., 1999. Sediment and water influxes into the Black Sea by Anatolian Rivers. *Zeitschrift fur Geomorphologie* 43: 61–79.

Allen, L.H., Sinclair, T.R. and Bennett, J.M., 1997. Evapotranspiration of vegetation of Florida: perpetual misconceptions versus mechanistic processes. *Soil and Crop Science Society of Florida Proceedings* 56: 1–10.

Allen, R., Pereira, L.S., Raes, D. and Smith, M., 1998. *Crop evaporation: guidelines for computing crop requirements.* Irrigation and Drainage Paper 56, FAO, Rome.

Allen, S.J., Wallace, J.S., Gash, J.H.C. and Sivakumar, M.K., 1994. Measurements of albedo variation over natural vegetation in the Sahel. *International Journal of Climatology* 14: 625–36.

Amarasekera, K.N., Lee, R.F., Williams, E.R. and Eltahir, E.A.B., 1997. ENSO and the natural variability in the flow of tropical rivers. *Journal of Hydrology* 200: 24–39.

Amorocho, J. and Wu, B., 1977. Mathematical models for the simulation of cyclonic storm sequences and precipitation fields. *Journal of Hydrology* 32: 329–45.

Amthor, J.S., 1995. Terrestrial higher-plant response to increasing atmospheric CO_2 in relation to the global carbon cycle. *Global Change Biology* 1: 243–74.

Anderson, M.G. and Burt, T.P., 1978. The role of topography in controlling through-flow generation. *Earth Surface Processes* 3: 331–4.

Andre, J.C. *et al.*, 1988. HAPEX-MOBILHY: first results from the special observing period. *Annales Geophysicae* 6: 477–92.

Andrews, J.E., Brimblecome, P., Jickells, T.D. and Liss, P.S., 1996. *An Introduction to Environmental Chemistry.* Blackwell, Oxford.

Andrieu, H., Creutin, J.D., Delrieu, G. and Faure, D., 1997. Use of a weather radar for the hydrology of a mountainous area. *Journal of Hydrology* 193: 1–25.

Arheimer, B., Andersson, L. and Lepisto, A., 1996. Variation of nitrogen concentration in forest streams – influences of flow, seasonality and catchment characteristics. *Journal of Hydrology* 179: 281–304.

Arkin, P.A. and Xie, P., 1994. The Global Precipitation Climatology Project: first algorithm intercomparison project. *Bulletin of the American Meteorological Society* 75: 401–19.

Arnell, N.W., 1994. Variations over time in European hydrological behaviour: a spatial perspective. In: *FRIEND – Flow Regimes from International Experimental and Network Data. Int. Ass. Hydrol. Sci. Publ.* 221, 179–84.

Arnell, N.W., 1996. *Global Warming, River Flows and Water Resources.* Wiley, Chichester.

Arnell, N.W., 1999a. A simple water balance model for the simulation of streamflow over a large geographic domain. *Journal of Hydrology* 217: 314–35.

Arnell, N.W., 1999b. Climate change and global water resources. *Global Environmental Change* 9: S31–S49.

Arnell, N.W., 1999c. The effect of climate change on hydrological regimes in Europe: a continental perspective. *Global Environmental Change* 9: 5–23.

Arnell, N.W., 2000. Thresholds and response to climate change forcing: the water sector. *Climatic Change* 46: 305–16.

Arnell, N.W. and Reynard, N.S., 1996. The effects of climate change due to global warming on river flows in Great Britain. *Journal of Hydrology* 183: 397–424.

Arnell, N.W. and Reynard, N.S., 2000. Climate change and UK hydrology. In: M.C. Acreman (ed.), *The Hydrology of the UK: A Study of Change.* Routledge, London, pp. 3–29.

Arnell, N.W., Krasovskaia, I. and Gottschalk, L., 1993. River flow regimes in Europe. In: A. Gustard (ed.), *Flow Regimes from International Experimental and Network Data (FRIEND).* Institute of Hydrology, Wallingford, Oxfordshire, pp. 112–21.

Arnell, N.W. *et al.*, 2001. Hydrology and water resources. In: J. McCarthy, O. Canziani, N. Leary, D. Dokken and K. White (eds), *Climate Change 2001: Impacts, Adaptation and Vulnerability. Contribution of Working Group II to the Third Assessment*

Report of the Intergovernmental Panel on Climate Change. Cambridge University Press, Cambridge, pp. 191–233.

Asdak, C., Jarvis, P.G., van Gardingen, P. and Fraser, A., 1998a. Rainfall interception loss in unlogged and logged forest areas of Central Kalimantan, Indonesia. *Journal of Hydrology* 206: 237–44.

Asdak, C., Jarvis, P.G. and van Gardingen, P., 1998b. Modelling rainfall interception in unlogged and logged forest areas of Central Kalimantan, Indonesia. *Hydrology and Earth System Sciences* 2: 211–20.

Ashridge, D., 1995. Processes of river bank erosion and their contribution to the suspended sediment load of the River Culm, Devon. In: I.D.L. Foster, A.M. Gurnell and B.W. Webb (eds), *Sediment and Water Quality in River Catchments*. Wiley, Chichester, pp. 229–45.

Aston, A.R., 1979. Rainfall interception by eight small trees. *Journal of Hydrology* 42: 383–96.

Atkinson, L.P. and Grosch, C.E., 1999. Continental runoff and effects on the North Atlantic Ocean Subtropical Mode Water. *Geophysical Research Letters* 26: 2977–80.

Austin, P.M. and Houze, M.A., 1972. Analysis of the structure of precipitation patterns in New England. *Journal of Applied Meteorology* 11: 926–35.

Avila, A., Neal, C. and Terradas, J., 1996. Climate change implications for streamflow and streamwater chemistry in a Mediterranean catchment. *Journal of Hydrology* 177: 99–116.

Baird, A.J. and Wilby, R.L. (eds), 1999. *Eco-hydrology: Plants and Water in Terrestrial and Aquatic Environments*. Routledge, London.

Bajracharya, R.M. and Lal, R., 1999. Land use effects on soil crusting and hydraulic response of surface crusts on a tropical Alfisol. *Hydrological Processes* 13: 59–72.

Baker, V.R., 1987. Palaeoflood hydrology of extraordinary flood events. *Journal of Hydrology* 96: 79–99.

Baker, V.R., 1998. Palaeohydrology and the hydrological sciences. In: G. Benito, V.R. Baker and K.J. Gregory (eds), *Palaeohydrology and Environmental Change*. Wiley, Chichester, pp. 1–10.

Bari, M.A., Smith, N., Rpurecht, J.K. and Boyd, B.W., 1996. Changes in streamflow components following logging and regeneration in the southern forest of Western Australia. *Hydrological Processes* 10: 447–61.

Barnston, A.G. and Livesey, R.E., 1987. Classification, seasonality and persistence of low-frequency atmospheric circulation patterns. *Monthly Weather Review* 115: 1083–126.

Barnston, A.G. and Schickedanz, P.T., 1984. The effect of irrigation on warm season precipitation in the Southern Great Plains. *Journal of Climate and Applied Meteorology* 23: 865–88.

Barriendos Vallve, M. and Martin-Vide, J., 1998. Secular climatic oscillation as indicated by catastrophic floods in the Spanish Mediterranean coastal area (14th–19th centuries). *Climatic Change* 38: 473–91.

Barry, R.G. and Chorley, R.J., 1998. *Atmosphere, Weather and Climate*. 7th edition. Routledge, London.

Bartlein, P.J., 1982. Streamflow anomaly patterns in the USA and southern Canada. *Journal of Hydrology* 57: 49–63.

Bastiaanssen, W.G.M., 2000. SEBAL-based sensible and latent heat fluxes in the irrigated Gediz Basin, Turkey. *Journal of Hydrology* 229: 87–100.

Bates, B.C., Jakeman, A.J., Charles, S.P., Sumner, N.R. and Fleming, P.M., 1996. Impact of climate change on Australia's water resources. In: W.J. Bouma, G.I.

Pearman and M.R. Manning (eds), *Greenhouse: Coping with Climate Change.* CSIRO, Collingwood, Victoria, Australia, pp. 248–62.

Bates, B.C., Rahman, A., Mein, R.G. and Weinmann, P.E., 1998. Climatic and physical factors that influence the homogeneity of regional floods in southeastern Australia. *Water Resources Research* 34: 3369–82.

Baumgartner, F. and Reichel, E., 1975. *The World Water Balance: Mean Annual Global, Continental and Maritime Precipitation, Evaporation and Runoff.* Ordenbourg, Munich.

Beckinsale, R.P., 1969. River regimes. In: R.J. Chorley (ed.), *Water, Earth and Man.* Methuen, London, pp. 455–71.

Beldring, S., Gottschalk, L., Seibert, J. and Tallaksen, L.M., 1999. Distribution of soil moisture and groundwater levels at patch and catchment scales. *Agricultural and Forest Meteorology* 98–9: 305–24.

Beljaars, A.C.M., Viterbo, P., Miller, M.J. and Betts, A.K., 1996. The anomalous rainfall over the United States during July 1993: sensitivity to land surface parameterization and soil moisture. *Monthly Weather Review* 124: 362–83.

Benton, G.S., Blackburn, R.T. and Snead, V.O., 1950. The role of the atmosphere in the hydrologic cycle. *Transactions of the American Geophysical Union* 31: 61–73.

Bergstrom, S. and Carlsson, B., 1994. River runoff to the Baltic Sea: 1950–1990. *Ambio* 23: 280–7.

Berner, E.K. and Berner, R.A., 1987. *The Global Water Cycle.* Prentice Hall, Englewood Cliffs, NJ.

Berner, E.K. and Berner, R.A., 1996. *Global Environment: Water, Air and Geochemical Cycles.* Prentice Hall, Upper Saddle River, NJ.

Bernier, M. *et al.*, 1999. Determination of snow water equivalent using RADARSAT SAR data in eastern Canada. *Hydrological Processes* 13: 3041–51.

Berri, G.J. and Flamenco, E.A., 1999. Seasonal volume forecast of the Diamante River, Argentina, based on El Niño observations and predictions. *Water Resources Research* 35: 3803–10.

Betts, R.A., Cox, P.M. and Woodward, F.I., 2000. Simulated response of potential vegetation to doubled-CO_2 climate change and feedbacks on near-surface temperatures. *Global Ecology and Biogeography* 9: 171–80.

Beven, K.J., 1995. Linking parameters across scales: subgrid parameterisations and scale dependent hydrological models. *Hydrological Processes* 9: 507–25.

Beven, K.J., 1997. TOPMODEL: a critique. *Hydrological Processes* 11: 1069–86.

Beven, K.J., 2001. *Hydrological Modelling: The Primer.* Wiley, Chichester.

Beven, K.J. and Germann, P., 1982. Macropores and water flow in soils. *Water Resources Research* 18: 1311–25.

Beyazgul, M., Kayam, Y. and Engelsman, F., 2000. Estimation methods for crop water requirements in the Gediz basin of Western Turkey. *Journal of Hydrology* 229: 19–26.

Beyerle, U. *et al.*, 1999. Infiltration of river water to a shallow aquifer investigated with $^3H/^3He$ noble gases and CFCs. *Journal of Hydrology* 220: 169–85.

Biodiversity Challenge Group, 1996. *High and Dry: The Impacts of Overabstraction of Water on Wildlife.* Biodiversity Challenge Group.

Birkett, C., Murtugudde, R. and Allan, T., 1999. Indian Ocean climate event brings floods to East Africa's lakes and the Sudd Marsh. *Geophysical Research Letters* 26: 1031–4.

Black, A.R., 1996. Major flooding and increased flood frequency in Scotland since 1988. *Physics and Chemistry of the Earth* 20: 463–8.

Black, A.R. and Werrity, A., 1997. Seasonaility of flooding: a case study of north Britain. *Journal of Hydrology* 195: 1–23.

Bloomfield, J., 1996. Characteristics of hydrogeologically significant fracture distributions in the Chalk: examples from the Upper Chalk of southern England. *Journal of Hydrology* 184: 355–79.

Blöschl, G. and Sivapalan, M., 1997. Process controls on regional flood frequency: coefficient of variation and basin scale. *Water Resources Research* 33: 2967–80.

Blöschl, G., Grayson, R.B. and Sivapalan, M., 1995. On the representative elementary area (REA) concept and its utility for distributed runoff modelling. *Hydrological Processes* 9: 313–30.

Blyth, K., 1997. An assessment of the capabilities of the ERS satellite's active microwave instruments for monitoring soil moisture change. *Hydrology and Earth System Sciences* 1: 159–74.

Bolle, H.J. *et al.*, 1993. EFEDA: European field experiment in a desertification-threatened area. *Annales Geophysicae* 11: 173–89.

Bonan, G.B., 1995. Sensitivity of a GCM simulation to inclusion of inland water surfaces. *Journal of Climate* 8: 2691–704.

Bonell, M., Cassells, D.S. and Gilmour, D.A., 1983. Vertical soil water movement in a tropical rainforest catchment in northeast Queensland. *Earth Surface Processes and Landforms* 8: 253–72.

Bonell, M., Hendricks, M.R., Imeson, A.C. and Hazelhoff, L., 1984. The generation of storm runoff in a forested clayey drainage basin in Luxembourg. *Journal of Hydrology* 71: 53–77.

Bosch, J.M. and Hewlett, J.D., 1982. A review of catchment experiments to determine the effects of vegetation changes on water yield and evapotranspiration. *Journal of Hydrology* 55: 3–23.

Boucher, K., 1997. Hydrological monitoring and measurement methods. In: R.L. Wilby (ed.), *Contemporary Hydrology*. Wiley, Chichester, pp. 107–49.

Bouten, W., Swarts, P.J.F. and de Water, E., 1991. Microwave transmission, a new tool in forest hydrological research. *Journal of Hydrology* 124: 119–30.

Boyer, D.G. and Pasquarell, G.C., 1999. Agricultural land use impacts on bacterial water quality in a karst groundwater aquifer. *Journal of the American Water Resources Association* 35: 291–300.

Bradbury, C.G. and Rushton, K.R., 1998. Estimating runoff-recharge in the Southern Lincolnshire Limestone catchment, UK. *Journal of Hydrology* 211: 86–99.

Bras, R.F. and Rodriguez-Iturbe, I., 1976. Network design for the estimation of areal rainfall events. *Water Resources Research* 21: 1185–95.

Brisco, B. *et al.*, 1992. Soil moisture measurement using portable dielectric probes and time domain reflectometry. *Water Resources Research* 28: 1339–46.

Broecker, W.S., 1987. Unpleasant surprises in the greenhouse? *Nature* 328: 123–6.

Bromley, J., Brouwer, J., Barker, A.P., Gaze, S. and Valentin, C., 1997. The role of surface water redistribution in an area of patterned vegetation in a semi-arid environment, south-west Niger. *Journal of Hydrology* 198: 1–29.

Brookes, A., 1987. River channel adjustments downstream of channelisation works in England and Wales. *Earth Surface Processes and Landforms* 12: 337–51.

Brooks, R.H. and Corey, A.T., 1964. *Hydraulic properties of porous media.* Hydrology Paper 3, Colorado State University, Fort Collins, Colorado.

Brostrom, A. *et al.*, 1998. Land surface feedbacks and palaeomonsoons in northern Africa. *Geophysical Research Letters* 25: 3615–18.

Brown, A.G., 1998. Fluvial evidence of the Medieval Warm Period and the Late Medieval Climatic Deterioration in Europe. In: G. Benito, V.R. Baker and K.J. Gregory (eds), *Palaeohydrology and Environmental Change*. Wiley, Chichester, pp. 43–52.

Brown, M.E. and Arnold, D.L., 1998. Land-surface-atmosphere interactions associated with deep convection in Illinois. *International Journal of Climatology* 18: 1637–53.

Browning, K.A. and Hill, F.F., 1981. Orographic rain. *Weather* 36: 326–9.

Brubaker, K.L., Entekhabi, D. and Eagleson, P.S., 1993. Estimation of continental precipitation recycling. *Journal of Climate* 6: 1077–89.

Brubaker, K.L., Entekhabi, D. and Eagleson, P.S., 1994. Atmospheric water vapor transport and continental hydrology over the Americas. *Journal of Hydrology* 155: 407–28.

Bryan, F. and Oort, A., 1984. Seasonal variation of the global water balance based on aerological data. *Journal of Geophysical Research* 89: 11717–30.

Bucher, B. and Demuth, S., 1985. Vergleichende Wasserbilanz eines flurbereinigten und eines nicht flurbereinigten Einzugsgebietes in Ostkaisersthul fur den Zeitraum 1977–1980. *Deutsche Gewasserkundliche Mitteilungen* 29: 1–4.

Budyko, M.I., 1974. *Climate and Life*. Academic Press, New York.

Bultot, F., Dupriez, G.L. and Gellens, D., 1990. Simulation of land use changes and impacts on the water balance – a case study for Belgium. *Journal of Hydrology* 114: 327–48.

Bultot, F., Gellens, D., Spreafico, M. and Schadler, B., 1992. Repercussions of a CO_2-doubling on the water balance – a case study in Switzerland. *Journal of Hydrology* 137: 199–208.

Burba, G.G., Verma, S.B. and Kim, J., 1999. A comparative study of surface energy fluxes of three communities (*Phragmites australis, Scirpus acutus* and open water) in a prairie wetland ecosystem. *Wetlands* 19: 451–7.

Burn, D.H. and Arnell, N.W., 1993. Synchronicity in global flood response. *Journal of Hydrology* 144: 381–404.

Burt, T.P. and Johnes, P.J., 1997. Managing water quality in agricultural catchments. *Transactions of the Institute of British Geographers* NS 22: 61–8.

Burt, T.P. and Shahgedanova, M., 1998. An historical record of evaporation losses since 1815 calculated using long-term observations from the Radcliffe Meteorological Station, Oxford, England. *Journal of Hydrology* 205: 101–11.

Buttle, J.M. and Peters, D.L., 1997. Inferring hydrological processes in a temperate basin using isotopic and geochemical hydrograph separation: a re-evaluation. *Hydrological Processes* 11: 557–74.

Cabrera, F., Clemente, L., Barrientos, E.D., Lopez, R. and Murillo, J.M., 1999. Heavy metal pollution of soils affected by the Guadiamar toxic flood. *Science of the Total Environment* 242: 117–29.

Cain, J.D., Batchelor, C.H., Gash, J.H.C. and Harding, R.J., 1998. Comment on the paper "Towards a rational definition of potential evaporation" by J.P. Lhomme. *Hydrology and Earth System Sciences* 2: 137.

Calder, I.R., 1990. *Evaporation in the Uplands*. Wiley, Chichester.

Calder, I.R. and Rosier, P.T.W., 1976. The design of large plastic sheet net-rainfall gauges. *Journal of Hydrology* 30: 403–6.

Calder, I.R. and Wright, I.R., 1986. Gamma ray attenuation studies of interception from Sitka spruce – some evidence for an additional transport mechanism. *Water Resources Research* 22: 409–17.

Calder, I.R., Wright, I.R. and Murdiyaso, D., 1986a. A study of evaporation from tropical rain forest – West Java. *Journal of Hydrology* 89: 13–33.

Calder, I.R., Kariyappa, G.S., Srinivasulu, N.V., Darling, W.G. and Lardner, A.J., 1986b. Investigation into the use of deuterium as a tracer for measuring transpiration from eucalypts. *Journal of Hydrology* 84: 345–51.

Calder, I.R., Hall, R.L. and Bastable, H.G., 1995. The impact of land use change on water resources in sub-Saharan Africa: a modelling study of Lake Malawi. *Journal of Hydrology* 170: 123–36.

Callender, G.S., 1938. The artificial production of carbon dioxide and its influence on temperature. *Quarterly Journal of the Royal Meteorological Society* 64: 223–37.

Cameron, C.S. *et al.*, 1997. Fog deposition in tall tussock grassland, South Island, New Zealand. *Journal of Hydrology* 193: 363–76.

Cameron, E.M., 1996. Hydrochemistry of the Fraser River, British Columbia: seasonal variation in major and minor components. *Journal of Hydrology* 182: 209–25.

Campbell, D.I. and Williamson, J.L., 1997. Evaporation from a raised peat bog. *Journal of Hydrology* 193: 142–60.

Carey, S.K. and Woo, M.K., 1999. Hydrology of two slopes in subarctic Yukon, Canada. *Hydrological Processes* 13: 2549–62.

Carlyle-Moses, D.E. and Price, A.G., 1999. An evaluation of the Gash interception model in a northern hardwood stand. *Journal of Hydrology* 214: 103–10.

Carpenter, S.R. *et al.*, 1998. Nonpoint pollution of surface waters with phosphorus and nitrogen. *Ecological Applications* 8: 559–68.

Carriquiry, J.D. and Sanchez, A., 1999. Sedimentation in the Colorado River delta and Upper Gulf of California after nearly a century of discharge loss. *Marine Geology* 158: 125–45.

Carroll, S.S. and Cressie, N., 1996. A comparison of geostatistical methodologies used to estimate snow water equivalent. *Water Resources Bulletin* 32: 267–78.

Casenave, A. and Valentin, C., 1992. A runoff capability classification system based on surface feature criteria. *Journal of Hydrology* 130: 231–49.

Cassells, D.S., Gilmour, D.A. and Bonell, M., 1985. Catchment response and watershed management in the tropical rainforests in north-eastern Australia. *Forest Ecology and Management* 10: 155–75.

Caudier, J., Saramillo, M., Solis, D. and de Leon, D., 1997. Water balance and nutrient inputs in bulk precipitation in tropical montane cloud forest in Panama. *Journal of Hydrology* 193: 83–96.

Cayan, D.R., Redmond, K.T. and Riddle, L.G., 1999. ENSO and hydrologic extremes in the western United States. *Journal of Climate* 12: 2881–93.

Cerda, A., 1997. Seasonal changes of the infiltration rates in a Mediterranean scrubland on limestone. *Journal of Hydrology* 198: 209–25.

Chahine, M.T., 1992. The hydrological cycle and its influence on climate. *Nature* 359: 373–80.

Chang, S.C. and Matzner, E., 2000. The effect of beech stemflow on spatial patterns of soil solution chemistry and seepage fluxes in a mixed beech/oak stand. *Hydrological Processes* 14: 135–44.

Chanzy, A. and Bruckler, L., 1993. Significance of soil surface moisture with respect to daily bare soil evaporation. *Water Resources Research* 29: 1113–25.

Charlesworth, S.M. and Foster, I.D.L., 1999. Sediment budgets and metal fluxes in two contrasting urban lake catchments in Coventry, UK. *Applied Geography* 19: 199–210.

Charney, J.G., 1975. Dynamics of deserts and droughts in the Sahel. *Quarterly Journal of the Royal Meteorological Society* 101: 193–202.

Chase, T.N., Pielke, R.A., Kittel, T.G.F., Baron, J.S. and Stohlgren, T.J., 1999. Potential impacts on Colorado Rocky Mountain weather due to land use changes on the adjacent Great Plains. *Journal of Geophysical Research – Atmospheres* 104: 16673–90.

Chase, T.N., Pielke, R.A., Kittel, T.G.F., Nemani, R.R. and Running, S.W., 2000. Simulated impacts of historical land cover changes on global climate in northern winter. *Climate Dynamics* 16: 93–105.

Chen, T.-C. and Pfaentdner, J., 1993. On the atmospheric branch of the hydrologic cycle. *Journal of Climate* 6: 161–7.

Chen, T.-C., Pfaendtner, J. and Weng, S.-P., 1994. Aspects of the hydrological cycle of the ocean-atmosphere system. *Journal of Physical Oceanography* 24: 1827–33.

Chiew, F.H.S., Piechota, T.C., Dracup, J.A. and McMahon, T.A., 1998. El Niño/Southern Oscillation and Australian rainfall, streamflow and drought: links and potential for forecasting. *Journal of Hydrology* 204: 138–49.

Chua, S.H. and Bras, R.L., 1982. Optimal estimation of mean areal precipitation in regions of orographic influence. *Journal of Hydrology* 57: 713–28.

Clair, T.A., Pollock, T.L. and Ehrman, J.M., 1994. Exports of carbon and nitrogen from river basins in Canada's Atlantic provinces. *Global Biogeochemical Cycles* 8: 441–50.

Clapp, R.B. and Hornberger, G.M., 1978. Empirical equations for some soil hydraulic properties. *Water Resources Research* 14: 601–4.

Clarke, R.T., Leese, M.N. and Newson, A.J., 1973. *Analysis of data from Plynlimon gauge networks, April 1971 to March 1973*. Report 27, Institute of Hydrology, Wallingford, Oxon.

Cleaveland, M.K., 2000. A 963-year reconstruction of summer (JJA) streamflow in the White River, Arkansas, USA, from tree rings. *The Holocene* 10: 33–41.

CLIVAR, 1997. *CLIVAR Science plan*. World Climate Research Programme, WMO, Geneva.

Coe, M.T., 2000. Modeling terrestrial hydrological systems at the continental scale: testing the accuracy of an atmospheric GCM. *Journal of Climate* 13: 686–704.

Coe, M.T. and Bonan, G.B., 1997. Feedbacks between climate and surface water in northern Africa during the middle Holocene. *Journal of Geophysical Research – Atmospheres* 102: 11087–101.

Cogley, J.G., 1994. *GGHYDRO – Global Hydrographic Data, Release 2.1*. Department of Geography, Trent University, Ontario, Canada.

Collins, D.N., 1998. Outburst and rainfall-induced peak runoff events in highly glacierized Alpine basins. *Hydrological Processes* 12: 2369–81.

Collins, R., Jenkins, A. and Harrow, M., 2000. The contribution of old and new water to a storm hydrograph determined by tracer addition to a whole catchment. *Hydrological Processes* 14: 701–11.

Compagnucci, R. and Vargas, W.M., 1998. Inter-annual variability of the Cuyo river's streamflow in the Argentinean Andean mountains and ENSO events. *International Journal of Climatology* 18: 1593–609.

Conway, D. and Hulme, M., 1996. The impacts of climate variability and future climate change in the Nile Basin on water resources in Egypt. *Water Resources Development* 12: 277–96.

Cook, P.G. and Kilty, S., 1992. A helicopter-borne electromagnetic survey to delineate groundwater recharge rates. *Water Resources Research* 28: 2953–61.

Cooper, D.M., Wilkinson, W.B. and Arnell, N.W., 1995. The effect of climate change on aquifer storage and river baseflow. *Hydrological Sciences Journal* 40: 615–31.

Cornell, S., Rendell, A. and Jickells, T., 1995. Atmospheric inputs of dissolved organic nitrogen to the oceans. *Nature* 376: 243–6.

Cosby, B.J., Hornberger, G.M., Rastetter, E.B. and Galloway, J.N., 1985. Modelling the effects of acid deposition: estimation of long-term water quality responses in a small forested catchment. *Water Resources Research* 21: 1591–601.

Cosgrove, W.J. and Rijsberman, F.R., 2000. *World Water Vision: Making Water Everybody's Business. World Water Council.* Earthscan, London.

Coughlan, M. and Avissar, R., 1996. The Global Energy and Water Cycle Experiment (GEWEX) Continental-Scale International Project (GCIP): an overview. *Journal of Geophysical Research – Atmospheres* 101(D3): 7139–48.

Coulthard, T.J., Kirkby, M.J. and Macklin, M.G., 2000. Modelling geomorphic response to environmental change in an upland catchment. *Hydrological Processes* 14: 2031–45.

Cowpertwait, P.S.P., 1991. Further development of the Neyman–Scott clustered point process for modelling rainfall. *Water Resources Research* 27: 1431–8.

Crabtree, R.E. and Trudgill, S.T., 1985. Hill slope hydrochemistry and stream response on wooded, permeable bedrock: the role of stemflow. *Journal of Hydrology* 80: 161–78.

Crayosky, T.W., DeWalle, D.R., Seybert, T.A. and Johnson, T.E., 1999. Channel precipitation dynamics in a forested Pennsylvania headwater catchment (USA). *Hydrological Processes* 13: 1303–14.

Crockford, R.H. and Richardson, D.P., 1990a. Partitioning of rainfall in a Eucalypt forest and pine plantation in south-eastern Australia: I. Throughfall measurement in a eucalypt forest: effect of method and species composition. *Hydrological Processes* 4: 131–44.

Crockford, R.H. and Richardson, D.P., 1990b. Partitioning of rainfall in a Eucalypt forest and pine plantation in south-eastern Australia: II. Stemflow and factors affecting stemflow in a dry sclerophyll eucalypt forest and *Pinus radiata* plantation. *Hydrological Processes* 4: 145–55.

Crockford, R.H. and Richardson, D.P., 1990c. Partitioning of rainfall in a Eucalypt forest and pine plantation in south-eastern Australia: III. Determination of the canopy storage capacity of a dry sclerophyll Eucalypt forest. *Hydrological Processes* 4: 157–67.

Crockford, R.H. and Richardson, D.P., 1990d. Partitioning of rainfall in a Eucalypt forest and pine plantation in south-eastern Australia: IV. The role of interception and canopy storage capacity on the interception of these forests, and the effect on interception of thinning the pine plantation. *Hydrological Processes* 4: 169–88.

Croke, J., Hairsine, P. and Fogarty, P., 1999. Runoff generation and re-distribution in logged eucalyptus forests, south-eastern Australia. *Journal of Hydrology* 216: 56–74.

Cullen, H.M. and de Menocal, P.B., 2000. North Atlantic influences on Tigris-Euphrates streamflow. *International Journal of Climatology* 20: 853–63.

Curry, J.A., Schramm, J.L. and Ebert, E.E., 1995. Sea ice-albedo climate feedback mechanism. *Journal of Climate* 8: 240–7.

Dalton, J., 1802. Experimental essays on the constitution of mixed gases. *Manchester Literary and Philosophical Society Memorandum* 5: 535–602.

Damant, C., Austin, G.L., Bellon, A. and Broughton, R.S., 1983. Errors in the Thiessen technique for estimating areal rain amounts using weather radar data. *Journal of Hydrology* 62: 81–94.

Darnell, W.L., Staylor, W.F., Gupta, S.K., Ritchey, N.A. and Wilber, A.C., 1992. Seasonal variation of surface radiation budget derived from ISCCP-C1 data. *Journal of Geophysical Research* 97: 15741–60.

Davis, R.A., 1997. Regional coastal morphodynamics along the United States Gulf of Mexico. *Journal of Coastal Research* 13: 595–604.

Dawdy, D.R., 1961. *Variation of flood ratios with size of drainage area.* Professional Paper 424-C, U.S. Geological Survey.

de Laine, R.J., 1969. Measuring rainfall on forest catchments. *Journal of Hydrology* 9: 103–12.

Delworth, T. and Manabe, S., 1989. The influence of soil wetness on near-surface atmospheric variability. *Journal of Climate* 2: 1447–62.

Demuth, S. and Heinrich, B., 1997. Temporal and spatial behaviour of drought in south Germany. In: *FRIEND'97 – Regional Hydrology: Concepts and Models for Sustainable Water Resources. Int. Ass. Hydrol. Sci. Publ.* 246, 151–7.

Desconnets, J.C., Taupin, J.D., Lebel, T. and Leduc, C., 1997. Hydrology of the HAPEX-Sahel central supersite: surface water drainage and aquifer recharge through the pool system. *Journal of Hydrology* 188–9: 155–78.

Desjardins, R.L., Schuepp, P.H., MacPherson, J.I. and Buckley, D.J., 1992. Spatial and temporal variations of the fluxes of carbon dioxide and sensible and latent heat over the FIFE site. *Journal of Geophysical Research – Atmospheres* 97(D17): 18467–75.

Devol, A.H., Forsberg, B.R., Richey, J.E. and Pimentel, T.P., 1995. Seasonal variation in chemical distributions in the Amazon (Solimoes) River: a multiyear time series. *Global Biogeochemical Cycles* 9: 307–28.

DeWalle, D.R., Swistock, B.R. and Sharpe, W.E., 1988. Three component tracer models for stormflow on a small Appalachian forested catchment. *Journal of Hydrology* 104: 301–10.

DeWalle, D.R., Edwards, P.J., Swistock, B.R., Aravena, R. and Drimmie, R.J., 1998. Seasonal isotope hydrology of three Appalachian forest catchments. *Hydrological Processes* 11: 1895–906.

Dillon, P.J. and Molot, L.A., 1997. Effect of landscape form on export of dissolved organic carbon, iron, and phosphorus from forested stream catchments. *Water Resources Research* 33: 2591–600.

Dillon, P.J., Molot, L.A. and Scheider, W.A., 1991. Phosphorus and nitrogen export from forested stream catchments in central Ontario. *Journal of Environmental Quality* 20: 857–64.

Dingman, S.L., 1993. *Physical Hydrology*. Prentice Hall, Englewood Cliffs, New Jersey.

Dingman, S.L., Seely-Reynolds, D. and Reynolds, R.C., 1988. Application of kriging to estimating mean annual precipitation in a region of orographic influence. *Water Resources Bulletin* 24: 329–39.

Dirks, K.N., Hay, J.E., Stow, C.D. and Harris, D., 1998. High-resolution studies of rainfall on Norfolk Island. Part II: Interpolation of rainfall data. *Journal of Hydrology* 208: 187–93.

Dolman, A.J., 1987. Summer and winter rainfall interception in an oak forest. Prediction with an analytical and a numerical simulation model. *Journal of Hydrology* 90: 1–9.

Dolman, A.J. and Blyth, E.M., 1997. Patch scale aggregation of heterogeneous land surface cover for mesoscale meteorological models. *Journal of Hydrology* 190: 252–68.

Dooge, J.C.I., 1986. Looking for hydrologic laws. *Water Resources Research* 22: 465–85.

Downing, R.A., 1993. Groundwater resources, their development and management in the UK: an historical perspective. *Quarterly Journal of Engineering Geology* 26: 335–58.

Dracup, J.A. and Kahya, E., 1994. The relationships between U.S. streamflow and La Niña events. *Water Resources Research* 30: 2133–41.

Drake, F., 2000. *Global Warming: the Science of Climate Change*. Arnold, London.

Dubicki, A., 1994. Changes in catchment discharge associated with forest dieback in regions of Poland affected by long-range transported air pollutants. *Ecological Engineering* 3: 291–8.

Dugdale, G., Hardy, S. and Milford, J.R., 1991. Daily catchment rainfall estimated from Meteosat. *Hydrological Processes* 5: 261–70.

Dunne, T. and Black, R.D., 1970. Partial area contributions to storm runoff in a small New England watershed. *Water Resources Research* 6: 1296–311.

Dunne, T., Moore, T.R. and Taylor, C.H., 1975. Recognition and prediction of runoff-producing zones in humid regions. *Hydrological Sciences Bulletin* 20: 305–27.

Durocher, M.G., 1990. Monitoring spatial variability of forest interception. *Hydrological Processes* 4: 215–29.

Dvorak, V., Hladny, J. and Kasparek, L., 1997. Climate change hydrology and water resources impact and adaptation for selected river basins in the Czech Republic. *Climatic Change* 36: 93–106.

Dykes, A.P., 1997. Rainfall interception from a lowland tropical rainforest in Brunei. *Journal of Hydrology* 200: 260–79.

Dykes, A.P. and Thornes, J.B., 2000. Hillslope hydrology in tropical rainforest steeplands in Brunei. *Hydrological Processes* 14: 215–35.

Earle, C.J., 1993. Asynchronous droughts in California streamflow as reconstructed from tree rings. *Quaternary Research* 39: 290–9.

Ellis, J.B., 1979. The nature and sources of urban sediments and their relation to water quality: a case study from north-west London. In: G.E. Hollis (ed.), *Man's Impact on the Hydrological Cycle in the United Kingdom*. GeoBooks, Norwich, pp. 199–216.

Elsenbeer, H. and Lack, A., 1996. Hydrometric and hydrochemical evidence for fast flowpaths at La Cuenca, Western Amazonia. *Journal of Hydrology* 180: 237–50.

Eltahir, E.A.B., 1996. El Niño and the natural variability in the flow of the Nile River. *Water Resources Research* 32: 131–7.

Eltahir, E.A.B., 1998. A soil moisture rainfall feedback mechanism. 1: Theory and observation. *Water Resources Research* 34: 765–76.

Ely, L.L., 1997. Response of extreme floods in the southwestern United States to climatic variations in the late Holocene. *Geomorphology* 19: 175–201.

Ely, L.L., Enzel, Y. and Cayan, D.R., 1994. Anomalous North Pacific atmospheric circulation and large winter floods in the southwestern United States. *Journal of Climate* 7: 977–87.

Engman, E.T. and Gurney, R.J., 1991. *Remote Sensing in Hydrology*. Chapman and Hall, London.

Environment Agency, 1998a. *The State of the Environment of England and Wales: Fresh Waters*. The Stationery Office, London.

Environment Agency, 1998b. *Soar Action Plan: Local Environment Agency Plan*, Environment Agency, Bristol.

Essery, C.I. and Wilcock, D.N., 1991. The variation in rainfall catch from standard UK Meteorological Office raingauges: a twelve year case study. *Hydrological Sciences Journal* 36: 23–34.

European Environment Agency, 1999. *Environment in the European Union at the Turn of the Century*. European Environment Agency, Copenhagen.

Evans, C.D., Jenkins, A., Helliwell, R.C. and Ferrier, R.C., 1998. Predicting regional recovery from acidification: the MAGIC model applied to Scotland, England and Wales. *Hydrology and Earth System Sciences* 2: 543–54.

Evans, C., Davies, T.D. and Murdoch, P.S., 1999. Component flow processes at four streams in the Catskill Mountains, New York, analysed using episodic concentration/discharge relationships. *Hydrological Processes* 13: 563–75.

Farquharson, F.A.K., Meigh, J.R. and Sutcliffe, J.V., 1992. Regional flood frequency analysis in arid and semi-arid area. *Journal of Hydrology* 138: 487–501.

Favis-Mortlock, D.T. and Guerra, S.J.T., 1999. The implications of general circulation model estimates of rainfall for future erosion: a case study from Brazil. *Catena* 37: 329–54.

Federer, C.A., Vörösmarty, C.J. and Fekete, B., 1996. Intercomparison of methods for potential evapotranspiration in regional or global water balance models. *Water Resources Research* 32: 2315–21.

Felzer, B. and Heard, P., 1999. Precipitation differences amongst GCMs used for the US national assessment. *Journal of the American Water Resources Association* 35: 1327–39.

Fennessey, N.M. and Vogel, R.M., 1996. Regional models of potential evaporation and reference evaporation for the north east USA. *Journal of Hydrology* 184: 337–54.

Field, C.B., Jackson, R.B. and Mooney, H.A., 1995. Stomatal responses to increased CO_2 – implications from the plant to the global scale. *Plant, Cell and Environment* 18: 1214–25.

Finch, J.W., 1998. Estimating direct groundwater recharge using a simple water balance model – sensitivity to land surface parameters. *Journal of Hydrology* 211: 112–25.

Folland, C.K., Palmer, T.N. and Parker, D.E., 1986. Sahel rainfall and worldwide sea temperatures, 1901–1985. *Nature* 320: 602–7.

Fontes, J.C. and Gasse, F., 1991. PALHYDAF (Palaeohydrology in Africa) program – objectives, methods, major results. *Palaeogeography, Palaeoclimatology, Palaeoecology* 84: 191–215.

Fowler, D., Cape, J.N. and Unsworth, M.H., 1989. Deposition of atmospheric pollutants on forests. *Philosophical Transactions of the Royal Society of London, Series B* 324: 247–65.

Fraedrich, K., 1994. An ENSO impact on Europe? *Tellus* 46A: 541–52.

Frech, M. and Jochum, A., 1999. The evaluation of flux aggregation methods using aircraft measurements in the surface layer. *Agricultural and Forest Meteorology* 98–99: 121–43.

Freer, J., Beven, K. and Ambroise, B., 1996. Bayesian estimation of uncertainty in runoff prediction and the value of data: an application of the GLUE approach. *Water Resources Research* 32: 2161–73.

Frihy, O.E., Dewidar, K.M. and El Banna, M.M., 1998. Natural and human impact on the northeastern Nile delta coast of Egypt. *Journal of Coastal Research* 14: 1109–18.

Fujieda, M., Kudoh, T., de Cicco, V. and de Calvarcho, J.L., 1997. Hydrological processes at two subtropical forest catchments: the Serra do Mar, São Paulo, Brazil. *Journal of Hydrology* 196: 26–46.

Fuller, I.C., Macklin, M.G., Lewin, J., Passmore, D.G. and Wintle, A.G., 1998. River response to high-frequency climate oscillations in southern Europe over the past 200 kyr. *Geology* 26: 275–8.

Gao, X.G., Sorooshian, S. and Gupta, H.V., 1996. Sensitivity analysis of the Biosphere-Atmosphere Transfer Scheme. *Journal of Geophysical Research – Atmospheres* 101(D3): 7279–89.

Garmouna, M., Teil, M.J., Blanchard, M. and Chevreuil, M., 1998. Spatial and temporal variations of herbicide (triazines and phenylureas) concentrations in the catchment basin of the Marne River (France). *Science of the Total Environment* 224: 93–107.

Garrels, R.M., Mackenzie, F.T. and Hunt, C., 1975. *Chemical Cycles and the Global Environment.* Kaufmann, Los Altos.

Gash, J.H.C., 1979. An analytical model of rainfall interception in forests. *Quarterly Journal of the Royal Meteorological Society* 105: 43–55.

Gash, J.H.C. and Morton, A.J., 1978. An application of the Rutter model to the estimation of the interception loss from Thetford Forest. *Journal of Hydrology* 38: 49–58.

Gash, J.H.C. and Nobre, C.A., 1997. Climatic effects of Amazonian deforestation: some results from ABRACOS. *Bulletin of the American Meteorological Society* 78: 823–30.

Gash, J.H.C. and Stewart, J.B., 1977. The evaporation from Thetford Forest during 1975. *Journal of Hydrology* 35: 385–96.

Gash, J.H.C., Wright, I.R. and Lloyd, C.R., 1980. Comparative estimates of interception loss from three coniferous forests in Great Britain. *Journal of Hydrology* 48: 89–105.

Gash, J.H.C., Lloyd, C.R. and Lachaud, G., 1995. Estimating sparse forest rainfall interception with an analytical model. *Journal of Hydrology* 170: 79–86.

Gash, J.H.C. *et al.*, 1997. The variability of evaporation during the HAPEX-Sahel Intensive Observation Period. *Journal of Hydrology* 188–9: 385–99.

Gates, W.L. *et al.*, 1999. An overview of the results of the Atmospheric Model Intercomparison Project (AMIP I). *Bulletin of the American Meteorological Society* 80: 29–55.

Gburek, W.J. and Folmar, G.J., 1999. A ground water recharge field study: site characterization and initial results. *Hydrological Processes* 13: 2813–31.

Gellens, D. and Roulin, E., 1998. Streamflow response of Belgian catchments to IPCC climate change scenarios. *Journal of Hydrology* 210: 242–58.

GEMS-Water, 1999. *Annotated Digital Atlas of Global Water Quality.* www.cciw.ca/gems/atlas-gwq/gems1.htm.

Genta, J.L., Perez-Iribarren, G. and Mechoso, C.R., 1998. A recent increasing trend in the streamflow of rivers in southeastern South America. *Journal of Climate* 11: 2858–62.

George, R.K., Waylen, P. and Laporte, S., 1998. Interannual variability of annual streamflow and the Southern Oscillation in Costa Rica. *Hydrological Sciences Journal* 43: 409–24.

Gilman, K., 1994. *Hydrology and Wetland Conservation.* Wiley, Chichester.

Gineste, P., Puech, C. and Merot, P., 1998. Radar remote sensing of the source areas from the Coet-Dan catchment. *Hydrological Processes* 12: 267–84.

Glacken, C.J., 1967. *Traces on the Rhodian Shore.* University of California Press, Berkeley.

Glantz, M., 1996. *Currents of Change.* Cambridge University Press, Cambridge.

Glantz, M.H., Rubinstein, A.Z. and Zonn, I., 1993. Tragedy in the Aral Sea basin. *Global Environmental Change* 3: 174–98.

Gleick, P.H. (ed.), 1993. *Water in Crisis: A Guide to the World's Fresh Water Resources*. Oxford University Press, New York.

Gleick, P.H., 1998. *The World's Water: The Biennial Report on Freshwater Resources 1998–1999*. Island Press, Washington D.C.

Goodrich, D.C., Faures, J.-M., Woolhiser, D.A., Lane, L.J. and Sorooshian, S., 1995. Measurement and analysis of small-scale convective storm rainfall variability. *Journal of Hydrology* 173: 283–308.

Goodrich, D.C. *et al.*, 1997. Linearity of basin response as a function of scale in a semiarid watershed. *Water Resources Research* 33: 2951–65.

Goovaerts, P., 2000. Geostatistical approaches for incorporating elevation into the spatial interpolation of rainfall. *Journal of Hydrology* 228: 113–29.

Gornitz, V., Rosenzweig, C. and Hillel, D., 1997. Effects of anthropogenic intervention in the land hydrologic cycle on global sea level rise. *Global and Planetary Change* 14: 147–61.

Gottlein, A. and Manderscheid, B., 1998. Spatial heterogeneity and temporal dynamics of soil water tension in a mature Norway spruce stand. *Hydrological Processes* 12: 417–28.

Gottschalk, L., 1985. Hydrological regionalisation of Sweden. *Hydrological Sciences Journal* 30: 65–83.

Gottschalk, L., Jensen, J.L., Lundquist, D., Solantie, R. and Tollan, A., 1979. Hydrological regimes in Nordic countries. *Nordic Hydrology* 10: 273–86.

Goutourbe, J.P. *et al.*, 1997. An overview of HAPEX-Sahel: a study in climate and desertification. *Journal of Hydrology* 188–9: 4–17.

Graham, S.T., Famiglietti, J.S. and Maidment, D.R., 1999. Five-minute, 1/2 degree, and 1 degree data sets of continental watersheds and river networks for use in regional and global hydrologic and climate system modelling studies. *Water Resources Research* 35: 583–7.

Granger, R.J., 2000. Satellite-derived estimates of evapotranspiration in the Gediz Basin. *Journal of Hydrology* 229: 70–6.

Granier, A., Biron, P. and Lemoine, D., 2000. Water balance, transpiration and canopy conductance in two beech stands. *Agricultural and Forest Meteorology* 100: 291–308.

Gray, D.M. and Prowse, T.D., 1993. Snow and floating ice. In: D.R. Maidment (ed.), *Handbook of Hydrology*. McGraw Hill, New York, pp. 7.1–7.58.

Green, W.H. and Ampt, G.A., 1911. Studies on soil physics: 1. Flow of air and water through soils. *Journal of Agricultural Science* 4: 1–24.

Greene, E.M., Liston, G.E. and Pielke, R.A., 1999. Relationships between landscape, snowcover depletion, and regional weather and climate. *Hydrological Processes* 13: 2453–66.

Gregory, J.M., Mitchell, J.F.B. and Brady, A.J., 1997. Summer drought in northern mid-latitudes in a time-dependent CO_2 climate experiment. *Journal of Climate* 10: 662–86.

Gregory, K.J. and Walling, D.E., 1973. *Drainage Basin Form and Process*. Edward Arnold, London.

Gregory, K.J., Starkel, L. and Baker, V.R. (eds), 1995. *Global Continental Palaeohydrology*. Wiley, Chichester.

Grelle, A., Lundberg, A., Lindroth, A., Moren, A.-S. and Cienciala, E., 1997. Evaporation components of a boreal forest: variations during the growing season. *Journal of Hydrology* 197: 70–87.

Gremillion, P., Gonyeau, A. and Wanelista, M., 2000. Application of alternative hydrograph separation models to detect change in flow paths in a watershed undergoing urban development. *Hydrological Processes* 14: 1485–501.

Grimes, D.I.F., Pardo-Iguzquiza, E. and Bonaficio, R., 1999. Optimal areal rainfall estimation using raingauges and satellite data. *Journal of Hydrology* 222: 93–108.

Guetter, A.K. and Georgakakos, K.P., 1993. River outflow of the conterminous United States, 1939–1988. *Bulletin of the American Meteorological Society* 74: 1873–91.

Guetter, A.K. and Georgakakos, K.P., 1996. Are the El Niño and La Niña predictors of the Iowa River seasonal flow? *Journal of Applied Meteorology* 35: 690–705.

Gupta, A.D. and Paudyal, G.N., 1988. Estimating aquifer recharge and parameters from water level observations. *Journal of Hydrology* 99: 103–16.

Gupta, V.K. and Dawdy, D.R., 1995. Physical interpretation of regional variations in the scaling exponents of flood quantiles. *Hydrological Processes* 9: 347–62.

Gurnell, A.M., 1976. A note on the contribution of fog drip to streamflow. *Weather* 31: 121–6.

Gurnell, A.M., 1978. The dynamics of a drainage network. *Nordic Hydrology* 9: 293–306.

Gurnell, A.M. and Petts, G.E., 2000. Causes of catchment scale hydrological changes. In: M. Acreman (ed.), *The Hydrology of the UK: A Study of Change.* Routledge, London, pp. 82–98.

Gurnell, A.M., Gregory, K.J., Hollis, S. and Hill, C.T., 1985. Detrended correspondence analysis of heathland vegetation: the identification of runoff contributing areas. *Earth Surface Processes and Landforms* 10: 343–51.

Gustard, A., Roald, L.A., Demuth, S., Lumadjeng, H.S. and Gross, R., 1989. *Flow Regimes from Experimental and Network Data (FREND).* Institute of Hydrology, Wallingford, Oxfordshire.

Gustard, A., Bullock, A. and Dixon, J.M., 1992. *Low Flow Estimation in the United Kingdom.* Report 108, Institute of Hydrology, Wallingford, UK.

Hailemariam, K., 1999. Impact of climate change on the water resources of Awash River Basin, Ethiopia. *Climate Research* 6: 91–6.

Haines, A.T., Finlayson, B.L. and McMahon, T.A., 1988. A global classification of river regimes. *Applied Geography* 8: 255–72.

Hall, F.G., 1999. Introduction to the special section: BOREAS in 1999: Experiment and science overview. *Journal of Geophysical Research – Atmospheres* 104(D22): 27627–39.

Hall, R.J. and Allen, S.J., 1997. Water use of poplar clones grown as short-rotation coppice at two sites in the UK. *Aspects of Applied Biology* 44: 1–8.

Hall, R.L., 1985. Further interception studies of heather using a wet surface weighing lysimeter system. *Journal of Hydrology* 81: 193–210.

Halldin, S. *et al.*, 1998. NOPEX: a northern hemisphere climate processes land surface experiment. *Journal of Hydrology* 212: 188–97.

Hamlet, A.F. and Lettenmaier, D.P., 1999a. Columbia River streamflow forecasting based on ENSO and PDO climate signals. *Journal of Water Resources Planning and Management – ASCE* 125: 333–41.

Hamlet, A.F. and Lettenmaier, D.P., 1999b. Effects of climate change on hydrology and water resources in the Columbia River basin. *Journal of the American Water Resources Association* 35: 1597–623.

Hamon, W.R., 1963. Computation of direct runoff amounts from storm rainfall. *Int. Ass. Hydrol. Sci. Publ.* 63, 52–62.

Hanan, N.P. and Prince, S.D., 1997. Stomatal conductance of west-central supersite vegetation in HAPEX-Sahel: measurements and empirical models. *Journal of Hydrology* 188–9: 536–62.

Hannah, D.M., Gurnell, A.M. and McGregor, G.R., 1999. A methodology for investigation of the seasonal evolution in proglacial hydrograph form. *Hydrological Processes* 13: 2603–21.

Hanratty, M.P. and Stefan, H.G., 1998. Simulating climate change effects in a Minnesota agricultural watershed. *Journal of Environmental Quality* 27: 1524–32.

Harding, R.J. and Lloyd, C., 1998. Fluxes of water and energy from three high latitude tundra sites in Svalbard. *Nordic Hydrology* 29: 267–84.

Harding, R.J. *et al.*, 1992. *Hydrological Impacts of Broadleaf Woodlands: Implications for Water Use and Water Quality.* Report 115/03/ST, National Rivers Authority.

Harrison, S.P. and Digerfeldt, G., 1993. European lakes as palaeohydrological and palaeoclimatic indicators. *Quaternary Science Reviews* 12: 233–48.

Harvey, L.D.D., 2000. *Climate and Global Environmental Change.* Prentice Hall, Harlow.

Hassan, H., Aramaki, T., Hanaki, K., Matsuo, T. and Wilby, R., 1998. Lake stratification and temperature profiles simulated using downscaled GCM output. *Water Science and Technology* 11: 217–26.

Havel, M., Peters, N.E. and Cerny, J., 1999. Longitudinal patterns of stream chemistry in a catchment with forest dieback, Czech Republic. *Environmental Pollution* 104: 157–67.

Headworth, H.G., 1972. The analysis of natural groundwater level fluctuations in the chalk of Hampshire. *Journal of the Institute of Water Engineers and Scientists* 26: 107–24.

Hedstrom, N.R. and Pomeroy, J.W., 1998. Measurements and modelling of snow interception in the boreal forest. *Hydrological Processes* 12: 1611–25.

Heikenheimo, M., Kargas, M., Tourula, T., Venalainen, A. and Tattari, S., 1999. Momentum and heat fluxes over Lakes Tamnaren and Raksjo determined by the bulk-aerodynamic and eddy-correlation methods. *Agricultural and Forest Meteorology* 98–99: 521–34.

Hennessy, K.J., Gregory, J.M. and Mitchell, J.F.B., 1997. Changes in daily precipitation under enhanced greenhouse conditions. *Climate Dynamics* 13: 667–80.

Heppell, C.M., Burt, T.P. and Williams, R.J., 2000. Variations in the hydrology of an underdrained clay hillslope. *Journal of Hydrology* 227: 236–56.

Herczeg, A.L., Leaney, F.W.J., Stadter, M.F., Allan, G.L. and Fifield, L.K., 1997. Chemical and isotopic indicators of point-source recharge to a karst aquifer, South Australia. *Journal of Hydrology* 192: 271–99.

Herwitz, S.T., 1986. Infiltration-excess caused by stemflow in a cyclone-prone tropical rainforest. *Earth Surface Processes and Landforms* 11: 401–12.

Hewlett, J.D. and Hibbert, A.R., 1967. Factors affecting the response of small watersheds to precipitation in humid areas. In: W.E. Sopper and H.W. Lull (eds), *Forest Hydrology.* Pergamon, Oxford, pp. 275–90.

Hill, T. and Neal, C., 1997. Spatial and temporal variation in pH, alkalinity and conductivity in surface runoff and groundwater for the Upper River Severn catchment. *Hydrology and Earth System Sciences* 1: 697–715.

Hillel, D., 1980. *Fundamentals of Soil Physics.* Academic Press, New York.

Hirsch, R.M., Slack, J.R. and Smith, R.A., 1982. Techniques of trend analysis for monthly water quality data. *Water Resources Research* 18: 107–21.

Hodson, A.J. *et al.*, 1998. Meteorological and runoff time-series characteristics in a small, high-Arctic glaciated basin, Svalbard. *Hydrological Processes* 12: 509–26.

Hoeg, S., Uhlenbrook, S. and Leibundgut, C., 2000. Hydrograph separation in a mountainous catchment – combining hydrochemical and isotopic tracers. *Hydrological Processes* 14: 1199–216.

Hoffert, M.I., Frei, A. and Narayanan, V.K., 1988. Application of solar max ACRIM data to analysis of solar-driven climatic variability on earth. *Climatic Change* 13: 267–86.

Hollis, G.E., 1975. The effect of urbanisation on floods of different recurrence intervals. *Water Resources Research* 11: 431–4.

Hongve, D., 1999. Production of dissolved organic carbon in forested catchments. *Journal of Hydrology* 224: 91–9.

Hope, D., Billett, M.F. and Cresser, M.S., 1997. Exports of organic carbon in two river systems in NE Scotland. *Journal of Hydrology* 193: 61–82.

Hopkinson, C. and Young, G.J., 1998. The effect of glacier wastage on the flow of the Bow River at Banff, Alberta, 1951–93. *Hydrological Processes* 12: 1745–62.

Hornberger, G., Raffensperger, J.P., Wiberg, P.L. and Eshleman, K.N., 1998. *Elements of Physical Hydrology*. Johns Hopkins University Press, Baltimore.

Horton, J.E., 1933. The role of infiltration in the hydrological cycle. *Transactions of the American Geophysical Union* 14: 446–60.

Hostetler, S.W. and Small, E.E., 1999. Response of North American freshwater lakes to simulated future climates. *Journal of the American Water Resources Association* 35: 1625–37.

Hough, M.N. and Jones, R.J.A., 1997. The UK Meteorological Office rainfall and evaporation calculation system: MORECS version 2.0 – and overview. *Hydrology and Earth System Sciences* 1: 227–39.

Houghton, R.A., 1999. The annual net flux of carbon to the atmosphere from changes in land use. *Tellus* 51B: 298–313.

House, W.A. *et al.*, 1997. Nutrient transport in the Humber rivers. *Science of the Total Environment* 194: 303–20.

Howard, A.G., 1998. *Aquatic Environmental Chemistry*. Oxford University Press, Oxford.

Howarth, R.W. *et al.*, 1996. Regional nitrogen budgets and riverine N&P fluxes for the drainages to the North Atlantic Ocean: natural and human influences. *Biogeochemistry* 35: 75–139.

Howells, G., 1990. *Acid Rain and Acid Waters*. Ellis Horwood, New York.

Hudson, J.A., Crane, S.B. and Blackie, J.R., 1997a. The Plynlimon water balance 1969–1995: the impact of forest and moorland vegetation on evaporation and streamflow in upland catchments. *Hydrology and Earth System Sciences* 1: 409–27.

Hudson, J.A., Gilman, K. and Calder, I.R., 1997b. Land use and water quality issues in the uplands with reference to the Plynlimon study. *Hydrology and Earth System Sciences* 1: 389–97.

Hulme, M. and Barrow, E., 1997. Introducing climate change. In: M. Hulme and E. Barrow (eds), *Climates of the British Isles: Present, Past and Future*. Routledge, London, pp. 1–7.

Hulme, M. and Jenkins, G., 1998. *Climate change scenarios for the United Kingdom: scientific report*. UKCIP Technical Report No. 1, Climatic Research Unit, University of East Anglia.

Hulme, M. *et al.*, 1999. Relative impacts of human-induced climate change and natural climatic variability. *Nature* 397: 688–91.

Humborg, C., Ittekkot, V., Cociasu, A. and Von Bodungen, B., 1997. Effect of Danube River dam on Black Sea biogeochemistry and ecosystem structure. *Nature* 386: 385–8.

Hunte, W. and Wittenberg, M., 1992. Effects of eutrophication and sedimentation on juvenile corals. *Marine Biology* 114: 625–31.

Hunzinger, H., 1997. Hydrology of montane forests in the Sierra de San Javier, Tucuman, Argentina. *Mountain Research and Development* 17: 299–317.

Hurrell, J.W., 1995. Decadal trends in the North Atlantic Oscillation: regional temperatures and precipitation. *Science* 269: 676–9.

Hutjes, R.W.A., Wierda, A. and Veen, A.W.L., 1990. Rainfall interception in the Tai Forest, Ivory Coast: application of two simulation models to a humid tropical system. *Journal of Hydrology* 114: 259–75.

Ibanez, C., Prat, N. and Canicio, A., 1996. Changes in the hydrology and sediment transport produced by large dams on the lower Ebro River and its estuary. *Regulated Rivers* 12: 51–62.

Ibanez, C., Pont, D. and Prat, N., 1997. Characterization of the Ebro and Rhône estuaries: a basis for defining and classifying salt-wedge estuaries. *Limnology and Oceanography* 42: 89–101.

Idso, S.B., Jackson, R.D., Reginato, R.J., Kimball, B.A. and Nakayama, F.S., 1975. The dependence of bare soil albedo on soil water content. *Journal of Applied Meteorology* 14: 109–13.

Inbar, M., Tamir, M. and Wittenberg, L., 1998. Runoff and erosion processes after a forest fire in Mount Carmel, a Mediterranean area. *Geomorphology* 24: 17–33.

Inman, D.L. and Jenkins, S.A., 1999. Climate change and the episodicity of sediment flux of small Californian rivers. *Journal of Geology* 107: 251–70.

Institute of Hydrology, 1999. *Flood Estimation Handbook*. 5 vols. Institute of Hydrology, Wallingford, Oxon.

IPCC (Intergovernmental Panel on Climate Change), 1990. *Climate Change: The IPCC Scientific Assessment*. Cambridge University Press, Cambridge.

IPCC (Intergovernmental Panel on Climate Change), 1996. *Climate Change 1995: The Science of Climate Change*. Cambridge University Press, Cambridge.

IPCC (Intergovernmental Panel on Climate Change), 2000. *Special Report on Emissions Scenarios*. Cambridge University Press, Cambridge.

IPCC (Intergovernmental Panel on Climate Change), 2001, *Climate Change 2001: The Scientific Base*. Cambridge University Press, Cambridge.

Isdale, P.J., Stewart, B.J., Tickle, K.S. and Lough, J.M., 1998. Palaeohydrological variation in a tropical river catchment: a reconstruction using fluorescent bands in corals of the Great Barrier Reef, Australia. *The Holocene* 8: 1–8.

Jackson, I.J., 1975. Relationship between rainfall parameters and interception by tropical forests. *Journal of Hydrology* 24: 215–38.

Jackson, T.J., 1993. Measuring surface soil moisture using passive microwave remote sensing. *Hydrological Processes* 7: 139–52.

Jackson, T.J. and Le Vine, D.E., 1996. Mapping surface soil moisture using an aircraft-based passive microwave instrument: algorithm and examples. *Journal of Hydrology* 184: 85–100.

Jackson, T.J., Schmugge, J. and Engman, E.T., 1996. Remote sensing applications to hydrology: soil moisture. *Hydrological Sciences Journal* 41: 517–30.

Jarvis, A.J., Mansfield, T.A. and Davies, W.J., 1999. Stomatal behaviour, photosynthesis and transpiration under rising CO_2. *Plant, Cell and Environment* 22: 639–48.

Jenkins, A., Ferrier, R.C. and Cosby, B.J., 1997. A dynamic model for assessing the impact of coupled sulphur and nitrogen deposition scenarios on surface water acidification. *Journal of Hydrology* 197: 111–27.

Jensen, M.E., Burman, R.D. and Allen, R.G., 1990. *Evaporation and irrigation water requirements.* Manuals and Reports on Engineering Practice 70, American Society of Civil Engineers, New York.

Johnson, R.C., 1990. The interception, throughfall and stemflow in a forest in Highland Scotland and the comparison with other upland forests in the UK. *Journal of Hydrology* 118: 281–7.

Jones, J.A. and Grant, G.E., 1996. Peak flow responses to clear-cutting and roads in small and large basins, western Cascades, Oregon. *Water Resources Research* 32: 959–74.

Jones, J.A.A., 1997a. Pipeflow contributing areas and runoff response. *Hydrological Processes* 11: 35–41.

Jones, J.A.A., 1997b. *Global Hydrology*. Longman, Harlow.

Jose, A.M. and Cruz, N.A., 1999. Climate change impacts and responses in the Philippines: water resources. *Climate Research* 12: 77–84.

Jouzel, J. *et al.*, 1987. Vostok ice core: a continuous isotope temperature record over the last climatic cycle (160,000 years). *Nature* 329: 402–8.

Kabat, P., Dolman, A.J. and Elbers, J.A., 1997. Evaporation, sensible heat and canopy conductance of fallow savannah and patterned woodland in the Sahel. *Journal of Hydrology* 188–9: 494–515.

Kahya, E. and Dracup, J.A., 1993. U.S. streamflow patterns in relation to the El Niño/Southern Oscillation. *Water Resources Research* 29: 2491–503.

Kahya, E. and Dracup, J.A., 1994. The influences of type 1 El Niño and La Niña events on streamflows in the Pacific southwest of the United States. *Journal of Climate* 7: 965–76.

Kallio, K. *et al.*, 1997. Impacts of climatic change on agricultural nutrient losses in Finland. *Boreal Environment Research* 2: 33–52.

Kamari, J., Rankinen, K., Finer, L., Piirainen, S. and Posch, M., 1998. Modelling the response of soil and runoff chemistry to forest harvesting in a low deposition area (Kangasvaara, Eastern Finland). *Hydrology and Earth System Sciences* 2: 485–95.

Karl, T.R. and Knight, R.W., 1998. Secular trends of precipitation amount, frequency and intensity in the United States. *Bulletin of the American Meteorological Society* 79: 231–42.

Kattner, G. *et al.*, 1999. Tracing dissolved organic substances and nutrients from the Lena River through Laptev Sea (Arctic). *Marine Chemistry* 65: 25–39.

Kauppi, P.E., Meilikainen, K. and Kuusela, K., 1992. Biomass and carbon budget of European forests, 1700 to 1990. *Science* 256: 70–2.

Keller, E.A., Valentine, D.W. and Gibbs, D.R., 1997. Hydrological response of small watersheds following the Southern California Painted Cave fire of June 1990. *Hydrological Processes* 11: 401–26.

Kelliher, F.M., Leuning, R., Raupach, M. and Schulze, E.-D., 1995. Maximum conductances for evaporation from global vegetation types. *Agricultural and Forest Meteorology* 73: 1–16.

Kelliher, F.M. *et al.*, 1997. Evaporation from an eastern Siberian larch forest. *Agricultural and Forest Meteorology* 85: 135–47.

Kelliher, F.M. *et al.*, 1998. Evaporation from a central Siberian pine forest. *Journal of Hydrology* 205: 279–96.

Khaliq, M.N. and Cunnane, C., 1996. Modelling point rainfall occurrences with the Modified Bartlett–Lewis Rectangular Pulses model. *Journal of Hydrology* 180: 109–37.

Kirby, C., Newson, M.D. and Gilman, K., 1990. *Plynlimon research: the first two decades.* Report 109, Institute of Hydrology, Wallingford, UK.

Kirby, C., Neal, C., Turner, H. and Moorhouse, P., 1997. A bibliography of hydrological, geomorphological, sedimentological, biological and hydrochemical references to the Institute of Hydrology experimental catchment studies in Plynlimon. *Hydrology and Earth System Sciences* 1: 755–63.

Kitchen, M. and Blackall, R.M., 1992. Orographic rainfall over low hills and associated corrections to radar measurements. *Journal of Hydrology* 139: 115–34.

Kite, G.W. and Droogers, P., 2000. Comparing evapotranspiration estimates from satellites, hydrological models and field data. *Journal of Hydrology* 229: 3–18.

Klaassen, W., Lankreijer, H.J.M. and Veen, A.W.L., 1996. Rainfall interception near a forest edge. *Journal of Hydrology* 185: 349–61.

Klaassen, W., Bosveld, F. and de Water, E., 1998. Water storage and evaporation as constituents of rainfall interception. *Journal of Hydrology* 212–13: 36–50.

Klemes, V., 1988. A hydrological perspective. *Journal of Hydrology* 100: 3–28.

Knox, J.C., 1993. Large increases in flood magnitude in response to modest changes in climate. *Nature* 361: 430–2.

Kobayashi, D., Ishii, Y. and Kodama, Y., 1999. Stream temperature, specific conductance and runoff process in mountain watersheds. *Hydrological Processes* 13: 865–76.

Kolpin, D.W., Barbash, J.E. and Gilliom, R.J., 1998. Occurrence of pesticides in shallow groundwater of the United States: initial results from the National Water Quality Assessment Program. *Environmental Science and Technology* 32: 558–66.

Korzun, V.I., 1978. *World Water Balance and Water Resources of the Earth.* Studies and Reports in Hydrology 25. Unesco, Paris.

Koster, R.D., Eagleson, P.S. and Broecker, W.S., 1988. *Tracer Water Transport and Subgrid Precipitation Variation within Atmospheric General Circulation Models.* Report 317, Massachusetts Institute of Technology, Ralph M. Parsons Laboratory, Department of Civil Engineering.

Krasovskaia, I., 1995. Quantification of the stability of river flow regimes. *Hydrological Sciences Journal* 40: 587–97.

Krasovskaia, I., Arnell, N.W. and Gottschalk, L., 1994. Flow regimes in northern and western Europe: development and application of procedures for classifying flow regimes. In: *FRIEND – Flow Regimes from International Experimental and Network Data. Int. Ass. Hydrol. Sci. Publ.* 221, 185–92.

Kreuger, J., 1998. Pesticides in stream water within an agricultural catchment in southern Sweden, 1990–1996. *Science of the Total Environment* 216: 227–51.

Kruijt, B., Barton, C., Rey, A. and Jarvis, P.G., 1999. The sensitivity of stand-scale photosynthesis and transpiration to changes in atmospheric CO_2 concentration and climate. *Hydrology and Earth System Sciences* 3: 55–69.

Kuhnel, I. *et al.*, 1990. Climatic influences on streamflow variability: a comparison between southeastern Australia and southeastern United States of America. *Water Resources Research* 26: 2483–96.

Kundzewicz, Z., 1997. Water resources for sustainable development. *Hydrological Sciences Journal* 42: 467–80.

Kustas, W.P. and Norman, J.M., 1996. Use of remote sensing for evapotranspiration monitoring over land surfaces. *Hydrological Sciences Journal* 41: 495–516.

Kuusisto, E., 1986. The energy balance of a melting snow cover in different environments. In: *Modeling snowmelt-induced processes. Int. Ass. Hydrol. Sci. Publ.* 155.

Lacey, G.C. and Grayson, R.B., 1998. Relating baseflow to catchment properties in south-eastern Australia. *Journal of Hydrology* 204: 231–50.

La Fleur, P.M. and Roulet, N.T., 1992. A comparison of evaporation rates from two fens of the Hudson Bay lowland. *Aquatic Botany* 44: 59–69.

Lal, R., 1996. Deforestation and land-use effects on soil degradation and rehabilitation in western Nigeria. 3: Runoff, soil erosion and nutrient loss. *Land Degradation and Development* 7: 99–119.

Latif, M. and Grötzner, A., 2000. The equatorial Atlantic oscillation and its response to ENSO. *Climate Dynamics* 16: 213–18.

Laudon, H. and Slaymaker, O., 1997. Hydrograph separation using stable isotopes, silica and electrical conductivity: an alpine example. *Journal of Hydrology* 201: 82–101.

Lavabre, J., Torres, D.S. and Cernesson, F., 1993. Changes in the hydrological response of a small Mediterranean basin a year after a wildfire. *Journal of Hydrology* 142: 273–99.

Lavee, H. and Poesen, J.W., 1991. Overland flow generation and continuity on stone-covered surfaces. *Hydrological Processes* 5: 345–60.

Law, F., 1956. The effect of afforestation upon the yield of water catchment areas. *Journal of the British Waterworks Association* 38: 484–94.

Lean, J., Bunton, C.B., Nobre, C.A. and Rowntree, P.R., 1996. The simulated impact of Amazonian deforestation on climate using measured ABRACOS vegetation characteristics. In: J.H.C. Gash, C.A. Nobre, J.M. Roberts and R.L. Victoria (eds), *Amazonian Deforestation and Climate*. Wiley, Chichester, pp. 549–76.

Leduc, C., Bromley, J. and Schroeter, P., 1997. Water table fluctuation and recharge in semi-arid climate: some results of the HAPEX-Sahel hydrodynamic survey (Niger). *Journal of Hydrology* 188–9: 123–38.

Leeks, G.J.L. and Jarvie, H.P., 1998. Introduction to the Land-Ocean Interaction Study (LOIS): rationale and international context. *Science of the Total Environment* 210/211: 5–20.

Leeks, G.J.L. and Marks, S.D., 1997. Dynamics of river sediments in forested headwater streams: Plynlimon. *Hydrology and Earth System Sciences* 1: 483–97.

Le Maitre, D.C. and Versfeld, D.B., 1997. Forest evaporation models: relationships between stand growth and evaporation. *Journal of Hydrology* 193: 240–57.

Lentz, S.J. and Limeburner, R., 1995. The Amazon River plume during AMASSEDS – spatial characteristics and saline variability. *Journal of Geophysical Research – Oceans* 100(C2): 2355–75.

Lettenmaier, D.P., Wood, A.W., Palmer, R.N., Wood, E.F. and Stakhiv, E.Z., 1999. Water resources implications of global warming: a US regional perspective. *Climatic Change* 43: 537–79.

Leung, L.R. and Wigmosta, M.S., 1999. Potential climate change impacts on mountain watersheds in the Pacific Northwest. *Journal of the American Water Resources Association* 35: 1463–71.

Leung, L.R., Hamlet, A.F., Lettenmaier, D.P. and Kumar, A., 1999. Simulations of the ENSO hydroclimate signals in the Pacific northwest Columbia River basin. *Bulletin of the American Meteorological Society* 80: 2313–29.

Lhomme, J.P., 1997. Towards a rational definition of potential evaporation. *Hydrology and Earth System Sciences* 1: 257–64.

Lins, H., 1985. Interannual streamflow variability in the United States based on principal components. *Water Resources Research* 21: 691–707.

Lins, H.F., 1997. Regional streamflow regimes and hydroclimatology of the United States. *Water Resources Research* 33: 1655–68.

Lins, H.F. and Slack, J.R., 1999. Streamflow trends in the United States. *Geophysical Research Letters* 26: 227–30.

Liu, S., 1998. Estimation of rainfall storage capacity in the canopies of cypress wetlands and slash pine uplands in North-Central Florida. *Journal of Hydrology* 207: 32–41.

Llorens, P., Poch, R., Latron, J. and Gallart, F., 1997. Rainfall interception by a *Pinus sylvestris* forest patch overgrown in a Mediterranean mountainous abandoned area. *Journal of Hydrology* 199: 331–45.

Lloyd, C.R. and Marques, A., 1988. Spatial variability of throughfall and stemflow measurements in Amazonian rain forest. *Agricultural and Forest Meteorology* 42: 63–73.

Lloyd, C.R., Gash, J.H.C., Shuttleworth, W.J. and Marques, A.D.O., 1988. The measurement and modelling of rainfall interception by Amazonian rain forest. *Agricultural and Forest Meteorology* 43: 277–97.

Loaiciga, H.A., Haston, L. and Michaelsen, J., 1993. Dendrochronology and long-term hydrologic phenomena. *Reviews of Geophysics* 31: 151–71.

Long, J.L.A., House, W.A., Parker, A. and Rae, J.E., 1998. Micro-organic compounds associated with sediments in the Humber rivers. *Science of the Total Environment* 210/211: 229–54.

Lorup, J.K., Refsgaard, J.C. and Mazvimavi, D., 1998. Assessing the effects of land use change on catchment runoff by combined use of statistical tests and hydrological modelling: case studies from Zimbabwe. *Journal of Hydrology* 205: 147–63.

Loukas, A. and Quick, M.C., 1996. Spatial and temporal distribution of storm precipitation in southwestern British Columbia. *Journal of Hydrology* 174: 37–56.

Lundberg, A., 1993. Evaporation of intercepted snow – review of existing and new measurement methods. *Journal of Hydrology* 151: 267–90.

Lundberg, A., Calder, I. and Harding, R., 1998. Evaporation of intercepted snow: measurement and modelling. *Journal of Hydrology* 206: 151–63.

Lvovich, M.I., 1938. *Opyt Klassifikatsii Rek USSR.* Trudy GGI, Leningrad.

Macdonald, R.W., Carmack, E.C., McLaughlin, F.A., Falkner, K.K. and Swift, J.H., 1999. Connections among ice, runoff and atmospheric forcing in the Beaufort Gyre. *Geophysical Research Letters* 26: 2223–6.

Mackenzie, F.T., 1998. *Our Changing Planet: An Introduction to Earth System Science and Global Environmental Change.* Prentice Hall, Upper Saddle River, New Jersey.

Maizels, J.K., 1983. Palaeovelocity and palaeodischarge determination for coarse gravel deposits. In: K.J. Gregory (ed.), *Background to Palaeohydrology.* Wiley, Chichester, pp. 101–39.

Malmer, A., 1996. Hydrological effects and nutrient losses of forest plantation establishment on tropical rainforest land in Sabah, Malaysia. *Journal of Hydrology* 174: 129–48.

Manabe, S. and Bryan, K., 1985. CO_2-induced change in a coupled ocean-atmosphere model and its paleoclimatic implications. *Journal of Geophysical Research* 90: 11689–707.

Manak, D.K. and Mysak, L.A., 1989. On the relationship between Arctic sea-ice anomalies and fluctuations in northern Canadian air temperature and river discharge. *Atmosphere and Ocean* 27: 682–91.

Mandelbrot, B.B. and Wallis, J.R., 1969. Computer experiments with fractional Gaussian noises. Part 1: Averages and variances. *Water Resources Research* 5: 228–41.

Maneux, E. *et al.*, 1999. Assessment of suspended matter input into the oceans by small mountainous coastal rivers: the case of the Bay of Biscay. *Comptes Rendus de l'Académie des Sciences, Serie II Fascicule A: Sciences de la Terre et des Planètes* 329: 413–20.

Mann, M.E., Bradley, R.S. and Hughes, M.K., 1999. Northern hemisphere temperatures during the past millennium: inferences, uncertainties and limitations. *Geophysical Research Letters* 26: 759–62.

Mantua, N.J., Hare, S.R., Zhang, Y., Wallace, J.M. and Francis, R.C., 1997. A Pacific inter-decadal climate oscillation with impact on salmon production. *Bulletin of the American Meteorological Society* 78: 1069–79.

Marengo, J.A., Tomasella, J. and Uvo, C.R., 1998. Trends in streamflow and rainfall in tropical South America: Amazonia, eastern Brazil and northwestern Peru. *Journal of Geophysical Research – Atmospheres* 103: 1775–83.

Marks, D., Kimball, J., Tingey, D. and Link, T., 1998. The sensitivity of snowmelt processes to climate conditions and forest cover during rain-on-snow: a case study of the 1996 Pacific Northwest flood. *Hydrological Processes* 12: 1569–87.

Marland, G., Boden, T.A., Andres, R.J., Brenkert, A.L. and Johnston, C., 1998. Global, regional, and national CO_2 emissions. *Trends: A Compendium of Data on Global Change*. Carbon Dioxide Information Analysis Center, Oak Ridge National Laboratory, Oak Ridge, Tennessee.

Marsh, T.J., Black, A., Acreman, M.C. and Elliot, C., 2000. River flows. In: M.C. Acreman (ed.), *The Hydrology of the UK: A Study of Change*. Routledge, London, pp. 101–34.

Marshall, S.J. and Clarke, G.K.C., 1999. Modeling North American freshwater runoff through the last glacial cycle. *Quaternary Research* 52: 300–15.

Martinez-Mena, M., Albaladejo, J. and Castillo, V.M., 1998. Factors influencing surface runoff generation in a Mediterranean semi-arid environment: Chicamo watershed, SE Spain. *Hydrological Processes* 12: 741–54.

Mason, B.J., 1992. *Acid Rain: Its Causes and its Effects on Inland Waters.* Clarendon Press, Oxford.

Matheussen, B., Kirschbaum, R.L., Goodman, I.A., O'Donnell, G.M. and Lettenmaier, D.P., 2000. Effects of land cover change on streamflow in the interior Columbia River basin (USA and Canada). *Hydrological Processes* 14: 867–85.

McCrutcheon, S.C., Martin, J.L. and Barnwell, T.O., 1993. Water quality. In: D.R. Maidment (ed.), *Handbook of Hydrology*. McGraw Hill, New York, pp. 11.1–11.73.

McGuffie, K., Henderson-Sellers, A., Zhang, H., Durbridge, T.B. and Pitman, A.J., 1995. Global climate sensitivity to tropical deforestation. *Global and Planetary Change* 10: 97–128.

McHale, M.R., Mitchell, M.J., McDonnell, J.J. and Cirmo, C.P., 2000. Nitrogen solutes in an Adirondack forested watershed: importance of dissolved organic nitrogen. *Biogeochemistry* 48: 165–84.

McKee, L., Eyre, B. and Hossain, S., 2000. Intra- and interannual export of nitrogen and phosphorus in the subtropical Richmond River catchment, Australia. *Hydrological Processes* 14: 1787–809.

McMahon, T.A., Finlayson, B.L., Haines, A.T. and Srikanthan, R., 1992. *Global Runoff: Continental Comparisons of Annual Flows and Peak Discharges.* Catena, Cremlingen-Drestedt.

Mechoso, C.R. and Iribarren, G.P., 1992. Streamflow in southeastern South America and the Southern Oscillation. *Journal of Climate* 5: 1535–9.

Meijninger, W.M.L. and de Bruin, H.A.R., 2000. The sensible heat fluxes over irrigated areas in western Turkey determined with a large aperture scintillometer. *Journal of Hydrology* 229: 42–9.

Meybeck, M., 1982. Carbon, nitrogen and phosphorus transport by world rivers. *American Journal of Science* 282: 401–50.

Meybeck, M., 1983. Riverine transport of atmospheric carbon – sources, global typology and budget. *Water, Air and Soil Pollution* 70: 443–63.

Meybeck, M., Chapman, D. and Helmer, R., 1989. *Global Freshwater Quality: A First Assessment*. Blackwell, Oxford.

Meyer, W.B., 1996. *Human Impact on the Earth*. Cambridge University Press, Cambridge.

Micklin, P.P., 1988. Desiccation of the Aral Sea – a water management disaster in the Soviet Union. *Science* 241: 1170–5.

Miles, E.L., Snover, A.K., Hamlet, A.F., Callahan, B. and Fluharty, D., 2000. Pacific northwest regional assessment: the impacts of climate variability and climate change on the water resources of the Columbia River basin. *Journal of the American Water Resources Association* 36: 399–420.

Miller, J.R. and Russell, G.L., 2000. Projected impact of climate change on the freshwater and salt budgets of the Arctic Ocean by a global climate model. *Geophysical Research Letters* 27: 1183–6.

Miller, J.R., Russell, G.L. and Caliri, G., 1994. Continental-scale river flow in climate models. *Journal of Climate* 7: 914–28.

Miller, N.L., Kim, J., Hartman, R.K. and Farrara, J., 1999. Downscaled climate and streamflow study of the southwestern United States. *Journal of the American Water Resources Association* 35: 1525–38.

Milliman, J.D. and Meade, R.H., 1983. World-wide delivery of river sediment to the oceans. *Journal of Geology* 91: 1–21.

Milliman, J.D. and Syvitski, P.M., 1992. Geomorphic/tectonic control of sediment discharge to the ocean: the importance of small mountainous rivers. *Journal of Geology* 100: 525–44.

Milliman, J.D., Farnsworth, K.L. and Albertin, C.S., 1999. Flux and fate of fluvial sediments leaving large islands in the East Indies. *Journal of Sea Research* 41: 97–107.

Milly, P.C.D. and Dunne, K.A., 1994. Sensitivity of the global water cycle to the water-holding capacity of land. *Journal of Climate* 7: 506–26.

Mitchell, J.F.B., Johns, T.C., Eagles, M., Ingram, W.J. and Davies, R.A., 1999. Towards the development of climate change scenarios. *Climatic Change* 41: 547–81.

Mkhandi, S. and Kachroo, R., 1997. Regional flood frequency analysis for Southern Africa. *Southern African FRIEND*. Technical Documents in Hydrology No. 15. Unesco, Paris, pp. 130–50.

Monteith, D.T., Evans, C.D. and Reynolds, B., 2000. Are temporal variations in the nitrate content of UK upland freshwaters linked to the North Atlantic Oscillation? *Hydrological Processes* 14: 1745–9.

Monteith, J.L., 1965. Evaporation and the environment. *Proceedings of the Symposium of the Society for Experimental Biology* 19: 205–34.

Moore, R.J., 1985. The probability-distributed principle and runoff production at point and basin scales. *Hydrological Sciences Journal* 30: 273–97.

Moore, R.J., Bell, V.A., Austin, R.M. and Harding, R.J., 1999. Methods for snowmelt forecasting in upland Britain. *Hydrology and Earth System Sciences* 3: 233–46.

Moore, W.S., 1996. Large groundwater inputs to coastal waters revealed by ^{226}Ra enrichments. *Nature* 380: 612–14.

Mosley, M.P., 2000. Regional differences in the effects of El Niño and La Niña on low flows and floods. *Hydrological Sciences Journal* 45: 249–67.

Moss, M.E., Pearson, C.P. and McKerchar, A.I., 1994. The Southern Oscillation Index as a predictor of the probability of low streamflows in New Zealand. *Water Resources Research* 30: 2717–23.

Munchow, A., Weingartner, T.J. and Cooper, L.W., 1999. The summer hydrography and surface circulation of the east Siberian shelf sea. *Journal of Physical Oceanography* 29: 2167–82.

Murdoch, P.S., Baron, J.S. and Miller, T.L., 2000. Potential effects of climate change on surface water quality in North America. *Journal of the American Water Resources Association* 36: 347–66.

Murukami, S., Tsuboyama, Y., Shimuzu, T., Fujicda, M. and Noguchi, S., 2000. Variation of evapotranspiration with stand age and climate in a small Japanese forested catchment. *Journal of Hydrology* 227: 114–27.

Naden, P.S. and Macdonald, A.T., 1987. Statistical modelling of water colour in the uplands: the Upper Nidd catchment 1979–1987. *Environmental Pollution* 60: 141–63.

Naden, P.S., Blyth, E.M., Broadhurst, P., Watts, C.D. and Wright, I.R., 2000. Modelling the spatial variation in soil moisture at the landscape scale: an application to five areas of ecological interest in the UK. *Hydrological Processes* 14: 785–809.

Nakai, Y., Sakamoto, T., Terajima, T., Kitamura, K. and Shirai, T., 1999. Energy balance above a boreal coniferous forest: a difference in turbulent fluxes between snow-covered and snow-free canopies. *Hydrological Processes* 13: 515–29

Nandakumar, N. and Mein, R.G., 1997. Uncertainty in rainfall-runoff model simulations and the implications for predicting the hydrologic effects of land use change. *Journal of Hydrology* 192: 211–32.

National Research Council, 1991. *Opportunities in the Hydrologic Sciences*. National Academy Press, Washington D.C.

National Rivers Authority, 1993. *Low Flows and Water Resources*. National Rivers Authority, Bristol.

Navar, J., 1993. The causes of stemflow variation in three semi-arid growing species of northeastern Mexico. *Journal of Hydrology* 145: 175–90.

Neal, C., 1997. Introduction to the Special Issue of Hydrology and Earth System Sciences: the water quality of the Plynlimon catchments. *Hydrology and Earth System Sciences* 1: 385–8.

Neal, C. *et al.*, 1993. Relationships between precipitation, streamflow and throughfall for a lowland beech plantation, Black Wood, Hampshire, Southern England. *Journal of Hydrology* 146: 221–33.

Neal, C., Hill, T., Hill, S. and Reynolds, B., 1997a. Acid neutralization capacity measurements in surface and ground waters in the Upper River Severn, Plynlimon: from hydrograph splitting to water flow pathways. *Hydrology and Earth System Sciences* 1: 687–96.

Neal, C. *et al.*, 1997b. The hydrochemistry of the headwaters of the River Severn, Plynlimon. *Hydrology and Earth System Sciences* 1: 583–617.

Neal, C. *et al.*, 2000. The water quality of the River Thames at a rural site downstream of Oxford. *Science of the Total Environment* 251: 441–57.

NERC, 1975. *Flood Studies Report*. Stationery Office, London.

New, M.G., Hulme, M. and Jones, P.D., 1999. Representing twentieth century space-time climate variability. Part 1: Development of a 1961–1990 mean monthly terrestrial climatology. *Journal of Climate* 12: 829–56.

Newbold, J.D., 1996. Cycles and spirals of nutrients. In: G.E. Petts and P. Calow (eds), *River Flows and Channel Flows*. Blackwell, Oxford, pp. 130–59.

Newson, M.D., 1997. *Water, Land and Development*. 2nd edition, Routledge, London.

Newton, R.M., Burns, D.A., Blette, V.L. and Driscoll, C.T., 1996. Effect of whole catchment liming on the episodic acidification of two Adirondack streams. *Biogeochemistry* 32: 299–322.

Nicholls, N., Gruza, G.V., Jouzel, J., Karl, T.R. and Ogallo, L.A., 1996. Observed climate variability and change. In: J.T. Houghton *et al.* (eds), *Climate Change 1995: The Science of Climate Change*. Cambridge University Press, Cambridge, pp. 137–92.

Nicholls, W.D., 1993. Estimating discharge of shallow groundwater by transpiration from greasewood in the northern Great Basin. *Water Resources Research* 29: 2771–8.

Nicholson, S., 2000. Land surface processes and Sahel climate. *Reviews of Geophysics* 38: 117–39.

Nijssen, B., Lettenmaier, D.P., Liang, X., Wetzel, S.W. and Wood, E.F., 1997. Streamflow simulation for continental-scale river basins. *Water Resources Research* 33: 711–24.

Njoku, E. and Entekhabi, D., 1996. Passive microwave remote sensing of soil moisture. *Journal of Hydrology* 184: 101–30.

Nobre, C., Sellers, P.J. and Shukla, J., 1991. Amazonian deforestation and regional climate change. *Journal of Climatology* 10: 957–88.

Nyberg, L., Rodhe, A. and Bishop, K., 1999. Water transit times and flow paths from two line injections of H-3 and Cl-36 in a microcatchment at Gardsjon, Sweden. *Hydrological Processes* 13: 1557–75.

Oki, T., 1999. The global water cycle. In: K.A. Browning and R.J. Gurney (eds), *Global Energy and Water Cycles*. Cambridge University Press, Cambridge, pp. 10–29.

Oki, T., Musiake, K., Matsuyama, H. and Masuda, K., 1995. Global atmospheric water balance and runoff from large river basins. *Hydrological Processes* 9: 655–78.

Olsen, J.R., Stedinger, J.R., Matalas, N.C. and Stakhiv, E.Z., 1999. Climate variability and flood frequency estimation for the Upper Mississipi and Lower Missouri rivers. *Journal of the American Water Resources Association* 35: 1509–23.

Onof, C. and Wheater, H.S., 1996. Analysis of the spatial coverage of British rainfall fields. *Journal of Hydrology* 176: 97–113.

Osborn, T.J., Hulme, M., Jones, P.D. and Basnett, T.A., 2000. Observed trends in the daily intensity of United Kingdom precipitation. *International Journal of Climatology* 20: 347–64.

Otter, L.B. and Scholes, M.C., 2000. Methane sources and sinks in a periodically-flooded South African savanna. *Global Biogeochemical Cycles* 14: 97–111.

Palanques, A., Plana, F. and Maldonado, A., 1990. Recent influences of man on the Ebro margin sedimentation system, northwestern Mediterranean Sea. *Marine Geology* 95: 247–63.

Pardé, M., 1955. *Fleuves et Rivières*. Colin, Paris.

Pardo-Iguzquiza, E., 1998. Optimal selection of number and location of rainfall gauges for areal rainfall estimation using geostatistics and simulated annealing. *Journal of Hydrology* 210: 206–20.

Parkhurst, R.S., Winter, T.C., Rosenberry, D.O. and Shurrock, A.M., 1998. Evaporation from a small prairie wetland in the Cottonwood Lake area, North Dakota – an energy-budget study. *Wetlands* 18: 272–87.

Penman, H.L., 1948. Natural evaporation from open water, bare soil and grass. *Proceedings of the Royal Society, Series A* 193: 120–45.

Penman, H.L., 1956. Evaporation: an introductory survey. *Netherlands Journal of Agricultural Science* 1: 9–29.

Peugeot, C., Esteves, M., Galle, S., Rajot, J.L. and Vandervaere, J.P., 1997. Runoff generation processes: results and analysis of field data collected at the east central supersite of the HAPEX-Sahel experiment. *Journal of Hydrology* 188–9: 179–202.

Pfister, C. and Brazdil, R., 1999. Climatic variability in sixteenth-century Europe and its social dimension: a synthesis. *Climatic Change* 43: 5–53.

Phillips, D.L., Dolph, J. and Marks, D., 1992. A comparison of geostatistical procedures for spatial analysis of precipitation in mountainous terrain. *Agricultural and Forest Meteorology* 58: 119–41.

Piechota, T.C. and Dracup, J.A., 1996. Drought and regional hydrologic variation in the United States: associations with El Niño Southern Oscillation. *Water Resources Research* 32.

Piechota, T.C., Dracup, J.A. and Fovell, R.G., 1997. Western U.S. streamflow and atmospheric circulation patterns during El Niño-Southern Oscillation. *Journal of Hydrology* 201: 249–71.

Piechota, T.C., Chiew, F.H.S., Dracup, J.A. and McMahon, T.A., 1998. Seasonal streamflow forecasting in eastern Australia and the El Niño Southern Oscillation. *Water Resources Research* 34: 3035–44.

Piexoto, J.P., de Aleida, M., Rosen, R.D. and Salstein, D.A., 1982. Atmospheric moisture transport and the water balance of the Mediterranean Sea. *Water Resources Research* 18: 83–90.

Pilgrim, D.H., Chapman, T.G. and Doran, D.G., 1988. Problem of rainfall-runoff modelling in arid and semi-arid regions. *Hydrological Sciences Journal* 33: 379–400.

Pilgrim, J.M., Fang, X. and Stefan, H.G., 1998. Stream temperature correlations with air temperatures in Minnesota: implications for climate warming. *Journal of the American Water Resources Association* 34: 1109–21.

Pilling, C. and Jones, J.A.A., 1999. High resolution climate change scenarios: implications for British runoff. *Hydrological Processes* 13: 2877–95.

Pitman, A.J. and Zhao, M., 2000. The relative impact of observed changes in land cover and carbon dioxide as simulated by a climate model. *Geophysical Research Letters* 27: 1267–70.

Pitman, A.J., Henderson-Sellers, A. and Zhang, Z.-L., 1990. Sensitivity of regional climates to localized precipitation in global models. *Nature* 346: 734–7.

Poesen, J., Ingelmo-Sanchez, F. and Mocher, H., 1990. The hydrological response of soil surfaces to rainfall as affected by cover and position of rock fragments in the top layer. *Earth Surface Processes and Landforms* 15: 653–71.

Ponce, V.M., Pandey, R.P. and Kumar, S., 1999. Groundwater recharge by channel infiltration in El Barbon basin, Baja California, Mexico. *Journal of Hydrology* 214: 1–7.

Potter, K.W., 1991. Hydrological impacts of changing land management practices in a moderate-sized agricultural catchment. *Water Resources Research* 27: 845–55.

Poulos, S.E., Collins, M. and Evans, G., 1996. Water-sediment fluxes of Greek rivers, southeastern Alpine Europe: annual yields, seasonal variability, delta formation and human impact. *Zeitschrift für Geomorphologie* 40: 243–61.

Poveda, G. and Mesa, O.J., 1997. Feedbacks between hydrological processes in tropical South America and large-scale ocean-atmospheric phenomena. *Journal of Climate* 10: 2690–702.

Price, A.G., Dunham, K., Carleton, T. and Band, L., 1997. Variability of water fluxes through the black spruce (*Picea mariana*) canopy and feather moss (*Pleurozium schreberi*) carpet in the boreal forest of Northern Manitoba. *Journal of Hydrology* 196: 310–23.

Price, M., 1996. *Introducing Groundwater*. Stanley Thornes, Cheltenham.

Priestley, C.H.N. and Taylor, R.J., 1972. On the assessment of surface heat flux and evaporation using large scale parameters. *Monthly Weather Review* 100: 81–92.

Prieto, M.D.R., Herrera, R. and Dursel, P., 1999. Historical evidence of streamflow fluctuations in the Mendoza River, Argentina, and their relationship with ENSO. *The Holocene* 9: 473–81.

Probst, J.L. and Tardy, Y., 1987. Long range streamflow and world continental runoff fluctuations since the beginning of this century. *Journal of Hydrology* 94: 289–311.

Putuhena, W.M. and Cordery, I., 1996. Estimation of the interception capacity of the forest floor. *Journal of Hydrology* 180: 283–99.

Quinton, W.L. and Marsh, P., 1999. A conceptual framework for runoff generation in a permafrost environment. *Hydrological Processes* 13: 2563–81.

Raddatz, R.L., 1998. Anthropogenic vegetation transformation and the potential for deep convection on the Canadian prairies. *Canadian Journal of Soil Science* 78: 657–66.

Rahmstorf, S., 1995. Bifurcation of the Atlantic thermohaline circulation in response to changes in the hydrological cycle. *Nature* 378: 145–9.

Ranzi, R., Grossi, G. and Bacchi, B., 1999. Ten years of monitoring areal snowpack in the Southern Alps using NOAA-AVHRR imagery, ground measurements and hydrological data. *Hydrological Processes* 13: 2079–95.

Rasmussen, E.M., 1977. *Hydrological Applications of Atmospheric Vapor-Flux Analyses*. Hydrology Reports, 11. World Meteorological Organization, Geneva.

Rawls, W.J., Ahuja, L.R., Brakensiek, D.L. and Shirmohammadi, A., 1993. Infiltration and soil water movement. In: D.R. Maidment (ed.), *Handbook of Hydrology*. McGraw Hill, New York, pp. 5.1–5.51.

Reale, O. and Shukla, J., 2000. Modeling the effects of vegetation on Mediterranean climate during the Roman Classical Period: Part II. Model simulation. *Global and Planetary Change* 25: 185–214.

Redmond, K.T. and Koch, R.W., 1991. Surface climate and streamflow variability in the western United States and their relationship to large-scale circulation indices. *Water Resources Research* 27: 2381–99.

Reinfelds, I. and Bishop, P., 1998. Palaeohydrology, palaeodischarges and palaeochannel dimensions: research strategies for meandering alluvial rivers. In: G. Benito, V.R. Baker and K.J. Gregory (eds), *Palaeohydrology and Environmental Change*. Wiley, Chichester, pp. 27–42.

Restrepo, J.D. and Kjerfve, B., 2000. Magdalena River: interannual variability (1975–1995) and revised water discharge and sediment load estimates. *Journal of Hydrology* 235: 137–49.

Reuss, J.O., Stottlemyer, R. and Troendle, C.A., 1997. Effect of clear cutting on nutrient fluxes in a subalpine forest at Fraser, Colorado. *Hydrology and Earth System Sciences* 1: 333–43.

Revelle, P. and Suess, H.E., 1957. Carbon dioxide exchange between atmosphere and ocean and the question of an increase in atmospheric CO_2 during the past decades. *Tellus* 9: 18–27.

Reynolds, W.D. and Black, D.E., 1991. Determination of hydraulic conductivity using a tension infiltrometer. *Soil Science Society of America Journal* 55: 633–9.

Riebsame, W.E., 1990. The United States Great Plains. In: B.L. Turner *et al.* (eds), *The Earth as Transformed by Human Action*. Cambridge University Press, Cambridge, pp. 561–75.

Risbey, J.S. and Entekhabi, D., 1996. Observed Sacremento Basin streamflow response to precipitation and temperature changes and its relevance to climate impact studies. *Journal of Hydrology* 184: 209–23.

Roberts, G. and Crane, S.B., 1997. The effects of clear-felling established forestry on streamflow losses from the Hore sub-catchment. *Hydrology and Earth System Sciences* 1: 477–82.

Roberts, J.M., 1977. The use of tree cutting techniques in the study of the water relations of mature *Pinus selvestris L. Journal of Experimental Biology* 28: 751–67.

Roberts, J.M., 1983. Forest transpiration: a conservative hydrological process? *Journal of Hydrology* 66: 133–41.

Roberts, N., 1994. *The Holocene: an Environmental History*. Blackwell, Oxford.

Robertson, A.W. and Mechoso, C.R., 1998. Interannual and decadal cycles in river flows of southeastern South America. *Journal of Climate* 11: 2570–81.

Robinson, J.S. and Sivapalan, M., 1997. An investigation into the physical causes of scaling and heterogeneity of regional flood frequency. *Water Resources Research* 33: 1045–60.

Robinson, M., 1986. Changes in catchment runoff following drainage and afforestation. *Journal of Hydrology* 86: 71–84.

Robinson, M., 1990. *Impact of improved land drainage on river flows*. Report 113, Institute of Hydrology, Wallingford, Oxon.

Robinson, M., 1998. Thirty years of forest hydrology changes at Coalburn: water balance and extreme flows. *Hydrology and Earth System Sciences* 2: 233–8.

Robinson, M. and Dean, T.J., 1993. Measurement of near surface soil water content using a capacitance probe. *Hydrological Processes* 7: 77–86.

Robinson, M. *et al.*, 2000. Land use change. In: M. Acreman (ed.), *The Hydrology of the UK: A Study of Change*. Routledge, London, pp. 30–54.

Robson, A.J., Jones, T.K., Reed, D.W. and Bayliss, A.C., 1998. A study of national trend and variation in UK floods. *International Journal of Climatology* 18: 165–82.

Rodda, J.C., 1967. The systematic error in rainfall measurement. *Journal of the Institution of Water Engineers* 21: 173–7.

Rodda, J.C. and Smith, S.W., 1986. The significance of the systematic error in rainfall measurement for assessing atmospheric deposition. *Atmospheric Environment* 20: 1059–64.

Rodhe, H., Langner, J., Gallardo, L. and Kjellstrom, E., 1995. Global transport of acidifying pollutants. *Water, Air and Soil Pollution* 85: 37–50.

Rodwell, M.J., Rowell, D.P. and Folland, C.K., 1999. Oceanic forcing of the wintertime North Atlantic Oscillation and European climate. *Nature* 398: 320–3.

Rondeau, B., Cossa, D., Gagnon, P. and Bilodeau, L., 2000. Budget and sources of suspended sediment transported in the St Lawrence River, Canada. *Hydrological Processes* 14: 21–36.

Ropelewski, C.F., 1992. Predicting El Niño events. *Nature* 356: 476–7.

Ropelewski, C.F. and Halpert, M.S., 1987. Global and regional scale precipitation patterns associated with El Niño/Southern Oscillation. *Monthly Weather Review* 115: 1606–26.

Rosen, M.R., Coshell, L., Turner, J.V. and Woodbury, R.J., 1996. Hydrochemistry and nutrient cycling in Yalgorup National Park, Western Australia. *Journal of Hydrology* 185: 241–74.

Rossi, F., Fiorentino, M. and Versace, P., 1984. Two component extreme value distribution for flood frequency analysis. *Water Resources Research* 20: 847–56.

Rutter, A.J., 1967. An analysis of evaporation from a stand of Scots pine. In: W.E. Sopper and H.W. Lull (eds), *Forest Hydrology*. Pergamon, Oxford, pp. 403–17.

Rutter, A.J., Morton, A.J. and Robins, P.C., 1975. A predictive model of rainfall interception in forests. II. Generalization of the model and comparison with observations in some coniferous and hardwood stands. *Journal of Applied Ecology* 12: 367–80.

Sahagian, D., 2000. Global physical effects of anthropogenic hydrological alterations: sea level and water redistribution. *Global and Planetary Change* 25: 39–48.

Sahin, V. and Hall, M.J., 1996. The effects of afforestation and deforestation on water yields. *Journal of Hydrology* 178: 293–310.

Said, F. *et al.*, 1997. Spatial variability in airborne surface flux measurements during HAPEX-Sahel. *Journal of Hydrology* 188–9: 878–911.

Salas, J.D., 1993. Analysis and modeling of hydrologic time series. In: D.R. Maidment (ed.), *Handbook of Hydrology*. McGraw Hill, New York, pp. 19.1–19.72.

Sami, K. and Hughes, D.A., 1996. A comparison of recharge estimates to a fractured sedimentary aquifer in South Africa from a chloride mass balance and an integrated surface-subsurface model. *Journal of Hydrology* 179: 111–36.

San Jose, J.J. and Montes, R., 1992. Rainfall partitioning by a semi-deciduous forest grove in the savannas of the Orinoco Llanas, Venezuela. *Journal of Hydrology* 132: 249–62.

Sandstrom, K., 1995a. The recent Lake Babati floods in semi-arid Tanzania: a response to changes in land cover. *Geografiska Annaler Series A – Physical Geography* 77: 35–44.

Sandstrom, K., 1995b. Modelling the effects of rainfall variability on groundwater recharge in semi-arid Tanzania. *Nordic Hydrology* 26: 313–30.

Sausen, R., Schubert, S. and Dümenil, L., 1994. A model of river runoff for use in coupled atmosphere-ocean models. *Journal of Hydrology* 155: 337–52.

Saxton, K.E., Rawls, W.J., Romberger, J.S. and Papendick, R.I., 1986. Estimating generalized soil water characteristics from texture. *Transactions of the American Society of Agricultural Engineers* 50: 1031–5.

Scatena, F.N., 1990. Watershed scale rainfall interception on two forested watersheds in the Luguillo Mountains of Puerto Rico. *Journal of Hydrology* 113: 89–102.

Schaap, M.G., Bouten, W. and Verstraten, J.M., 1997. Forest floor water content dynamics in a Douglas fir stand. *Journal of Hydrology* 201: 367–83.

Schar, C., Luthi, D., Beyerle, U. and Heise, E., 1999. The soil-precipitation feedback: a process study with a regional climate model. *Journal of Climate* 12: 722–41.

Schellekens, J., Scatema, F.N., Bruijnzeel, L.A. and Wickel, A.J., 1999. Modelling rainfall interception by a lowland tropical rain forest in northeastern Puerto Rico. *Journal of Hydrology* 225: 168–84.

Schinke, H. and Matthaus, W., 1998. On the causes of major Baltic inflows – an analysis of long time series. *Continental Shelf Research* 18: 67–97.

Schmugge, T.J., Jackson, T.J. and McKim, H.L., 1980. Survey of methods for soil moisture determination. *Water Resources Research* 16: 961–79.

Schnurr, R. and Lettenmaier, D.P., 1998. A case study of statistical downscaling in Australia using weather classification by recursive partitioning. *Journal of Hydrology* 212–13: 362–79.

Schofield, N.J. and Ruprecht, J.K., 1989. Regional analysis of stream salinisation in south-west Western Australia. *Journal of Hydrology* 112: 19–39.

Schreiber, P. and Demuth, S., 1997. Regionalisation of low flows in southwest Germany. *Hydrological Sciences Journal* 42: 845–58.

Scott, D.F., 1997. The contrasting effects of wildfire and clearfelling on the hydrology of a small catchment. *Hydrological Processes* 11: 543–56.

Scott, D.F. and Lesch, W., 1997. Streamflow responses to afforestation with *Eucalyptus grandis* and *Pinus patula* and to felling in the Mokobulaan experimental catchments, South Africa. *Journal of Hydrology* 199: 360–77.

Scribner, E.A., Battaglin, W.A., Goolsby, D.A. and Thurman, E.M., 2000. Changes in herbicide concentrations in Midwestern streams in relation to changes in use, 1989–1998. *Science of the Total Environment* 248: 255–63.

Sear, D.A., Briggs, A.R. and Brookes, A., 1998. A preliminary analysis of the morphological adjustment within and downstream of a lowland river subject to river restoration. *Aquatic Conservation* 8: 167–84.

Sear, D.A., Wilcock, D., Robinson, M. and Fisher, K., 2000. River channel modification in the UK. In: M. Acreman (ed.), *The Hydrology of the UK: A Study of Change.* Routledge, London, pp. 55–81.

Searcy, C., Dean, K. and Stringer, W., 1996. A river-coastal sea ice interaction model: Mackenzie River delta. *Journal of Geophysical Research – Oceans* 101(C4): 8885–94.

Sefton, C.E.M. and Howarth, S.M., 1998. Relationships between dynamic response characteristics and physical descriptors of catchments in England and Wales. *Journal of Hydrology* 211: 1–16.

Sellers. P.J., Shuttleworth, W.J., Dorman, J.L., Dalcher, A. and Roberts, J.M., 1989. Calibrating the simple biosphere model for Amazonian tropical forest using field and remote-sensing data. 1. Average calibration with field data. *Journal of Applied Meteorology* 28: 727–59.

Sellers, P., Hall, F.G., Asrar, G., Strebel, D.E. and Murphy, R.E., 1992. An overview of the First International Satellite Land Surface Climatology Project (ISLSCP) Field Experiment (FIFE). *Journal of Geophysical Research* 97, D17: 18345–71.

Sellers, P.J. *et al.*, 1997. The impact of using area-averaged land surface properties – topography, vegetation condition, soil wetness – in calculations of intermediate scale (approximately 10 km^2) surface-atmosphere heat and moisture fluxes. *Journal of Hydrology* 190: 269–301.

Sempere, R., Charriere, B., Van Wambeke, F. and Cauwet, G., 2000. Carbon inputs of the Rhône River to the Mediterranean Sea: biogeochemical implications. *Global Biogeochemical Cycles* 14: 669–81.

Semtner, A.J., 1984. The climate response of the Arctic Ocean to Soviet river diversions. *Climatic Change* 6: 109–30.

Sene, K.J. and Plinston, D.T., 1994. A review and update of the hydrology of Lake Victoria in East Africa. *Hydrological Sciences Journal* 39: 47–63.

Sene, K.J., Gash, J.H.C. and McNeil, D.D., 1991. Evaporation from a tropical lake: comparison of theory with direct measurements. *Journal of Hydrology* 127: 193–217.

Seo, D.-J., Breidenbach, J.P. and Johnson, E.R., 1999. Real-time estimation of mean field bias in radar rainfall data. *Journal of Hydrology* 223: 131–47.

Sevruk, B., 1987. Point precipitation measurements: why are they not corrected? In: *Water for the Future: Hydrology in Perspective. Int. Ass. Hydrol. Sci. Publ.* 164, 477–86.

Shanley, J.B. and Chalmers, A., 1999. The effect of frozen soil on snowmelt runoff at Sleepers River, Vermont. *Hydrological Processes* 13: 1843–57.

Shao, Y. and Henderson-Sellers, A., 1996. Modelling soil moisture: a Project for the Intercomparison of Land Surface Parameterisations (PILPS) Phase 2(b). *Journal of Geophysical Research* 101(D3): 7227–50.

Shaw, S.R., 1983. An investigation of the cellular structure of storms using correlation techniques. *Journal of Hydrology* 62: 63–79.

Shaw, T.J., Moore, W.S., Kloepfer, J. and Sochaski, M.A., 1998. The flux of barium to the coastal waters of the southeastern USA: the importance of submarine groundwater discharge. *Geochimica et Cosmochimica Acta* 62: 3047–54.

Shentsis, I., Meirovich, L., Ben-Zvi, A. and Rosenthal, E., 1999. Assessment of transmission losses and groundwater recharge from runoff events in a wadi under shortage of data on lateral inflow, Negev, Israel. *Hydrological Processes* 13: 1649–63.

Shiklomanov, I.A., 1993. World fresh water resources. In: P.H. Gleick (ed.), *Water in Crisis: A Guide to the World's Fresh Water Resources.* Oxford University Press, New York, pp. 13–24.

Shiklomanov, I.A., 1998. *Asssessment of water resources and water availability in the world. Background Report for the Comprehensive Assessment of the Freshwater Resources of the World.* Stockholm Environment Institute, Stockholm.

Sholkowitz, E.R., Elderfield, H., Szymczak, R. and Casey, K., 1999. Island weathering: river sources of rare earth elements to the Western Pacific Ocean. *Marine Chemistry* 68: 39–57.

Shorthouse, C.A. and Arnell, N.W., 1997. Spatial and temporal variability in European river flows and the North Atlantic Oscillation. In: *FRIEND'97 – Regional Hydrology: Concepts and Models for Sustainable Water Resources. Int. Ass. Hydrol. Sci. Publ.* 246, 77–85.

Shorthouse, C.A. and Arnell, N.W., 1998. Scales of spatial and temporal variability in European river flows. In: H. Wheater and C. Kirby (eds), *Hydrology in a Changing Environment.* Wiley, Chichester, pp. 215–24.

Shukla, J. and Mintz, Y., 1982. Influence of land surface evapo-transpiration on the earth's climate. *Science* 215: 1498–501.

Shuttleworth, W.J., 1977. The exchange of wind-driven fog and mist between vegetation and the atmosphere. *Boundary Layer Meteorology* 12: 463–89.

Shuttleworth, W.J., 1989. Micrometeorology of temperate and tropical forest. *Philosophical Transactions of the Royal Society, London, Series B* 324: 299–334.

Shuttleworth, W.J., 1993. Evaporation. In: D.R. Maidment (ed.), *Handbook of Hydrology*. McGraw Hill, New York, pp. 4.1–4.53.

Shuttleworth, W.J. and Calder, I.R., 1979. Has the Priestley–Taylor equation any relevance to forest evaporation? *Journal of Applied Meteorology* 18: 639–46.

Shuttleworth, W.J. *et al.*, 1988. An integrated micrometeorological system for evaporation measurement. *Agricultural and Forest Meteorology* 43: 295–317.

Sidle, R.C. *et al.*, 2000. Stormflow generation in steep forested headwaters: a linked hydrogeomorphic paradigm. *Hydrological Processes* 14: 369–85.

Silberstein, R.P. and Sivapalan, M., 1995. Estimation of terrestrial water and energy balances over heterogeneous catchments. *Hydrological Processes* 9: 613–30.

Simpson, H.J., Cane, M.A., Lin, S.K., Zebiak, S.E. and Herczeg, A.L., 1993. Forecasting annual discharge of River Murray, Australia, from a geophysical model of ENSO. *Journal of Climate* 6: 386–90.

Singh, B. and Szeicz, B., 1979. The effects of intercepted rainfall on the water balance of hardwood forest. *Water Resources Research* 15: 131–8.

Singh, P. and Kumar, N., 1997. Effect of orography on precipitation in the western Himalaya region. *Journal of Hydrology* 199: 183–206.

Sklash, M.G. and Farvolden, R.N., 1979. The role of groundwater in storm runoff. *Journal of Hydrology* 43: 45–65.

Slaymaker, O. and Spencer, T., 1998. *Physical Geography and Global Environmental Change*. Longman, Harlow.

Small, E.E., Sloan, L.C., Hostetler, S. and Giorgi, F., 1999. Simulating the water balance of the Aral Sea with a coupled regional climate-lake model. *Journal of Geophysical Research – Atmospheres* 104(D6): 6583–602.

Smith, D.I., 1998. *Water in Australia: Resources and Management*. Oxford University Press, Melbourne.

Smith, D.M. and Allen, S.J., 1996. Measurement of sap flow in plant stems. *Journal of Experimental Botany* 305: 1833–44.

Smith, J.A., 1992. Representation of basin scale in flood peak distributions. *Water Resources Research* 28: 2993–9.

Smith, S.J., Wigley, T.M.L., Nakicenovic, N. and Raper, S.C.B., 2000. Climate implications of greenhouse gas emissions scenarios. *Technological Forecasting and Social Change* 65: 195–204.

Solomon, D.K., Schiff, S.L., Poveda, R.J. and Clarke, W.B., 1993. A validation of the $^3H/^3He$ method for determining groundwater recharge. *Water Resources Research* 29: 2951–62.

Song, J., Willmott, C.J. and Hanson, B., 1997. Influence of heterogeneous land surfaces on surface energy and mass fluxes. *Theoretical and Applied Climatology* 58: 175–88.

Sorman, A.U., Abdul Razzak, M.J. and Morel-Seytoux, H.J., 1997. Groundwater recharge estimation from ephemeral streams. Case study: Wadi Tabalah, Saudi Arabia. *Hydrological Processes* 11: 1607–20.

Soto, B. and Diaz-Fierros, F., 1997. Soil water balance as affected by throughfall in gorse (*Ulex europaeus*, L.) shrubland after burning. *Journal of Hydrology* 195: 218–31.

Soulsby, C., 1995. Contrasts in storm event hydrochemistry in an acidic afforested catchment in upland Wales. *Journal of Hydrology* 170: 159–79.

Soulsby, C., 1997. Hydrochemical processes. In: R.L. Wilby (ed.), *Contemporary Hydrology*. Wiley, Chichester, pp. 59–106.

Soulsby, C., Helliwell, R.C., Ferrier, R.C., Jenkins, A. and Harriman, R., 1997. Seasonal snowpack influence on the hydrology of a sub-arctic catchment in Scotland. *Journal of Hydrology* 192: 17–32.

Soulsby, C., Malcolm, R., Helliwell, R., Ferrier, R.C. and Jenkins, A., 2000. Isotope hydrology of the Allt a' Mharcaidh catchment, Cairngorms, Scotland: implications for hydrological pathways and residence times. *Hydrological Processes* 14: 747–62.

Stamm, J.F., Wood, E.F. and Lettenmaier, D.P., 1994. Sensitivity of a GCM simulation of global climate to the representation of land-surface hydrology. *Journal of Climate* 7: 1218–39.

Stanescu, V.A. and Ungureanu, V., 1997. Hydrological regimes in the FRIEND-AMHY area: space variability and stability. In: *FRIEND'97 – Regional Hydrology: Concepts and Models for Sustainable Water Resources. Int. Ass. Hydrol. Sci. Publ.* 246, 67–75.

Stanners, D. and Bourdeau, P. (eds), 1995. *Europe's Environment: the Dobris Assessment.* European Environment Agency, Copenhagen.

Starkel, L., Gregory, K.J. and Thornes, J.B. (eds), 1991. *Temperate Palaeohydrology: Fluvial Processes in the Temperate Zone over the Last 15000 Years.* Wiley, Chichester.

Starr, V.P. and Piexoto, J., 1958. On the global balance of water vapour and the hydrology of deserts. *Tellus* 10: 189–94.

Stedinger, J.R., Vogel, R.M. and Foufoula-Georgiou, E., 1993. Frequency analysis of extreme events. In: D.R. Maidment (ed.), *Hydrology in Practice.* McGraw Hill, New York, pp. 18.1–18.66.

Stednick, J.D., 1996. Monitoring the effects of timber harvest on annual water yield. *Journal of Hydrology* 176: 79–95.

Steele, M., Thomas, D., Rothrock, D. and Martin, S., 1996. A simple model study of the Arctic Ocean freshwater balance, 1979–1985. *Journal of Geophysical Research – Oceans* 101(C9): 20833–48.

Stern, D.I. and Kaufmann, R.K., 1996. *Estimates of Global Anthropogenic Sulfate Emissions 1860–1993.* CEES Working Paper 9602, Center for Energy and Environmental Studies, Boston University, Boston MA.

Stevens, P.A., Reynolds, B., Hughes, S., Norris, D.A. and Dickinson, A.L., 1997. Relationships between spruce plantation age, solute and soil chemistry in Hafren forest. *Hydrology and Earth System Sciences* 1: 627–37.

Stewart, J.B., 1977. Evaporation from the wet canopy of a pine forest. *Water Resources Research* 13: 915–22.

Stewart, J.B., 1988. Modelling surface conductance of pine forest. *Agricultural and Forest Meteorology* 43: 19–35.

Stewart, J.B. and Thom, A.S., 1973. Energy budgets in pine forests. *Quarterly Journal of the Royal Meteorological Society* 99: 154–70.

Stewart, R.E. *et al.*, 1998. The Mackenzie GEWEX study: the water and energy cycles of a major North American river basin. *Bulletin of the American Meteorological Society* 79: 2665–83.

Stieglitz, M., Rind, D., Famiglietti, J. and Rosenzweig, C., 1997. An efficient approach to modelling the topographic control of surface hydrology for regional and global climate modeling. *Journal of Climate* 10: 118–37.

Stohlgren, T.J., Chase, T.N., Pielke, R.A., Kittel, T.G.F. and Baron, J.S., 1998. Evidence that local land use practices influence regional climate, vegetation and streamflow patterns in adjacent natural areas. *Global Change Biology* 4: 495–504.

Stonefelt, N.M.D., Fontaine, T.A. and Hotchkiss, R.H., 2000. Impacts of climate change on water yield in the Upper Wind Basin. *Journal of the American Water Resources Association* 36: 321–36.

Sud, Y.C., Yang, R. and Walker, G.R., 1996. Impact of in situ deforestation in Amazonia on the regional climate: general circulation model simulation study. *Journal of Geophysical Research – Atmospheres* 101: 7095–109.

Sullivan, T.J., McMartin, B. and Charles, D.F., 1996. Re-examination of the role of landscape change in the acidification of lakes in the Adirondack Mountains, New York. *Science of the Total Environment* 183: 231–48.

Sun, H. and Furbish, D.J., 1997. Annual precipitation and river discharges in Florida in response to El Niño- and La Niña-sea surface temperature anomalies. *Journal of Hydrology* 199: 74–87.

Tallaksen, L.M., 1995. A review of baseflow recession analysis. *Journal of Hydrology* 148: 349–70.

Tallaksen, L.M. and Hisdal, H., 1997. Regional analysis of extreme streamflow drought duration and deficit volume. In: *FRIEND'97 – Regional Hydrology: Concepts and Models for Sustainable Water Resources. Int. Ass. Hydrol. Sci. Publ.* 246, 141–50.

Tallaksen, L.M., Madsen, H. and Clausen, B., 1997. On the definition and modelling of streamflow drought duration and deficit volume. *Hydrological Sciences Journal* 42: 15–34.

Taniguchi, M., Tsujimura, M. and Tanaka, T., 1996. Significance of stemflow in groundwater recharge. 1. Evaluation of the stemflow contribution to recharge using a mass balance approach. *Hydrological Processes* 10: 71–80.

Taylor, C.M., 2000. The influence of antecedent rainfall on Sahelian surface evaporation. *Hydrological Processes* 14: 1245–59.

Taylor, R.G. and Howard, K.W.F., 1996. Groundwater recharge in the Victoria Nile basin of East Africa: support for the soil moisture balance approach using stable isotope tracers. *Journal of Hydrology* 180: 31–53.

Terajima, T., Sakamoto, T. and Shirai, T., 2000. Morphology, structure and flow phases in soil pipes developing in forested hillslopes underlain by a Quaternary sand-gravel formation, Hokkaido, northern main island in Japan. *Hydrological Processes* 14: 713–26.

Ternon, J.F., Oudot, C., Dessier, A. and Diverres, D., 2000. A seasonal tropical sink for atmospheric CO_2 in the Atlantic Ocean: the role of the Amazon River discharge. *Marine Chemistry* 68: 183–201.

Thompson, J.R. and Hollis, G.E., 1995. Hydrological modelling and the sustainable development of the Hadejia-Nguru Wetlands, Nigeria. *Hydrological Sciences Journal* 40: 97–116.

Thoms, M.C. and Sheldon, F., 2000. Water resource development and hydrological change in a large dryland river: the Barwon-Darling River, Australia. *Journal of Hydrology* 228: 10–21.

Thornthwaite, C.W., 1948. An approach towards a rational definition of climate. *Geographical Review* 38: 55–94.

Tipping, E., Lofts, S. and Lawlor, A.J., 1998. Modelling the chemical speciation of trace metals in the surface waters of the Humber system. *Science of the Total Environment* 210–11: 63–78.

Tobon Marin, C., Bouten, W. and Sevink, J., 2000. Gross rainfall and its partitioning into throughfall, stemflow and evaporation of intercepted water in four forest ecosystems in western Amazonia. *Journal of Hydrology* 237: 40–57.

Todini, E., 1996. The ARNO rainfall-runoff model. *Journal of Hydrology* 175: 339–82.

Traaen, T.S. *et al.*, 1997. Whole-catchment liming at Tjonnstrond, Norway: an 11-year record. *Water, Air and Soil Pollution* 94: 163–80.

Tremblay, L.B. and Mysak, L.A., 1998. On the origin and evolution of sea-ice anomalies in the Beaufort-Chukchi Sea. *Climate Dynamics* 14: 451–60.

Trenberth, K.E., 1998. Atmospheric moisture recycling: role of advection and local evaporation. *Journal of Climate* 12(3): 1368–81.

Trenberth, K. and Guillemot, C.J., 1995. Evaluation of the global atmospheric moisture budget as seen from analyses. *Journal of Climate* 8: 2255–72.

Trenberth, K.E. and Guillemot, C.J., 1996. Physical processes involved in the 1988 drought and 1993 floods in North America. *Journal of Climate* 9: 1288–98.

Trenberth, K.E. and Hurrell, J.W., 1994. Decadal atmosphere-ocean variations in the Pacific. *Climate Dynamics* 9: 303–19.

Trenberth, K.E., Houghton, J.T. and Meira Filho, L.G., 1996. The climate system: an overview. In: J.T. Houghton *et al.* (eds), *Climate Change 1995: The Science of Climate Change. Contribution of Working Group 1 to the Second Assessment Report of the Intergovernmental Panel on Climate Change.* Cambridge University Press, Cambridge, pp. 55–64.

Tricker, A.S., 1981. Spatial and temporal patterns of infiltration. *Journal of Hydrology* 49: 261–77.

Trimble, S.W., 1981. Changes in sediment storage in Coon Creek Basin, Driftless Area, Wisconsin, 1853–1975. *Science* 214: 181–3.

Troendle, C.A. and Reuss, J.O., 1997. Effect of clear cutting on snow accumulation and water outflow at Fraser, Colorado. *Hydrology and Earth System Sciences* 1: 325–32.

Tsintikis, D., Georgakakos, K.P., Artan, G.A. and Tsonis, A.A., 1999. A feasibility study on mean areal rainfall estimation and hydrologic response in the Blue Nile region using Meteosat. *Journal of Hydrology* 221: 97–116.

Turley, C.M., 1999. The changing Mediterranean Sea – a sensitive ecosystem? *Progress in Oceanography* 44: 387–400.

Turner, B.L. *et al.*, 1990. *The Earth as Transformed by Human Action.* Cambridge University Press, Cambridge.

Uchida, T., Kosugi, K. and Mizuyama, T., 1999. Runoff characteristics of pipeflow and effects of pipeflow on rainfall-runoff phenomena in a mountainous watershed. *Journal of Hydrology* 222: 18–36.

UNEP, 1999. *Global Environment Outlook 2000.* Earthscan, London.

Uvo, C.B. and Graham, N.E., 1998. Seasonal runoff forecasts for northern South America: a statistical model. *Water Resources Research* 34: 3515–24.

Valente, F., David, J.S. and Gash, J.H.C., 1997. Modelling interception loss for two sparse eucalypt and pine forests in central Portugal using reformulated Rutter and Gash analytical models. *Journal of Hydrology* 190: 141–62.

van de Giesen, N.C., Stomph, T.J. and de Ridder, N., 2000. Scale effects of Hortonian overland flow and rainfall-runoff dynamics in a West African catena landscape. *Hydrological Processes* 14: 165–75.

van Genuchten, M.T., 1980. A closed-form equation for predicting the hydraulic conductivity of undisturbed soils. *Soil Science Society of America Journal* 44: 892–8.

van Oldenbergh, G.J., Burgers, G. and Tank, A.K., 2000. On the El Niño teleconnection to spring precipitation. *International Journal of Climatology* 20: 565–74.

van Overmeeren, R.A., Sariowan, S.V. and Gehrels, J.C., 1997. Ground penetrating radar for determining volumetric soil water content: results of comparative measurements at two test sites. *Journal of Hydrology* 197: 316–38.

van Wesemael, B., Mulligan, M. and Poesen, J., 2000. Spatial patterns of soil water balance on intensively cultivated hillslopes in a semi-arid environment: the impact of rock fragments and soil thickness. *Hydrological Processes* 14: 1811–28.

vanLoon, G.W. and Duffy, S.J., 2000. *Environmental Chemistry: a Global Perspective.* Oxford University Press, Oxford.

Vehviläinen, B. and Huttunen, M., 1997. Climate change and water resources in Finland. *Boreal Environment Research* 2: 3–18.

Vidale, P.L., Pielke, R.A., Steyaert, l.T. and Barr, A., 1997. Case study modelling of turbulent and mesoscale fluxes over the BOREAS region. *Journal of Geophysical Research – Atmospheres* 102: 29167–88.

Viney, N.R. and Sivapalan, M., 1996. The hydrological response of catchments to simulated changes in climate. *Ecological Modelling* 86: 189–93.

Virtue, W.A. and Clayton, J.W., 1997. Sheep dip chemicals and water pollution. *Science of the Total Environment* 194/5: 207–17.

Vogel, R.M. and Kroll, C.N., 1992. Regional geohydrologic-geomorphic relationships for the estimation of low flow statistics. *Water Resources Research* 28: 2451–8.

Vogel, R.M. and Sankarasubramanian, A., 2000. Spatial scaling properties of annual streamflow in the United States. *Hydrological Sciences Journal* 45: 465–76.

Vörösmarty, C. *et al.*, 1996. Analysing the discharge regime of a large tropical river through remote sensing, ground-based climate data, and modelling. *Water Resources Research* 32: 3137–50.

Vörösmarty, C.J., Federer, C.A. and Schloss, A.L., 1998. Potential evaporation functions compared on US watersheds: possible implications for global-scale water balance and terrestrial ecosystem modeling. *Journal of Hydrology* 207: 147–69.

Vörösmarty, C.J., Fekete, B.M., Meybeck, M. and Lammers, R.B., 2000. Global system of rivers: its role in organising continental land mass and defining land-to-ocean linkages. *Global Biogeochemical Cycles* 14: 599–621.

Wai, M.M.K. and Smith, E.A., 1998. Linking boundary layer circulation and surface processes during FIFE 89, Part II: maintenance of secondary circulation. *Journal of the Atmospheric Sciences* 55: 1260–76.

Wai, M.M.-K. *et al.*, 1997. Variability in boundary layer structure during HAPEX-Sahel wet-dry season transition. *Journal of Hydrology* 188–9: 965–97.

Wallace, J.S. and Holwill, C.J., 1997. Soil evaporation from tiger bush in south west Niger. *Journal of Hydrology* 188: 426–42.

Walling, D.E., 1977. Suspended sediment and solute response characteristics of the River Exe, Devon, England. In: R. Davidson and A.W. Nickling (eds), *Research in Fluvial Geomorphology*. GeoAbstracts, Norwich, pp. 169–97.

Walling, D.E., 1996. Suspended sediment transport by rivers: a geomorphological and hydrological perspective. *Advances in Limnology* 47: 1–27.

Walling, D.E., 1997. The response of sediment yields to environmental change. In: *Human Impact on Erosion and Sedimentation. Int. Ass. Hydrol. Sci. Publ.* 245, 77–89.

Walling, D.E., 1999. Linking land use, erosion and sediment yields in river basins. *Hydrobiologia* 410: 223–40.

Walling, D.E. and Webb, B.W., 1981. Water quality. In: J. Lewin (ed.), *British Rivers*. George Allen & Unwin, London, pp. 126–69.

Walling, D.E. and Webb, B.W., 1986. Solutes in river systems. In: S.T. Trudgill (ed.), *Solute Processes*. Wiley, Chichester, pp. 251–327.

Walling, D.E. and Webb, B.W., 1996. Water quality: I. Physical characteristics. In: G. Petts and P. Calow (eds), *River Flows and Channel Forms*. Blackwell, Oxford, pp. 77–101.

Wang, Q.X. and Takahashi, H., 1998. Regional hydrological effects of grassland degradation in the Loess Plateau of China. *Hydrological Processes* 12: 2279–88.

Ward, R.C. and Robinson, M., 2000. *Principles of Hydrology*. 4th edition. McGraw Hill, London.

Waylen, P.R. and Caviedes, C.N., 1990. Annual and seasonal fluctuations of precipitation and streamflow in the Acongagua River basin, Chile. *Journal of Hydrology* 120: 79–102.

Weatherly, J.W. and Walsh, J.E., 1996. The effects of precipitation and river runoff in a coupled ice-ocean model of the Arctic. *Climate Dynamics* 12: 785–98.

Webb, B.W. and Walling, D.E., 1996. Water quality. II: Chemical characteristics. In: G.E. Petts and P. Calow (eds), *River Flows and Channel Forms*. Blackwell, Oxford, pp. 102–29.

Wells, N., 1997. *The Atmosphere and Ocean: A Physical Introduction*. 2nd edition. Wiley, Chichester.

Westmacott, J.R. and Burn, D.H., 1997. Climate change effects on the hydrologic regime within the Churchill-Nelson River basin. *Journal of Hydrology* 202: 263–79.

Weston, K.J. and Roy, M.G., 1994. The directional-dependence of the enhancement of rainfall over complex topography. *Meteorological Applications* 1: 267–75.

Whitehead, P.G. and Neal, C., 1987. Modelling the effect of acid deposition in upland Scotland. *Transactions of the Royal Society of Edinburgh, Earth Sciences* 78: 385–92.

Wigley, T.M.L. and Jones, P.D., 1985. Influence of precipitation changes with direct CO_2 effects on streamflow. *Nature* 314: 149–52.

Wijffels, S.E., Scmitt, R.W., Bryden, H.L. and Stigebrandt, A., 1992. Transport of freshwater by the oceans. *Journal of Physical Oceanography* 22: 155–62.

Wilby, R., 1993. Evidence of ENSO in the synoptic climate of the British Isles since 1880. *Weather* 48: 234–9.

Wilby, R.L. and Wigley, T.M.L., 1997. Downscaling general circulation model output: a review of methods and limitations. *Progress in Physical Geography* 21: 530–48.

Wilby, R.L., Hay, L.E. and Leavesley, G.H., 1999. A comparison of downscaled and raw GCM output: implications for climate change scenarios in the San Juan River basin, Colorado. *Journal of Hydrology* 225: 67–91.

Wilcock, D.N. and Essery, C.I., 1984. Infiltration measurements in a small lowland catchment. *Journal of Hydrology* 74: 191–204.

Wilkinson, J. *et al.*, 1997. Major, minor and trace element composition of cloudwater and rainwater at Plynlimon. *Hydrology and Earth System Sciences* 1: 557–69.

Wilks, D.S., 1989. Adapting stochastic weather generation algorithms for climate change studies. *Climatic Change* 22: 67–84.

Williams, K.W.J., DeWalle, D.R., Edwards, P.J. and Schnabel, R.R., 1997. Indicators of nitrate export from forested watersheds of the mid-Appalachians, United States of America. *Global Biogeochemical Cycles* 11: 649–56.

Williams, M.R. and Melack, J.M., 1997. Solute export from forested and partially deforested catchments in the central Amazon. *Biogeochemistry* 38: 67–102.

Williams, R., Burt, T. and Brighty, G., 2000. River water quality. In: M. Acreman (ed.), *The Hydrology of the UK: A Study of Change*. Routledge, London, pp. 134–49.

Wittenberg, H., 1999. Baseflow recession and recharge as non-linear storage processes. *Hydrological Processes* 13: 715–26.

Wittenberg, H. and Sivapalan, M., 1999. Watershed groundwater balance estimation using streamflow recession analyses and baseflow separation. *Journal of Hydrology* 219: 20–33.

Wolman, M.G., 1967. A cycle of sedimentation and erosion in urban river channels. *Geografiska Annaler* 49A: 385–95.

Wolman, M.G. and Miller, J.C., 1960. Magnitude and frequency of forces in geomorphic processes. *Journal of Geology* 68: 54–74.

Woo, M.K. and Waylen, P.R., 1984. Areal prediction of annual floods generated by two distinct processes. *Hydrological Sciences Journal* 29: 75–88.

Wood, E.F., 1997. Effects of soil moisture aggregation on surface evaporative fluxes. *Journal of Hydrology* 190: 397–412.

Wood, E.F., Sivapalan, M. and Beven, K.J., 1990. Similarity and scale in catchment storm response. *Reviews of Geophysics* 28: 1–18.

Wood, E.F., Lettenmaier, D.P. and Zartarian, V.G., 1992. A land surface hydrology parameterization with subgrid variability for general circulation models. *Journal of Geophysical Research – Atmospheres* 97(D3): 2717–28.

Woodruff, J.F. and Hewlett, J.D., 1970. Predicting and mapping the average hydrologic response for the eastern United States. *Water Resources Research* 6: 1312–26.

Woods, R.A., Sivapalan, M. and Duncan, M., 1995. Investigating the representative elementary area concept: an approach based on field data. *Hydrological Processes* 9: 291–312.

World Commission on Dams, 2000. *Dams and Development: A New Framework for Decision-Making*. Earthscan, London.

World Meteorological Organization, 1997. *Comprehensive Assessment of the Freshwater Resources of the World*. WMO, Geneva.

World Resources Institute, 1996. *World Resources: A Guide to the Global Environment 1996–1997*. Oxford University Press, New York.

Wright, K.A., Sendek, K.H., Rice, R.M. and Thomas, R.B., 1990. Logging effects on streamflow: storm runoff at Caspar Creek in north-western California. *Water Resources Research* 26: 1657–67.

Wright, R.F. *et al.*, 1994. Changes in acidification of lochs in Galloway, south-western Scotland, 1979–1988: the MAGIC model used to evaluate the role of afforestation, calculate critical loads, and predict fish status. *Journal of Hydrology* 161: 257–85.

Wright, R.F., Beier, C. and Cosby, B.J., 1998. Effects of nitrogen deposition and climate change on nitrogen runoff at Norwegian boreal forest catchments: the MERLIN model applied to Risdalsheia (RAIN and CLIMEX) catchments. *Hydrology and Earth System Sciences* 2: 399–414.

Xie, P. and Arkin, P.A., 1996. Analyses of global monthly precipitation using gauge observations, satellite estimates and numerical model predictions. *Journal of Climate* 9: 840–58.

Yair, A. and Lavee, H., 1976. Runoff generation processes and runoff yield from arid talus-mantled slopes. *Earth Surface Processes* 1: 235–47.

Yair, A. and Lavee, H., 1985. Runoff generation in arid and semi-arid zones. In: M.G. Anderson and T.P. Burt (eds), *Hydrological Forecasting*. Wiley, Chichester, pp. 182–220.

Yamanaka, T., Takeda, A. and Shimada, J., 1998. Evaporation beneath the soil surface: some observational evidence and numerical experiments. *Hydrological Processes* 12: 2193–203.

Yang, D., Yu, G., Xie, Y., Zhan, D. and Li, Z., 2000. Sedimentary records of large Holocene floods from the middle reaches of the Yellow River, China. *Geomorphology* 33: 73–88.

Yang, Y., Lerner, D.N., Barrett, M.H. and Tellam, J.H., 1999. Quantification of groundwater recharge in the city of Nottingham, UK. *Environmental Geology* 38: 183–98.

Yoshino, F., 1999. Studies on the characteristics of variation and spatial correlation of the long-term annual runoff in the world rivers. *Journal of the Japanese Society of Hydrology and Water Resources* 12: 109–20.

Young, A.R., Round, C.E. and Gustard, A., 2000. Spatial and temporal variations in the occurrence of low flow events in the UK. *Hydrology and Earth System Sciences* 4: 35–45.

Younger, P.L., 1997. The longevity of minewater pollution: a basis for decision-making. *Science of the Total Environment* 194–5: 457–66.

Yu, Z., Lakhtakia, M.N., Yarnal, B. *et al.*, 1999. Simulating the river-basin response to atmospheric forcing by linking a mesoscale meteorological model and hydrologic model system. *Journal of Hydrology* 218: 72–91.

Zheng, X.Y. and Eltahir, E.A.B., 1997. The response to deforestation and desertification in a model of West African monsoons. *Geophysical Research Letters* 24: 155–8.

Zhu, T.X., Cai, Q.C. and Zeng, B.Q., 1997. Runoff generation on a semi-arid agricultural catchment: field and experimental studies. *Journal of Hydrology* 196: 99–118.

Ziegler, A.D. and Giambelluca, T.W., 1997. Importance of rural roads as source areas for runoff in mountainous areas of northern Thailand. *Journal of Hydrology* 196: 204–29.

Index

Note – individual entries are organised in the sequence: text, figures (in ***bold italic***), tables (in **bold**) and boxed material (**bold Box**)